Lecture Notes in Mathematics 1974

Editors:
J.-M. Morel, Cachan
F. Takens, Groningen
B. Teissier, Paris

Subseries:
École d'Été de Probabilités de Saint-Flour

Saint-Flour Probability Summer School

The Saint-Flour volumes are reflections of the courses given at the Saint-Flour Probability Summer School. Founded in 1971, this school is organised every year by the Laboratoire de Mathématiques (CNRS and Université Blaise Pascal, Clermont-Ferrand, France). It is intended for PhD students, teachers and researchers who are interested in probability theory, statistics, and in their applications.

The duration of each school is 13 days (it was 17 days up to 2005), and up to 70 participants can attend it. The aim is to provide, in three high-level courses, a comprehensive study of some fields in probability theory or statistics. The lecturers are chosen by an international scientific board. The participants themselves also have the opportunity to give short lectures about their research work.

Participants are lodged and work in the same building, a former seminary built in the 18th century in the city of Saint-Flour, at an altitude of 900 m. The pleasant surroundings facilitate scientific discussion and exchange.

The Saint-Flour Probability Summer School is supported by:

– Université Blaise Pascal
– Centre National de la Recherche Scientifique (C.N.R.S.)
– Ministère délégué à l'Enseignement supérieur et à la Recherche

For more information, see back pages of the book and
http://math.univ-bpclermont.fr/stflour/

Jean Picard
Summer School Chairman
Laboratoire de Mathématiques
Université Blaise Pascal
63177 Aubière Cedex
France

Frank den Hollander

Random Polymers

École d'Été de Probabilités
de Saint-Flour XXXVII – 2007

 Springer

Frank den Hollander
University of Leiden
Department of Mathematics
Niels Bohrweg 1
2333 RA Leiden
The Netherlands
denholla@math.leidenuniv.nl

ISSN 0721-5363
ISBN 978-3-642-00332-5   e-ISBN 978-3-642-00333-2
DOI 10.1007/978-3-642-00333-2
Springer Dordrecht Heidelberg London New York

Lecture Notes in Mathematics ISSN print edition: 0075-8434
                    ISSN electronic edition: 1617-9692
                    ISSN École d'Été de Probabilités de St. Flour, print edition: 0721-5363

Library of Congress Control Number: 2009922219

Mathematics Subject Classification (2000): 60K35, 60F10, 60K37, 82B26, 82B27

Cover design: SPi Publisher Services

Printed on acid-free paper

Springer is part of Springer Science+Business Media (www.springer.com)

Preface

This monograph contains the worked out and expanded notes of the lecture series I presented at the 37–TH PROBABILITY SUMMER SCHOOL IN SAINT-FLOUR, 8–21 JULY 2007. The goal I had set myself for these lectures was to provide an up-to-date account of some key developments in the *mathematical* theory of polymer chains, focusing on a number of models that are at the heart of the subject. In order to achieve this goal, I decided to limit myself to single polymers living on a lattice, to consider only models for which a transparent picture has emerged, and from the latter select those that lead to challenging open questions capable of attracting future research. Needless to say, my choice was influenced by my personal taste and involvement.

Polymers are studied intensively in mathematics, physics, chemistry and biology. Our focus will lie at the interface between *probability theory* and *equilibrium statistical physics*. To fully appreciate the results to be described, the reader needs a basic knowledge of both these areas. No other background is required. We will look at a number of *paradigm* models that exhibit interesting phenomena. The key objects of interest will be free energies, phase transitions as a function of underlying parameters and associated critical behavior, scaling properties of path measures in the different phases and associated invariance principles, as well as effects of randomness in the interactions. The emphasis will be on techniques coming from large deviation theory, combinatorics, ergodic theory and variational calculus.

We start with TWO BASIC MODELS of polymer chains: *simple random walk* and *self-avoiding walk*. After having collected a few key properties of these models, which serve to set the stage, we turn to the main body of the monograph, which is divided into two parts.

In PART A, we look at four models of POLYMERS WITH SELF-INTERACTION: (1) *soft polymers*, where self-intersections are not forbidden but are penalized, resulting in a repulsive interaction modeling the effect of "steric hindrance"; (2) *elastic polymers*, where self-intersections are penalized in a way that depends on their distance along the chain, in such a way that long loops are less penalized than short loops; (3) *polymer collapse*, where due to attractive interactions the polymer may roll itself up to form a ball; (4) *polymer adsorption*, where the polymer interacts with a linear substrate to which it may be attracted, either moving on both sides of the substrate (pinning at an interface) or staying on one side of the substrate (wetting of a surface).

In PART B, we look at five models of POLYMERS IN RANDOM ENVIRONMENT: (1) *charged polymers*, where positive and negative charges are arranged randomly along the chain, resulting in a mixture of repulsive and attractive interactions; (2) *copolymers near a linear selective interface*, where the polymer consists of a random concatenation of two types of monomers that interact differently with two solvents separated by a linear interface; (3) *copolymers near a random selective interface*, where the linear interface is replaced by a percolation-type interface; (4) *random pinning and wetting of polymers*, where the polymer interacts with a linear interface or surface consisting of different types of atoms or molecules arranged randomly; (5) *polymers in a random potential*, where the polymer interacts with different types of atoms or molecules arranged randomly in space.

All the results that are presented come with a complete mathematical proof. Nonetheless, there are a few places where proofs are a bit sketchy and the reader is referred to the literature for further details. Not doing so would have meant lengthening the exposition considerably. Still, even where proofs are tight I have taken care that the reader can always hold on to the main line of the argument.

All chapters can be read *essentially independently*. Each chapter tells a story that is *self-contained*, both in terms of content and of notation. Each chapter ends with a brief description of a number of important *extensions* (added to further enlarge the panorama) and with a number of *challenges* for the future (ranging from "doable in principle" via "very tough indeed" to "almost beyond hope"). An index with key words is added after the references, to help the reader connect the terminology that is used in the different chapters. For the topics covered in Parts A and B, I believe to have caught most of the relevant *mathematical* literature. There is a huge literature in physics and chemistry, of which only a few snapshots are being offered.

The choice I made of what material to cover was not driven by content alone. I also wanted to exhibit a number of key techniques that are currently available in the area and are being developed further. Thus, the reader will encounter the *method of local times* (Chapters 3, 6 and 8), *large deviations* and *variational calculus* (Chapters 3, 6, 9 and 10), the *lace expansion* in combination with the *induction approach* (Chapters 4 and 5), *generating functions* (Chapters 6 and 7), the *method of excursions* (Chapters 7, 9 and 11), the *subadditive ergodic theorem* (Chapters 9, 10 and 11), *partial annealing estimates* (Chapters 9, 10 and 11), *coarse-graining* (Chapter 10), and *martingales* (Chapter 12).

I greatly benefited from reading overview works that address various *mathematical* aspects of polymers, in particular, the monographs by Barber and Ninham [12], Madras and Slade [230], Hughes [175], Vanderzande [300], van der Hofstad [154], Sznitman [288], Janse van Rensburg [188], Slade [280], and Giacomin [116], the review papers by van der Hofstad and König [165], Bolthausen [28], and Soteros and Whittington [283], as well as the PhD theses of Caravenna [51], Pétrélis [263], and Vargas [302]. If the present monograph contributes towards making the area more accessible to a broad mathematical readership, as the above works do, then I will consider my goal reached.

While planning this monograph, I decided not to touch upon combinatorial counting techniques, exact enumeration methods and power series analysis, which provide invaluable insight for models that are too hard to handle analytically. Nor will the reader find a description of knotted polymers, which have many fascinating properties and a broad range of applications, nor of branched polymers, which are related to superprocesses arising as scaling limit. These are rapidly growing subjects, which the reader is invited to explore. For overviews, see Dušek [93], Guttmann [137], Orlandini and Whittington [257], and Guttmann [138]. Similarly, there is no discussion of models of two or more polymers interacting with each other, like in a polymer melt, nor of models dealing with dynamical aspects of polymers, such as reptation in a polymer melt. For overviews, see de Gennes [114], and Doi and Edwards [92]. The literature offers plenty of possibilities for the latter two topics as well, but so far the mathematics is rather thin.

I am grateful to Marek Biskup, Erwin Bolthausen, Andreas Greven, Remco van der Hofstad, Wolfgang König, Nicolas Pétrélis, Gordon Slade, Stu Whittington and Mario Wüthrich for co-authoring the joint papers we wrote on random polymers and for the many interesting and enjoyable discussions we have shared over the years. I am further grateful to Anton Bovier, Matthias Birkner, Thierry Bodineau, Francesco Caravenna, Francis Comets, Giambattista Giacomin, Tony Guttmann, Neil O'Connell, Andrew Rechnitzer, Chris Soteros, Alain-Sol Sznitman, Fabio Toninelli, Ivan Velenik and Lorenzo Zambotti for fruitful exchange on various occasions.

Matthias Birkner, Giambattista Giacomin, Remco van der Hofstad and Gordon Slade read parts of the prefinal draft and offered a number of useful remarks. Stu Whittington commented on three drafts in various stages of development, patiently answered a long list of questions and provided many references. He generously offered his guidance, which has been both stimulating and reassuring. Nicolas Pétrélis helped me to prepare my lectures in Saint-Flour and assisted me afterwards to finish the present monograph. Not only did we spend many hours together discussing the content, Nicolas carefully went through the full text and drew many of the figures. He was an indispensable companion in bringing the whole enterprise to a good end.

It is a pleasure to thank the staff of EURANDOM in Eindhoven for providing so many opportunities to do quiet research in a stimulating environment. It continues to be an honor and a pleasure to be affiliated with the institute, where most of the above colleagues are at home. Over the years, my research has been amply supported by NWO (Netherlands Organization for Scientific Research), which I gratefully acknowledge as well.

Finally, I thank Jean Picard for the invitation to lecture at the Saint-Flour summer school and for the pleasant exchange we have had before, during and after the event.

Mathematical Institute   *Frank den Hollander*
Leiden University
P.O. Box 9512
2300 RA Leiden
The Netherlands
www.math.leidenuniv.nl/~denholla/
December 2008

Contents

1

Introduction

1.1 What is a Polymer?

A *polymer* is a large molecule consisting of *monomers* that are tied together
by chemical bonds. The monomers can be either small units (such as CH_2
in polyethylene; see Fig. 1.1) or larger units with an internal structure (such
as the adenine-thymine and cytosine-guanine base pairs in the DNA double
helix; see Fig. 1.2). Polymers abound in nature because of the *multivalency* of
atoms like carbon, silicon, oxygen, nitrogen, sulfur and phosphorus, which are
capable of forming long concatenated structures.

Polymer Classification. Polymers come in two varieties: (1) HOMOPOLY-
MERS, with all their monomers identical (such as polyethylene); (2) COPOLY-
MERS, with two or more different types of monomer (such as DNA). The order
of the monomer types in copolymers can be either periodic (e.g. in agar) or
random (e.g. in carrageenan).

An important classification of polymers is into SYNTHETIC POLYMERS
(such as nylon, polyethylene and polystyrene) and NATURAL POLYMERS (also
called biopolymers). Major subclasses of the latter are: (a) *proteins* (strings
of amino-acids), the chief constituents of all living objects, carrying out a
multitude of tasks; (b) *nucleic acids* (DNA, RNA), the building blocks of
genes that are the very core of life processes; (c) *polysaccharides* (e.g. agar,
amylose, carrageenan, cellulose), which form part of the structure of animals
and plants and provide an energy source; (d) *lignin* (plant cement), which
fills up the space between cellulose fibres; (e) *rubber*, occurring in the fluid of
latex cells in certain trees and shrubs. Apart from (a)–(e), which are *organic*
materials, clays and minerals are *inorganic* examples of natural polymers.
Synthetic polymers typically are homopolymers, natural polymers typically
are copolymers (with notable exceptions). Bacterial polysaccharides tend to
be periodic, plant polysaccharides tend to be random.

Yet another classification of polymers is into LINEAR and BRANCHED. In
the former, the monomers have one reactive group (such as CH_2), leading

F. den Hollander, *Random Polymers*,
Lecture Notes in Mathematics 1974, DOI: 10.1007/978-3-642-00333-2_1,
© Springer-Verlag Berlin Heidelberg 2009

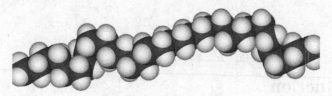

Fig. 1.1. Polyethylene.

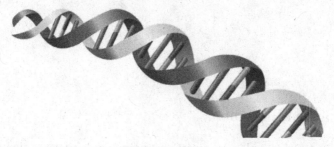

Fig. 1.2. DNA.

to a linear organization as a result of the polymerization process. In the latter, the monomers have two or more reactive groups (such as hydroxy acid), leading to an intricate network with multiple cross connections. Most natural polymers are linear, like DNA, RNA, proteins and the polysaccharides agar, alginate, amylose, carrageenan and cellulose. Some polysaccharides are branched, like amylopectin. Many synthetic polymers are linear, and many are branched. An example of a branched polymer is rubber, both natural and synthetic.

Polymerization. The chemical process of building a polymer from monomers is called *polymerization*. The size of a polymer, i.e., the number of constituent monomers (also called the degree of polymerization) may vary from 10^3 up to 10^{10} (smaller molecules do not qualify to be called polymers, larger molecules have not been recorded). Human DNA has $10^9 - 10^{10}$ base pairs, lignin consists of $10^6 - 10^7$ phenyl-propanes, while polysaccharides carry $10^3 - 10^4$ sugar units. Both in synthetic and in natural polymers, the size distribution may either be broad, with numbers varying significantly from polymer to polymer (e.g. in nylons and in polysaccharides) or be narrow (e.g. in proteins and in DNA). In synthetic polymers the size distribution can be made narrow through specific polymerization methods. The size of the monomer units varies from $1.5\,\text{Å}$ (for CH_2 in polyethylene) to $20\,\text{Å}$ (for the base pairs in DNA), with $1\,\text{Å} = 10^{-10}\,\text{m}$. For more background on the structure of polymers, the reader is referred to Green and Milne [129].

The chemical bonds in a polymer are flexible, so that the polymer can arrange itself in many different *spatial configurations*. The longer the chain, the more involved these configurations tend to be. For instance, the polymer can wind around itself to form knots, can be extended due to repulsive forces between the monomers as a result of excluded-volume, or can collapse to a ball due to attractive van der Waals forces between the monomers or repulsive forces between the monomers and a poor solvent. It can also interact with a surface on which it may or may not be adsorbed, or it can live in a slit between two confining surfaces.

As mentioned above, the polymer can be either homogeneous (homopolymer) or inhomogeneous (copolymer). A typical example of the latter is a copolymer whose monomers carry positive and negative charges, randomly arranged along the chain. Another example is a copolymer consisting of hydrophobic and hydrophilic monomers. If such a copolymer is placed near an interface separating oil and water, then it may try to wiggle around the interface in order to match the monomers as much as possible, and in doing so stay closely tied to the interface.

Targets. In the present monograph we will consider various different models of *linear* polymers, aimed at describing a variety of different physical settings, of the type alluded to above. Key quantities of interest will be the number of different spatial configurations, the typical end-to-end distance (subdiffusive, diffusive or superdiffusive), the space-time scaling limit, the fraction of monomers at an interface or adsorbed onto a surface, the average length and height of excursions away from an interface or a surface, all typically in the limit as the polymer gets long. We will pay special attention to the free energy of the polymer in this limit, and to the presence of *phase transitions* as a function of underlying model parameters, signalling drastic changes in behavior when these parameters cross critical values. We will also study the effect of randomness in the interactions.

Classical monographs on polymers with a physical and chemical orientation are Flory [102] and de Gennes [114]. (It is worthwhile to read their Nobel lectures: Flory [103] and de Gennes [115].) Monographs with a mathematical orientation are Barber and Ninham [12], Madras and Slade [230], Hughes [175], Vanderzande [300], Janse van Rensburg [188] and Giacomin [116].

1.2 What is the Model Setting?

Our polymers will live on the d-dimensional Euclidean lattice $\mathbb{Z}^d$, $d \geq 1$. They will be modelled as *random paths* on this lattice, where the monomers are the vertices in the path, and the chemical bonds connecting the monomers are the edges in the path.

The choice of model will depend on two key objects. Namely, for each $n \in \mathbb{N}_0 = \mathbb{N} \cup \{0\}$ we need to specify

$$\begin{aligned}
&\mathcal{W}_n = \text{a collection of allowed } n\text{-step } paths \text{ on } \mathbb{Z}^d, \\
&H_n = \text{a Hamiltonian that associates an } energy \text{ to each path in } \mathcal{W}_n.
\end{aligned} \tag{1.1}$$

As we will see, there is flexibility in the choice of $\mathcal{W}_n$, depending on the particular application we have in mind. We will be considering both undirected and directed paths (see Figs. 1.3 and 1.4). The choice of H_n will be driven by the underlying physics, and captures the interaction of the polymer with itself and/or its environment. Typically, H_n depends on one or two parameters, including temperature.

For each $n \in \mathbb{N}_0$, the *random polymer* of length n is defined by assigning to each $w \in \mathcal{W}_n$ a probability given by

$$P_n(w) = \frac{1}{Z_n} e^{-H_n(w)}, \qquad w \in \mathcal{W}_n, \tag{1.2}$$

where Z_n is the normalizing partition sum. This is called the *Gibbs measure* associated with the pair $(\mathcal{W}_n, H_n)$, and it describes the polymer in *equilibrium* with itself and/or its environment (at fixed temperature and fixed polymer

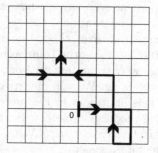

Fig. 1.3. A 19-step path on $\mathbb{Z}^2$, modeling a polymer in the plane with 20 monomers tied together by 19 chemical bonds. The steps in the path are: 3 east, 2 south, 1 west, 4 north, 5 west, 2 east and 2 north.

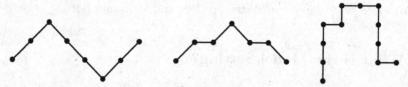

Fig. 1.4. Three examples of directed paths on $\mathbb{Z}^2$, in combinatorics commonly referred to as ballot paths (with allowed steps ↗ or ↘), generalized ballot paths (with allowed steps ↗, → and ↘), and partially directed self-avoiding walks (with allowed steps ↑, ↓ or →, subject to no self-intersections). Ballot paths and generalized ballot paths that begin and end at the same horizontal line and lie entirely on or above this line are called Dyck paths, repectively, Motzkin paths. Without the one-sidedness restriction they are called bilateral Dyck paths, respectively, bilateral Motzkin paths.

length). Under this Gibbs measure, paths with a low energy have a high probability, while paths with a high energy have a low probability.[1]

Note that $(P_n)_{n\in\mathbb{N}}$ in general is *not* a consistent family of probability distributions, i.e., P_n is not the projection of P_{n+1} obtained by summing out the position of the $(n+1)$-st monomer. Rather, for each n we have a different distribution, modeling a polymer chain of a fixed length.

The polymer measure in (1.2) will be our main focus in PART A. In PART B we will consider models in which the Hamiltonian also depends on a *random environment* (e.g. randomly ordered charges or monomer types). We will generically denote this random environment by ω and its probability distribution by $\mathbb{P}$. We will write H_n^ω to exhibit the ω-dependence of the Hamiltonian. Three types of Gibbs measures will make their appearance:

(1) The *quenched* Gibbs measure

$$P_n^\omega(w) = \frac{1}{Z_n^\omega}\, e^{-H_n^\omega(w)}, \qquad w \in \mathcal{W}_n. \tag{1.3}$$

(2) The *average quenched* Gibbs measure

$$\mathbb{E}(P_n^\omega(w)) = \int P_n^\omega(w)\,\mathbb{P}(d\omega), \qquad w \in \mathcal{W}_n. \tag{1.4}$$

(3) The *annealed* Gibbs measure

$$\mathbb{P}_n(w) = \frac{1}{\mathbb{Z}_n} \int e^{-H_n^\omega(w)}\,\mathbb{P}(d\omega), \qquad w \in \mathcal{W}_n. \tag{1.5}$$

The latter is used to model a polymer whose random environment is not frozen but takes part in the equilibration. We will mostly be interested in the behavior of the polymer in the limit as $n \to \infty$, typically after some appropriate scaling.[2]

We will not (!) consider models where the length or the configuration of the polymer changes with time (e.g. due to growing or shrinking, or to a Metropolis dynamics associated with the Hamiltonian for an appropriate choice of allowed transitions). These *non-equilibrium* situations are very interesting and challenging indeed, but so far the available mathematics is very thin.

---

[1] W. Kuhn, in the 1930's, seems to have been the first to put forward the "ensemble description" of polymers given by (1.2) (see Flory [100], Chapter I). The Gibbs measure in (1.2) maximizes the *entropy* $\sum_{w\in\mathcal{W}_n}[-P_n(w)\log P_n(w)]$ subject to the constraint of constant *energy* $\sum_{w\in\mathcal{W}_n} P_n(w)H_n(w)$, as required by equilibrium statistical physics.

[2] The reader must carefully distinguish between the upper index ω, labeling the random environment, and the argument w, labeling the path, even though these symbols look very much alike. In Part B they will appear side by side in many formulas.

1.3 The Central Role of Free Energy

The *free energy* of the polymer is defined as

$$f = \lim_{n \to \infty} \frac{1}{n} \log Z_n, \tag{1.6}$$

or, in the presence of a random environment,

$$f = \lim_{n \to \infty} \frac{1}{n} \log Z_n^{\omega} \qquad \omega - a.s. \tag{1.7}$$

The limit in (1.7) typically is constant ω-a.s. (i.e., constant on a set of ω's with $\mathbb{P}$-measure 1), a property referred to as "self-averaging w.r.t. ω".

Existence. It can frequently be shown that the partition sum Z_n satisfies the inequality

$$Z_n \leq Z_m \, Z_{n-m} \qquad \forall 0 \leq m \leq n. \tag{1.8}$$

For instance, this typically occurs when H_n assigns a repulsive self-interaction to the polymer, in which case the inequality follows by viewing the n-polymer as a concatenation of an m-polymer and an $(n-m)$-polymer. We will see examples in Part A. It follows from (1.8) that $n \mapsto \log Z_n$ is a *subadditive sequence*, i.e.,

$$\log Z_n \leq \log Z_m + \log Z_{n-m} \qquad \forall 0 \leq m \leq n, \tag{1.9}$$

and, with $f_n = \frac{1}{n} \log Z_n$, the latter implies that

$$f = \lim_{n \to \infty} f_n = \inf_{n \in \mathbb{N}} f_n \quad \text{exists in } [-\infty, \infty) \tag{1.10}$$

(for a proof see e.g. Madras and Slade [230], Lemma 1.2.2). If, moreover, $\inf_{w \in \mathcal{W}_n} H_n \leq Cn$ for some $C < \infty$, then $-\infty$ can be excluded as limiting value. The same is true when H_n assigns an attractive self-interaction to the polymer, in which case the inequalities in (1.8–1.9) are reversed, inf in (1.10) is replaced by sup, and ∞ is excluded as limiting value as soon as $|\mathcal{W}_n| \leq e^{Cn}$ and $\inf_{w \in \mathcal{W}_n} H_n \geq -Cn$ for some $C < \infty$. When H_n assigns both repulsive and attractive interactions to the polymer (as e.g. in a self-avoiding walk with self-attraction), then the above argument is generally not available, and the existence of the free energy either remains open or has to be established by other means.

In the presence of a random environment ω, the subadditivity property in (1.9) must be replaced by a random form of subadditivity. When applicable, $n \mapsto \log Z_n^{\omega}$ is called a *subadditive random process* (rather than a subadditive sequence), and Kingman's subadditive ergodic theorem [214] implies the ω-a.s. existence of the free energy in (1.7). We will see examples in Part B.

Convexity. Suppose that the Hamiltonian depends linearly on a single parameter $\beta \in \mathbb{R}$, and write $\beta H_n(w)$, $Z_n(\beta)$, P_n^β, $f_n(\beta)$ and $f(\beta)$ to exhibit this dependence. Then $\beta \mapsto f_n(\beta)$ is convex. Indeed, using Hölder's inequality, we have

$$
\begin{aligned}
Z_n\left(\tfrac{1}{2}\beta_1 + \tfrac{1}{2}\beta_2\right) &= E_n\left(e^{\left(\frac{1}{2}\beta_1 + \frac{1}{2}\beta_2\right)H_n}\right) \\
&\leq E_n\left(e^{\beta_1 H_n}\right)^{\frac{1}{2}} E_n\left(e^{\beta_2 H_n}\right)^{\frac{1}{2}} = Z_n(\beta_1)^{\frac{1}{2}} Z_n(\beta_2)^{\frac{1}{2}},
\end{aligned}
\tag{1.11}
$$

which yields

$$
f_n(\tfrac{1}{2}(\beta_1 + \beta_2)) \leq \tfrac{1}{2}f_n(\beta_1) + \tfrac{1}{2}f_n(\beta_2),
\tag{1.12}
$$

i.e., $\beta \mapsto f_n(\beta)$ is convex. Passing to the limit $n \to \infty$, we get that also $\beta \mapsto f(\beta)$ is convex.

Differentiability. Convexity implies continuity and, when combined with a sign, also monotonicity. Moreover, at those values of β where $f(\beta)$ is differentiable, convexity implies that

$$
f'(\beta) = \lim_{n \to \infty} f_n'(\beta).
\tag{1.13}
$$

The latter observation is important, because it tells us that

$$
\begin{aligned}
f'(\beta) &= \lim_{n \to \infty} \frac{1}{n} \log Z_n'(\beta) \\
&= \lim_{n \to \infty} \frac{1}{n} \sum_{w \in \mathcal{W}_n} [-H_n(w)] P_n^\beta(w).
\end{aligned}
\tag{1.14}
$$

What this says is that $-\beta f'(\beta)$ is the limiting energy per monomer under the Gibbs measure. Thus, at those values of β where the free energy fails to be differentiable this quantity is discontinuous, signalling the occurrence of a *phase transition*, i.e., a drastic change in the behavior of the typical path.

A similar argument applies when other parameters are in play. Derivatives of the free energy w.r.t. these parameters indicate how quantities related to these parameters change at a phase transition. We will see plenty of examples as we go along.

2

Two Basic Models

In this chapter we consider two basic models for a polymer chain: *simple random walk* (SRW), describing a polymer with no self-interaction, and *self-avoiding walk* (SAW), describing a polymer with a self-interaction given by the "excluded-volume effect", i.e., no site can be occupied by more than one monomer. Our goal is to give a *quick summary* of what is known for these models in order to set the stage for the physically more realistic models treated in Parts A and B. We will focus on those results that are relevant to later chapters and *omit proofs*, for which we refer the reader to the literature. A standard reference to SRW is the monograph by Spitzer [284]. A standard reference to SAW is the monograph by Madras and Slade [230]. Figs. 2.1 and 2.2 below are borrowed from Bill Casselman and Gordon Slade.

2.1 Simple Random Walk

Simple random walk (SRW) on $\mathbb{Z}^d$ is the random process $(S_n)_{n \in \mathbb{N}_0}$ defined by

$$S_0 = 0, \qquad S_n = \sum_{i=1}^{n} X_i, \quad n \in \mathbb{N}, \tag{2.1}$$

where

$$X = (X_i)_{i \in \mathbb{N}} \tag{2.2}$$

is an i.i.d. sequence of random variables taking values in $\mathbb{Z}^d$ with marginal law

$$P(X_1 = x) = \begin{cases} \frac{1}{2d}, & x \in \mathbb{Z}^d \text{ with } \|x\| = 1, \\ 0, & \text{otherwise,} \end{cases} \tag{2.3}$$

with $\| \cdot \|$ the Euclidean norm. If we think of the path of this process as modeling a polymer chain, then X_i (the step at time i) corresponds to the orientation of the chemical bond between the $(i-1)$-st and i-th monomer, while S_n (the position at time n) corresponds to the location of the n-th

F. den Hollander, *Random Polymers*,
Lecture Notes in Mathematics 1974, DOI: 10.1007/978-3-642-00333-2_2,
© Springer-Verlag Berlin Heidelberg 2009

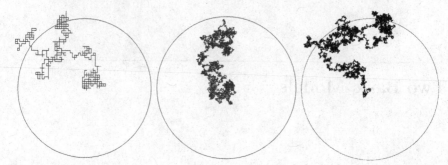

Fig. 2.1. Simulation of a SRW on $\mathbb{Z}^2$ taking $n = 10^3$, 10^4 and 10^5 steps. The circles have radius $\sqrt{n}$, in units of the step size.

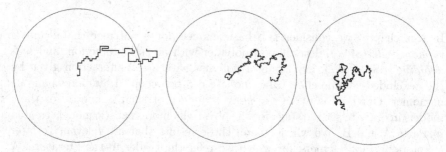

Fig. 2.2. Simulation of a SAW on $\mathbb{Z}^2$ with $n = 10^2$, 10^3 and 10^4 steps. The circles have radius $n^{3/4}$, in units of the step size.

monomer, the end-point of the polymer. We write P, E to denote probability and expectation w.r.t. SRW.

The SRW defined in (2.1–2.3) corresponds to choosing the set of paths and the Hamiltonian in (1.1) as

$$\mathcal{W}_n = \left\{ w = (w_i)_{i=0}^n \in (\mathbb{Z}^d)^{n+1} : w_0 = 0, \|w_{i+1} - w_i\| = 1 \; \forall 0 \le i < n \right\},$$
$$H_n \equiv 0,$$

$$(2.4)$$

and the path measure P_n in (1.2) as the uniform distribution on $\mathcal{W}_n$. This describes a *random polymer without interaction*, with $(S_i)_{i=0}^n$ taking values in $\mathcal{W}_n$. Clearly, $|\mathcal{W}_n| = (2d)^n$, and P_n is the projection of P onto $\mathcal{W}_n$. Consequently, $(P_n)_{n \in \mathbb{N}_0}$ is a consistent family. Below we collect a few key properties of SRW.

1. Transition probabilities. Let

$$\widehat{p}(k) = E\left(e^{i(k \cdot X_1)} \right) = \frac{1}{2d} \sum_{\substack{x \in \mathbb{Z}^d \\ \|x\|=1}} e^{i(k \cdot x)} = \frac{1}{d} \sum_{j=1}^d \cos(k_j),$$

$$k = (k_1, \ldots, k_d) \in [-\pi, \pi]^d,$$

$$(2.5)$$

denote the characteristic function of X_1, with $(\cdot, \cdot)$ the inner product on $\mathbb{R}^d$. Then

$$\widehat{p}_n(k) = E\left(e^{i(k \cdot S_n)}\right) = [\widehat{p}(k)]^n, \qquad k \in [-\pi, \pi)^d, \, n \in \mathbb{N}_0. \qquad (2.6)$$

Inversion gives

$$P(S_n = x) = \frac{1}{(2\pi)^d} \int_{[-\pi,\pi)^d} e^{-i(k \cdot x)} \, [\widehat{p}(k)]^n \, dk, \qquad x \in \mathbb{Z}^d, \, n \in \mathbb{N}_0, \qquad (2.7)$$

which is the Fourier representation of the two-point function. The $(r + 1)$-point functions follow from the two-point function via the Markov property (in accordance with (2.6)):

$$P\left(S_{n_1} = x_1, \ldots, S_{n_r} = x_r\right) = \prod_{s=1}^{r} P\left(S_{n_s - n_{s-1}} = x_s - x_{s-1}\right) \qquad (2.8)$$

(with $n_0 = x_0 = 0$). Note that SRW exhibits *diffusive behavior*:

$$E(S_n) = 0 \text{ and } E(\|S_n\|^2) = n \text{ for all } n \in \mathbb{N}_0. \qquad (2.9)$$

2. Local limit theorem. Using that $\widehat{p}(k) = 1$ if and only if $k = 0$ and $\widehat{p}(k) = 1 - \frac{1}{2d}\|k\|^2 + o(\|k\|^2)$ as $k \to 0$, we deduce from (2.7) that

$$P(S_{2n} = 0) = \frac{1}{(2\pi)^d} \int_{[-\pi,\pi)^d} [\widehat{p}(k)]^{2n} \, dk$$

$$\sim \frac{1}{(2\pi)^d} \int_{\mathbb{R}^d} e^{-\frac{1}{d}\|k\|^2 n} \, dk \qquad (2.10)$$

$$= 2\left(\frac{d}{4\pi n}\right)^{\frac{d}{2}} \qquad \text{as } n \to \infty.$$

A similar calculation gives

$$P(S_n = x) \sim 2\left(\frac{d}{2\pi n}\right)^{\frac{d}{2}} e^{-\frac{d}{2n}\|x\|^2} \qquad \text{as } n \to \infty \text{ and } \|x\|/n \to 0, \qquad (2.11)$$

provided x and n have the same parity.

3. Green function. Let

$$G(x) = \sum_{n \in \mathbb{N}_0} P(S_n = x), \qquad x \in \mathbb{Z}^d, \qquad (2.12)$$

denote the average number of visits to x by SRW. Then $G(x) = \infty$ for all $x \in \mathbb{Z}^d$ when $d = 1, 2$ (recurrence) and $G(x) < \infty$ for all $x \in \mathbb{Z}^d$ when $d \geq 3$ (transience). Inserting (2.7) into (2.12), we get

$$G(x) = \frac{1}{(2\pi)^d} \int_{[-\pi,\pi)^d} e^{-i(k \cdot x)} [1 - \widehat{p}(k)]^{-1} \, dk, \qquad x \in \mathbb{Z}^d. \qquad (2.13)$$

For $\|x\|$ large, the integral is dominated by the small k-values and so, similarly as in (2.10), a little calculation gives

$$G(x) \sim \frac{1}{(2\pi)^d} \int_{\mathbb{R}^d} e^{-i(k \cdot x)} \, 2d\|k\|^{-2} \, dk$$

$$= C_d \, \|x\|^{-d+2} \qquad \text{as } \|x\| \to \infty \text{ when } d \geq 3, \qquad (2.14)$$

where $C_d = \frac{d}{2}\Gamma(\frac{d}{2} - 1)\pi^{-d/2} = 2/(d-2)\omega_d$ with Γ the Gamma-function and ω_d the volume of the unit ball in $\mathbb{R}^d$.

4. Return probability. The probability that SRW returns to the origin, written F_d, equals $F_d = 1 - [G(0)]^{-1}$ (Spitzer [284], Section 1). For $d = 1, 2$, $F_d = 1$ (recurrence), while for $d \geq 3$, $0 < F_d < 1$ (transience), e.g. $F_3 \approx 0.340$, $F_4 \approx 0.193$ and $F_5 \approx 0.135$ (see Griffin [131]).

5. Functional central limit theorem. The scaling limit of SRW is *Brownian motion* on $\mathbb{R}^d$ (see Fig. 2.1), i.e., under the law P_n of SRW,

$$\left(\frac{1}{\sqrt{n}} S_{\lfloor nt \rfloor} \right)_{0 \leq t \leq 1} \Longrightarrow (B_t)_{0 \leq t \leq 1} \qquad \text{as } n \to \infty, \qquad (2.15)$$

with $\Longrightarrow$ denoting convergence in distribution on the space of càdlàg paths endowed with the Skorohod topology (Billingsley [14], Section 14). In Fourier language, the central limit theorem, i.e., (2.15) at $t = 1$, reads

$$\widehat{p}_n \left(\frac{k}{\sqrt{n}} \right) = e^{-\frac{1}{2d}\|k\|^2 \, [1+o(1)]} \qquad \text{as } n \to \infty \text{ and } \|k\|^2/n \to 0. \qquad (2.16)$$

2.2 Self-avoiding Walk

Self-avoiding walk (SAW) on $\mathbb{Z}^d$ corresponds to choosing the set of paths and the Hamiltonian in (1.1) as

$$\mathcal{W}_n = \{ w = (w_i)_{i=0}^n \in (\mathbb{Z}^d)^{n+1} \colon w_0 = 0, \, \|w_{i+1} - w_i\| = 1 \, \forall 0 \leq i < n,$$
$$w_i \neq w_j \, \forall 0 \leq i < j \leq n \},$$
$$H_n \equiv 0,$$

$$(2.17)$$

and the path measure P_n in (1.2) as the uniform distribution on $\mathcal{W}_n$. This describes a *random polymer with self-exclusion*. Because of the self-avoidance constraint, $(P_n)_{n \in \mathbb{N}_0}$ is not (!) a consistent family.

Let $c_n = |\mathcal{W}_n|$. For $d = 1$ we trivially have $c_n = 2$ for all $n \in \mathbb{N}$. For $d \geq 2$ no closed form expression is known for c_n, but for small and moderate n it can be computed by exact enumeration methods (see Guttmann [137]). The current record is $n = 71$ for $d = 2$ (Jensen [200], [201]), $n = 30$ for $d = 3$ and $n = 24$ for $d \geq 4$ (Clisby, Liang and Slade [68]). Larger n can be handled either via numerical simulation (presently up to $n = 2^{25} \approx 3.3 \times 10^7$ in $d = 3$) or with the help of extrapolation techniques. Some rigorous information on the asymptotic behavior of c_n as $n \to \infty$ is available. We collect the main results below.

1. Concatenation and subadditivity. Pick $m, n \in \mathbb{N}$. Any two SAWs of length m and n can be concatenated. This may or may not produce a SAW of length $m+n$, depending on whether the concatenation is or is not self-avoiding (see Fig. 2.3). Consequently,

$$c_{m+n} \leq c_m c_n, \qquad m, n \in \mathbb{N}, \tag{2.18}$$

i.e., $n \mapsto \log c_n$ is subadditive. From this in turn it follows that (see Madras and Slade [230], Lemma 1.2.2)

$$\lim_{n \to \infty} (c_n)^{1/n} = \inf_{n \in \mathbb{N}} (c_n)^{1/n} = \mu \in [1, \infty) \qquad \text{exists.} \tag{2.19}$$

The logarithm of the limit, $\log \mu$, is referred to as the *connective constant*. For SRW we have $\mu = 2d$.

Since $d^n \leq c_n \leq 2d(2d-1)^{n-1}$, it follows that $d \leq \mu \leq 2d-1$. The precise value of $\mu = \mu(d)$ is not known. Approximate values are $\mu(2) \approx 2.638$ and $\mu(3) \approx 4.684$, with many more decimal places known: see Jensen [200], [201], and Clisby, Liang and Slade [68]. In the latter paper, an asymptotic expansion of $\mu(d)$ in powers of $1/2d$ is derived containing 13 terms. The value of μ is *not* universal, i.e., it changes when we take a different lattice in the same dimension (e.g. the triangular lattice in $d = 2$ or the body-centered cubic lattice in $d = 3$).

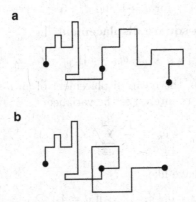

Fig. 2.3. Concatenation of two self-avoiding walks. The first concatenation is self-avoiding, the second is not.

2. Scaling for c_n. Exact enumeration and numerical simulation suggest that

$$c_n = \begin{cases} A\,\mu^n\,n^{\gamma-1}\,[1+o(1)], & \text{if } d \neq 4, \\ A\,\mu^n\,(\log n)^{\frac{1}{4}}\,[1+o(1)], & \text{if } d = 4, \end{cases} \qquad \text{as } n \to \infty, \qquad (2.20)$$

with A a non-universal amplitude and γ a universal *critical exponent*. The values of γ are

$$\gamma = 1 \ (d=1), \quad \tfrac{43}{32} \ (d=2), \quad 1.157\ldots \ (d=3), \quad 1 \ (d \geq 5). \qquad (2.21)$$

For $d = 1$, the scaling behavior in (2.20) is trivial: $A = 2$, $\mu = 1$, $\gamma = 1$. For $d \geq 5$, a proof based on the *lace expansion* was given by Hara and Slade [147], [148]; this proof is *computer-assisted*, and provides error bounds replacing $o(1)$. For $d = 2$ it is known that if (!) SAW has a conformally invariant scaling limit, then (!) this scaling limit is $\text{SLE}_{8/3}$, the *Schramm-Loewner Evolution* with parameter 8/3. The value $\gamma = 43/32$, which was first predicted by Nienhuis [248] based on non-rigorous arguments, would be implied by this scaling. Theoretical support for scaling to $\text{SLE}_{8/3}$ is provided in Lawler, Schramm and Werner [227], numerical support in Kennedy [209] (see Fig. 2.2 for a simulation). An overview on SLE is given in the Saint-Flour lectures by Werner [309]. For $d = 4$ work in progress by Brydges and Slade (private communication) based on *renormalization group theory* gives hope for a proof of the logarithmic correction. The case $d = 3$ remains open, but there are accurate estimates in Guida and Zinn-Justin [134] and in Clisby, Liang and Slade [68].

By noting that (2.20) implies

$$c_{2n}/(c_n)^2 \sim \begin{cases} A^{-1}(2/n)^{\gamma-1}, & \text{if } d \neq 4, \\ A^{-1}(\log n)^{-\frac{1}{4}}, & \text{if } d = 4, \end{cases} \qquad \text{as } n \to \infty, \qquad (2.22)$$

we see that γ has an interpretation as an *intersection* exponent, describing the rate of decay of the probability that two n-step SAWs starting at 0 have no site in common (apart from 0 itself). Note that this probability tends to 0 for $2 \leq d \leq 4$ and to A^{-1} for $d = 1$ and $d \geq 5$.

3. Scaling for mean-square displacement. Let

$$c_n(x) = |\{w \in \mathcal{W}_n \colon w_n = x\}|, \qquad x \in \mathbb{Z}^d, \ n \in \mathbb{N}_0. \qquad (2.23)$$

Then $\frac{1}{c_n}\sum_{x \in \mathbb{Z}^d} x c_n(x)$, the mean displacement of an n-step SAW, is zero by symmetry. A quantity of interest is the variance

$$v_n = \frac{1}{c_n}\sum_{x \in \mathbb{Z}^d} \|x\|^2 c_n(x). \qquad (2.24)$$

This is believed to behave like

$$v_n = \begin{cases} D\,n^{2\nu}\,[1+o(1)], & \text{if } d \neq 4, \\ D\,n(\log n)^{\frac{1}{4}}\,[1+o(1)], & \text{if } d = 4, \end{cases} \qquad \text{as } n \to \infty, \qquad (2.25)$$

with D a non-universal diffusion constant and ν a universal *critical exponent*. (Compare (2.25) with (2.9).) The values of ν are

$$\nu = 1 \ (d = 1), \quad \tfrac{3}{4} \ (d = 2), \quad 0.588\ldots \ (d = 3), \quad \tfrac{1}{2} \ (d \geq 5). \qquad (2.26)$$

As with (2.20), for $d = 1$ the scaling behavior in (2.25) is trivial, while for $d \geq 5$ a proof was given by Hara and Slade [147], [148] (see also Slade [277], [278], [279]). These two cases correspond to *ballistic*, respectively, *diffusive* behavior. The situation for $d = 2, 3, 4$ is the same as for (2.20). For a random walk on a hierarchical lattice that is effectively four-dimensional, the scaling in the second line of (2.25) was proved by Brydges and Imbrie [41] with the help of renormalization group theory. There is promise for these techniques to carry over to $\mathbb{Z}^4$.

4. Functional central limit theorem. It was proved in Hara and Slade [147], [148] (see also Slade [277], [278], [279]) that, for $d \geq 5$ under the law of P_n of the n-step SAW,

$$\left(\frac{1}{D\sqrt{n}} S_{\lfloor nt \rfloor} \right)_{0 \leq t \leq 1} \Longrightarrow (B_t)_{0 \leq t \leq 1} \qquad \text{as } n \to \infty, \qquad (2.27)$$

which is the analogue of (2.15). Let, as in (2.6),

$$\hat{c}_n(k) = \frac{1}{c_n} \sum_{w \in \mathcal{W}_n} e^{i(k \cdot w_n)}, \qquad k \in [-\pi, \pi]^d, \ n \in \mathbb{N}_0. \qquad (2.28)$$

Then (2.27) at $t = 1$ reads

$$\hat{c}_n \left(\frac{k}{D\sqrt{n}} \right) = e^{-\frac{1}{2d} \|k\|^2 \, [1 + o(1)]} \qquad \text{as } n \to \infty \text{ and } \|k\|^2 \leq C < \infty, \qquad (2.29)$$

which is the analogue of (2.16).

5. Green function. It was proved in Hara, van der Hofstad and Slade [146], and Hara [145] that there exists a $C > 0$ such that

$$G_\mu(x) = \sum_{n \in \mathbb{N}_0} c_n(x) \, \mu^{-n} \sim C \|x\|^{-d+2} \qquad \text{as } \|x\| \to \infty \text{ for } d \geq 5, \qquad (2.30)$$

which is the analogue of (2.14). Thus, the diffusive scaling manifests itself also at the level of the Green function. For SRW we have $\mu = 2d$ and $c_n(x)\mu^{-n} = P(S_n = x)$ (compare with (2.12)).

6. Flory argument. The following heuristic argument in support of (2.26) is due to Flory [101]. Suppose that SAW of length n has radius L. Then its density is $\asymp n/L^d$ monomers per site (where $\asymp$ means that the ratio of the two sides is bounded above and below by strictly positive and finite constants). Therefore its "repulsive energy" per site is $\asymp (n/L^d)^2$, and so its total repulsive energy is $\asymp n^2/L^d$. Now, the probability that SRW has radius L is roughly $\exp[-L^2/n]$. Hence, the probability that SAW has radius L is roughly $\exp[-\{n^2/L^d + L^2/n\}]$. Put $L = n^\nu$. Then the term between braces equals $n^{2-d\nu} + n^{2\nu-1}$, which is minimal when $2 - d\nu = 2\nu - 1$, or

$$\nu = \frac{3}{d+2}. \tag{2.31}$$

For this choice of ν the term between braces takes the value $n^{(4-d)/(d+2)}$, and so we need $1 \leq d \leq 4$ for the total repulsive energy not to vanish. For $d \geq 5$, the total repulsive energy vanishes and we are in the diffusive regime with $\nu = \frac{1}{2}$. (Madras and Slade [230], Section 2.2, gives a more refined version of the above argument.)

The value in (2.31) for the critical exponent in (2.25) when $1 \leq d \leq 4$ fits with (2.26), except for $d = 3$, where it is off by 2 percent. Thus, Flory's simple heuristics is remarkably accurate. Still, nobody has yet succeeded in turning it into a rigorous argument. For $d = 2, 3, 4$ it has not even been proved that v_n grows faster than $n^{1/2}$ and slower than n.

The fact that SAW has the same scaling behavior as SRW for $d \geq 5$ is expressed by saying that "SAW and SRW are in the same universality class". Correspondingly, $d = 4$ is called the *upper critical dimension*. In physical jargon, for $d \geq 5$ SAW has "mean-field behavior", meaning that its critical exponents cease to depend on dimension and are the same as for SRW (which has no interaction). The intuitive reason for the crossover is that in low dimension long loops of SRW are dominant, causing the excluded-volume effect in SAW to be long-ranged, whereas in high dimension short loops are dominant, causing it to be short-ranged. The behavior in (2.24–2.26) is called *ballistic* in $d = 1$, *subballistic* and *superdiffusive* in $d = 2, 3, 4$, and *diffusive* in $d \geq 5$.

Over the years, SAW has been a breeding ground for the development of methods that have been successfully applied to a variety of polymer models. Key examples are *concatenation arguments* (Hammersley [141]), *folding arguments* (Hammersley and Welsh [143], Hammersley and Whittington [144]), *pattern theorems* (Kesten [211]), and the *lace expansion* (Brydges and Spencer [44]). Also, SAW has led to the development of the so-called *pivot algorithm* to simulate self-avoiding lattice paths (Madras and Sokal [231]) and to advanced *exact numerations methods* (Guttman [137]), both of which have had considerable spin-off in other areas of probability theory and statistical physics as well. Guttmann [138] contains an extensive overview of rigorous, approximate and numerical results for lattice polygons, i.e., self-avoiding loops, including results obtained by exact enumeration methods.

Polymers with Self-interaction

OUTLINE:

In Part A we look at a four variations on the two "plain vanilla" models described in Chapter 2.

- In Chapters 3 and 4, we consider a polymer for which self-intersections are not forbidden (as in SAW) but are penalized, resulting in a repulsive interaction modeling the effect of "steric hindrance". This model, which we call *soft polymer*, may be viewed as an "interpolation" between SRW and SAW. Our main result will be that the soft polymer is ballistic in $d = 1$ and diffusive in $d \geq 5$, like SAW. We will obtain the full scaling behavior.

- In Chapter 5, we consider a polymer for which self-intersections are penalized in a way that depends on their distance along the chain, in such a way that long loops are less penalized than short loops. This model, which we call *elastic polymer*, provides a crude way of incorporating the effect of "stiffness" of a polymer (i.e., its resistance against making sharp bends). We will show that the smaller the stiffness, the lower the critical dimension for diffusive behavior.

- In Chapter 6, we add a reward for self-touchings (pairs of monomers occupying neighboring sites), which results in a mix of repulsive and attractive interactions. This is a model for a polymer in a poor solvent. If the attraction is strong enough to compete with the repulsion, then the polymer *collapses* from a "random coil" to a "compact ball". We will show that there are three phases – extended, collapsed and localized – with different scaling properties.

- In Chapter 7, finally, we consider a polymer that interacts with a linear substrate: monomers at the substrate receive a reward. If the attraction is strong enough, then the polymer *adsorbs* onto the substrate. We consider both the case where the substrate is penetrable ("pinning by an interface") and where it is impenetrable ("wetting of a surface"). The latter is a model for paint on a wall: at low temperature the paint sticks to the wall, at high temperature it does not. We also look at what happens when a *force* is applied to the right endpoint of an adsorbed polymer, pulling it away from and off the substrate with the help of optical tweezers. In addition, we look at a polymer in a slit between two impenetrable substrates and compute the effective force the polymer exerts on the substrates, which can be either attractive or repulsive. This is a model for colloidal dispersions of particles with polymers attached to them.

What is challenging about the models to be described below is that the polymers appear as random objects with a *long-range interaction*: monomers that are far apart along the chain can meet and can interact with each other. This places polymers in a league of their own, with specific questions driven by specific applications. As such, polymers are different from more classical probabilistic objects like Markov chains, percolation, the contact process or interacting diffusions. Consequently, their study requires the development of proper ideas and techniques.

3

Soft Polymers in Low Dimension

In Chapters 3 and 4 we consider a variation of the SAW in which self-intersections are not forbidden but are penalized. We refer to this as the *soft polymer*. In Chapter 3 will show that the soft polymer has *ballistic* behavior in $d = 1$. The proof uses a Markovian representation of the local times of one-dimensional SRW (a powerful technique that is useful also for other models), in combination with large deviation theory, variational calculus and spectral calculus. In Chapter 4 we will show that the soft polymer has *diffusive* behavior in $d \geq 5$. The proof there uses a combinatorial expansion technique called the lace expansion, and is based on the idea that in high dimension SAW can be viewed as a "perturbation" of SRW.

The above scaling says that in $d = 1$ and $d \geq 5$ the soft polymer is in the *same universality class* as SAW. This is expected to be true also for $2 \leq d \leq 4$, but a proof is missing.

In Section 3.1 we define the model, in Section 3.2 we state the main result, a large deviation principle (LDP) for the location of the right endpoint. In Section 3.3 we outline a five-step program to prove the LDP for bridge polymers, i.e., polymers confined between their endpoints. This program is carried out in Section 3.4. In Section 3.5 we remove the bridge condition and prove the full LDP. It will turn out that the rate function has an interesting *critical value strictly below the typical speed*. The main technique that is used is the *method of local times*.

3.1 A Polymer with Self-repellence

The soft polymer on $\mathbb{Z}^d$ treated in Chapters 3 and 4 is defined by choosing the set of paths and the Hamiltonian in (1.1) as

$$\mathcal{W}_n = \left\{ w = (w_i)_{i=0}^n \in (\mathbb{Z}^d)^{n+1} \colon w_0 = 0, \|w_{i+1} - w_i\| = 1 \ \forall \, 0 \leq i < n \right\},$$
$$H_n(w) = \beta I_n(w),$$

$$(3.1)$$

F. den Hollander, *Random Polymers*,
Lecture Notes in Mathematics 1974, DOI: 10.1007/978-3-642-00333-2_3,
© Springer-Verlag Berlin Heidelberg 2009

where $\beta \in [0, \infty)$ and

$$I_n(w) = \sum_{\substack{i,j=0 \\ i<j}}^{n} 1_{\{w_i = w_j\}} \tag{3.2}$$

is the *intersection local time* of w. This model goes under the name of *weakly self-avoiding random walk*: every self-intersection contributes an energy β to the Hamiltonian and is therefore penalized by a factor $e^{-\beta}$. Another name used in the literature is Domb-Joyce model. Think of β as a *strength of self-repellence* parameter.

We write P_n^β to denote the law of the soft polymer of length n with parameter β, as in (1.2). We add a factor $(1/2d)^n$ to P_n^β in order to be able to compare it with the law P_n of SRW, i.e., we put

$$P_n^\beta(w) = \frac{1}{Z_n^\beta} e^{-\beta I_n(w)} P_n(w), \qquad w \in \mathcal{W}_n, \tag{3.3}$$

so that we may think of P_n^β as an exponential tilting of P_n. Thus, P_n^β is the law of a *random process* $(S_i)_{i=0}^n$ with weak self-repellence, taking values in $\mathcal{W}_n$. Note that, like for SAW in Section 1.2, $(P_n^\beta)_{n \in \mathbb{N}_0}$ is not (!) a consistent family when $\beta \in (0, \infty)$. The case $\beta = 0$ corresponds to SRW, the case $\beta = \infty$ to SAW.

In what follows we focus on the case $d = 1$. In Chapter 4 we deal with the case $d \geq 5$.

3.2 Weakly Self-avoiding Walk in Dimension One

Intuitively, we expect that typical paths under the measure P_n^β hang around the origin for a while and then wander off to infinity at a strictly positive speed because of the self-repellence (there is a trivial symmetry between moving to the left and moving to the right). Ballistic behavior was first shown by Bolthausen [25], without existence and identification of the speed. Theorems 3.1 and 3.2 below, which are taken from Greven and den Hollander [130], establish existence and identify the speed in terms of a variational problem. See also den Hollander [168], Chapter IX.

Theorem 3.1. *For every $\beta \in (0, \infty)$ there exists a $\theta^*(\beta) \in (0, 1)$ such that*

$$\lim_{n \to \infty} P_n^\beta \left(\left| \frac{1}{n} S_n - \theta^*(\beta) \right| \leq \epsilon \, \middle| \, S_n \geq 0 \right) = 1 \text{ for all } \epsilon > 0. \tag{3.4}$$

Theorem 3.2. *The function $\beta \mapsto \theta^*(\beta)$ can be computed in terms of a variational problem. It follows from the solution of this variational problem that*

$$\beta \mapsto \theta^*(\beta) \text{ is analytic on } (0, \infty),$$

$$\lim_{\beta \downarrow 0} \theta^*(\beta) = 0, \qquad \lim_{\beta \to \infty} \theta^*(\beta) = 1. \tag{3.5}$$

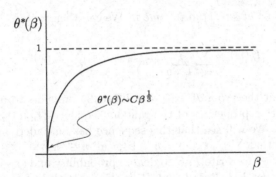

Fig. 3.1. The linear speed of the soft polymer.

The quantity $\theta^*(\beta)$ is *the speed of the soft polymer with strength of repellence* β. In Section 3.6 we will see that $\beta \mapsto \theta^*(\beta)$ looks like the curve in Fig. 3.1.

Theorems 3.1 and 3.2 will follow from the following *large deviation principle*, which is the main result of the present chapter.

Theorem 3.3. *For every $\beta \in (0, \infty)$ the family $(P_n^{+,\beta})_{n \in \mathbb{N}_0}$ defined by*

$$P_n^{+,\beta}(\cdot) = P_n^\beta \left(\frac{1}{n} S_n \in \cdot \,\middle|\, S_n \geq 0 \right) \tag{3.6}$$

satisfies a large deviation principle (LDP) on $[0,1]$ with rate n and with rate function I_β, identified in (3.66) below, having $\theta^*(\beta)$ as its unique zero.

The full proof of this LDP will have to wait until Section 3.5 (see Theorem 3.14 and Fig. 3.4). For the definition of LDP, we refer the reader to Dembo and Zeitouni [82], Chapter 1, and den Hollander [168], Chapter III. In essence, Theorem 3.3 says that

$$\lim_{\delta \downarrow 0} \lim_{n \to \infty} \frac{1}{n} \log P_n^{+,\beta}([\theta - \delta, \theta + \delta]) = -I_\beta(\theta). \tag{3.7}$$

Before we get going on the proof of Theorem 3.3, we first rewrite the definition of P_n^β in (3.3) in a way that is more convenient. Let

$$\widehat{I}_n(w) = \sum_{i,j=0}^{n} 1_{\{w_i = w_j\}}. \tag{3.8}$$

Then $\widehat{I}_n(w) = 2I_n(w) + (n+1)$. Hence, we may as well put $\widehat{I}_n(w)$ in the exponential weight factor (which only changes β to 2β). Henceforth we write $I_n(w)$ again, suppressing the overscript.

The following object is of paramount importance for the argument given below. Define

$$\ell_n(x) = \sum_{i=0}^{n} 1_{\{w_i = x\}}, \qquad x \in \mathbb{Z}, n \in \mathbb{N}_0, \tag{3.9}$$

i.e., the *local time at site x up to time n*. We can then write

$$I_n(w) = \sum_{x \in \mathbb{Z}} \sum_{i,j=0}^{n} 1_{\{w_i = w_j = x\}} = \sum_{x \in \mathbb{Z}} \ell_n(x)^2. \tag{3.10}$$

We thus see that the proof of Theorem 3.3 really amounts to understanding the large deviation properties of the random sequence $\{\ell_n(x)\}_{x \in \mathbb{Z}}$ under the law P_n of SRW. We will see that this sequence has an underlying *Markovian* structure. Note that $\sum_{x \in \mathbb{Z}} \ell_n(x) = n + 1$ for all $n \in \mathbb{N}$.

In what follows we write P, E to denote probability and expectation w.r.t. SRW (as in Section 2.1). Recall that P_n is the projection of P onto $\mathcal{W}_n$.

3.3 The Large Deviation Principle for Bridges

In order to obtain the desired LDP for $P_n^{+,\beta}(\frac{1}{n} S_n \in \cdot)$, we begin by deriving an LDP under the restriction that the path be a *bridge*, i.e., that it lies between its endpoints. This restriction will be crucial for the proof in Section 3.4, and will only be removed in Section 3.5.

Folding a path into a bridge. Our first lemma shows that the bridge condition does not change the normalizing constant.

Lemma 3.4. *For $n \to \infty$,*

$$E\left(e^{-\beta I_n} 1_{\{S_n \geq 0\}}\right) = e^{o(n)} E\left(e^{-\beta I_n} 1_{\circledast_n}\right), \tag{3.11}$$

with $I_n = I_n((S_i)_{i=0}^n)$ and

$$\circledast_n = \{S_0 \leq S_i \leq S_n \; \forall 0 \leq i \leq n\}. \tag{3.12}$$

Proof. The proof uses a folding argument due to Hammersley and Welsh [143]. Fix n.

First, suppose that the path is a *half-bridge* to the right, i.e., $S_i > S_0$ $\forall 0 < i \leq n$. We can then do a reflection procedure starting from the left endpoint of the path, as follows. Put $i_0 = 0$ and, for $j = 1, 2, \ldots$, define (R_j, i_j) recursively as

$$R_j = \max_{i_{j-1} < i \leq n} (-1)^j (S_{i_{j-1}} - S_i),$$
$$i_j = \text{the largest } i \text{ where the maximum is attained.}$$

The recursion is stopped at the smallest integer k such that $i_k = n$. What this definition says is that R_j is the span of the subwalk $(S_{i_{j-1}}, \ldots, S_n)$. Each subwalk $(S_{i_{j-1}}, \ldots, S_{i_j})$ lies strictly on one side of the point $S_{i_{j-1}}$, and

$$R_1 + \cdots + R_k \leq n \quad \text{and} \quad R_1 > R_2 > \cdots > R_k \geq 1. \tag{3.13}$$

If, for $j = 1, 2, \ldots, k - 1$, we reflect $(S_{i_j}, \ldots, S_n)$ around the point S_{i_j}, then we end up with a bridge, i.e., a path satisfying $S_0 < S_i \leq S_n \; \forall 0 < i \leq n$. Moreover, this bridge is less penalized than the original path because it has less self-intersections.

Next, we drop the assumption that the path be a half-bridge and only suppose that $S_n \geq 0$. Let

$$i_- = \min\left\{0 \leq i \leq n \colon S_i = \min_{0 \leq j \leq n} S_j\right\},$$

$$i_+ = \max\left\{0 \leq i \leq n \colon S_i = \max_{0 \leq j \leq n} S_j\right\}.$$

(3.14)

Then, when $i_- > 0$ and $i_+ < n$, both $(S_{i_-}, \ldots, S_0)$ and $(S_n, \ldots, S_{i_+})$ are half-bridges, and the above reflection procedure applies. If we fold both pieces outwards after the reflection procedure is through, then we end up with a bridge. (The cases $i_- = 0$ and $i_+ = n$ need no reflection.) Hence, we conclude that

$$E\left(e^{-\beta I_n} 1_{\{S_n \geq 0\}}\right) \leq N_n^2 \, E\left(e^{-\beta I_n} 1_{\circledast_n}\right),$$

(3.15)

where N_n is the number of solutions of (3.13) summed over k (which is the number of ordered partitions of $\{1, \ldots, n\}$). However, it is known that $N_n = \exp[O(\sqrt{n})]$ (see Madras and Slade [230], Theorem 3.1.4), so this factor is harmless and the claim follows. $\square$

The LDP for bridges. Our main result, whose proof will be given in Section 3.4, is the following LDP for the speed of the bridge soft polymer.

Theorem 3.5. *For every $\beta \in (0, \infty)$ the family $(P_n^{\beta, \text{bridge}})$, $n \in \mathbb{N}_0$, defined by*

$$P_n^{\beta, \text{bridge}}(\cdot) = P_n^\beta\left(\frac{1}{n} S_n \in \cdot \;\middle|\; \circledast_n\right)$$

(3.16)

satisfies the LDP on $(0, 1]$ with rate n and with rate function J_β identified in (3.21) and Lemma 3.12 below (see Fig. 3.3 below). The unique zero of J_β is $\theta^*(\beta)$ in Theorem 3.1.

To prove Theorem 3.5 we will carry out the following *program*:

(I) Pick $\theta \in (0, 1]$ and consider the quantity

$$P_n^\beta(S_n = \lceil \theta n \rceil \mid \circledast_n) = \frac{\widehat{K}_n(\theta)}{\int_{\theta \in (0,1]} d(\theta n) \widehat{K}_n(\theta)},$$

(3.17)

where

$$\widehat{K}_n(\theta) = E\left(e^{-\beta I_n} 1_{\{S_n = \lceil \theta n \rceil\}} 1_{\circledast_n}\right)$$

(3.18)

($\lceil \theta n \rceil$ and n must have the same parity). The value $\theta = 0$ is not relevant for bridges.

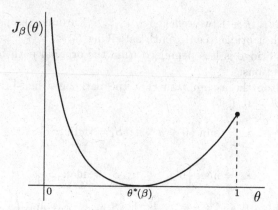

Fig. 3.2. The rate function J_β for bridge soft polymers.

(II) Show that there exists a function $\widehat{J}_\beta\colon (0,1] \to (0,\infty)$ such that

$$\lim_{n\to\infty} \frac{1}{n} \log \widehat{K}_n(\theta) = -\widehat{J}_\beta(\theta), \tag{3.19}$$

with the property that $\theta \mapsto \widehat{J}_\beta(\theta)$ is continuous, strictly convex and minimal at $\theta^*(\beta)$. Identify $\widehat{J}_\beta$ in terms of a variational problem.

(III) Combine (I) and (II), to obtain

$$\lim_{n\to\infty} \frac{1}{n} \log P_n^\beta\big(S_n = \lceil\theta n\rceil \,|\, \circledast_n\big) = -J_\beta(\theta), \tag{3.20}$$

with

$$J_\beta(\theta) = \widehat{J}_\beta(\theta) - \inf_{\theta\in(0,1]} \widehat{J}_\beta(\theta). \tag{3.21}$$

Evidently, $\theta \mapsto J_\beta(\theta)$ is also continuous, strictly convex and minimal at $\theta^*(\beta)$, which is its unique zero (see Fig. 3.2).

The argument in Section 3.4 will show that the same results as in (3.19–3.20) apply when θ is replaced by $\theta_n \to \theta$ as $n \to \infty$, which is why we get Theorem 3.5.

The above *program* will be carried out in Section 3.4, in five steps organized as Sections 3.4.1–3.4.5. The first two steps are a preparation that is needed to get the key quantities in the right format for applying large deviation theory. The actual application of large deviation theory and the analysis of the ensuing variational problem are carried out in the last three steps. The computation is technical but powerful, and can be carried over to other one-dimensional models as well.

After we have completed the proof of Theorem 3.5 we will show how to remove the bridge condition. This is done in Section 3.5 and leads to Theorem 3.3, the LDP we are actually after. It will turn out that the associated rate function is different from J_β but still has $\theta^*(\beta)$ as its unique zero (see Fig. 3.4 below), which is why Theorems 3.1–3.2 will follow as corollaries.

3.4 Program of Five Steps

3.4.1 Step 1: Adding Drift

We begin by going through a number of rewrites of the quantity $K_n(\theta)$ defined in (3.18).

Fix $\theta \in (0,1)$. (The case $\theta = 0$ is degenerate, the case $\theta = 1$ is trivial.) Let P_θ, E_θ denote probability and expectation for the random walk with drift θ (i.e., with probabilities $\frac{1}{2}(1+\theta)$ and $\frac{1}{2}(1-\theta)$ to step to the right and to the left, respectively). Then we can write (3.18) as

$$\widehat{K}_n(\theta) = (1-\theta)^{-\frac{n-\lceil \theta n \rceil}{2}}(1+\theta)^{-\frac{n+\lceil \theta n \rceil}{2}} \widetilde{K}_n(\theta), \tag{3.22}$$

with

$$\widetilde{K}_n(\theta) = E_\theta\big(e^{-\beta I_n} 1_{\{S_n = \lceil \theta n \rceil\}} 1_{\circledast_n}\big). \tag{3.23}$$

Indeed, every path from 0 to $\lceil \theta n \rceil$ makes the same number of steps to the left and to the right, so we pick up a simple Radon-Nikodym factor. Thus it suffices to study the asymptotics of $\widetilde{K}_n(\theta)$, i.e., *our task now is to relate the soft polymer with drift θ to the random walk with drift θ.*

The advantage of the reformulation in (3.22–3.23) is that the path does not care to return to $[0, S_n]$ after time n.

Lemma 3.6. *For every $\theta \in (0,1)$ and $n \in \mathbb{N}_0$,*

$$E_\theta\big(e^{-\beta I_n} 1_{\{S_n = \lceil \theta n \rceil\}} 1_{\circledast_n}\big) = \frac{1}{\theta}\, E_\theta\big(e^{-\beta I_n} 1_{\{S_n = \lceil \theta n \rceil\}} 1_{\circledast_n \cap \odot_n}\big), \tag{3.24}$$

with

$$\odot_n = \big\{S_i > S_n \ \forall i > n\big\}. \tag{3.25}$$

Proof. Simply use that I_n does not depend on S_i for $i > n$, and that $P_\theta(\odot_n) = \theta$ for all n (Spitzer [284], Section 1). $\square$

An important consequence of Lemma 3.6 is that on the event $\{S_n = \lceil \theta n \rceil\} \cap \circledast_n \cap \odot_n$ we may write

$$I_n = \sum_{x=0}^{\lceil \theta n \rceil} \ell(x)^2, \tag{3.26}$$

where

$$\ell(x) = \sum_{i \in \mathbb{N}_0} 1_{\{S_i = x\}}, \qquad x \in \mathbb{Z}, \tag{3.27}$$

is the *total local time at site x*. Indeed, this follows from the observation that on the event $\{S_n = \lceil \theta n \rceil\} \cap \circledast_n \cap \odot_n$ we have

$$\ell_n(x) = \begin{cases} \ell(x), & \text{if } 0 \le x \le \lceil \theta n \rceil, \\ 0, & \text{otherwise.} \end{cases}$$

Therefore we may rewrite (3.23) as

$$\widetilde{K}_n(\theta) = \frac{1}{\theta} E_\theta\left(e^{-\beta\sum_{x=0}^{\lceil\theta n\rceil}\ell^2(x)}1_{\{S_n=\lceil\theta n\rceil\}}1_{\circledast_n\cap\odot_n}\right). \tag{3.28}$$

Note that time n has been replaced by space $\lceil\theta n\rceil$ in (3.28). The total local times turn out to have a nice structure, as we show next.

3.4.2 Step 2: Markovian Nature of the Total Local Times

In this section we show that $\{\ell(x)\}_{x\in\mathbb{N}_0}$ admits a nice Markovian description. This will allow us to deduce the asymptotics of $\widetilde{K}_n(\theta)$ from an LDP for Markov chains. Let

$$m(x) = \sum_{i\in\mathbb{N}_0} 1_{\{S_i=x,\,S_{i+1}=x+1\}}, \qquad x\in\mathbb{Z}, \tag{3.29}$$

be the *total number of jumps from x to $x+1$*. Then, on the event $\{S_n = \lceil\theta n\rceil\}\cap\circledast_n\cap\odot_n$, the total number of jumps from $x+1$ to x equals

$$\sum_{i\in\mathbb{N}_0} 1_{\{S_i=x+1,\,S_{i+1}=x\}} = m(x)-1, \qquad 0\leq x\leq\lceil\theta n\rceil, \tag{3.30}$$

because the net number of jumps along the edge between x and $x+1$ must be $+1$. Since $\ell(x)$ is the sum of the number of jumps to x coming from the left and from the right, we have

$$\ell(x) = m(x-1)+m(x)-1_{\{x>0\}}, \qquad 0\leq x\leq\lceil\theta n\rceil. \tag{3.31}$$

Moreover, on the event $\{S_n = \lceil\theta n\rceil\}\cap\circledast_n\cap\odot_n$, the total time spent between 0 and $\lceil\theta n\rceil$ is $n+1$. Therefore

$$\{S_n = \lceil\theta n\rceil\}\cap\circledast_n\cap\odot_n$$
$$= \left\{\sum_{x=0}^{\lceil\theta n\rceil}[m(x-1)+m(x)-1_{\{x>0\}}] = n+1,\, m(-1)=0,\, m(\lceil\theta n\rceil)=1\right\}. \tag{3.32}$$

Therefore we may rewrite (3.28) as

$$\widetilde{K}_n(\theta) = \frac{1}{\theta} E_\theta\left(e^{-\beta\sum_{x=0}^{\lceil\theta n\rceil}[m(x-1)+m(x)-1_{\{x>0\}}]^2}\right.$$
$$\left.\times 1_{\left\{\sum_{x=0}^{\lceil\theta n\rceil}[m(x-1)+m(x)-1_{\{x>0\}}]=n+1\right\}}1_{\{m(-1)=0,\,m(\lceil\theta n\rceil)=1\}}\right). \tag{3.33}$$

The indicator $1_{\{x>0\}}$ and the restrictions $m(-1)=0$ and $m(\lceil\theta n\rceil)=1$ are to be thought of as harmless boundary terms.

The main reason for the reformulation in (3.33) is the following fact, which goes back to Knight [215].

Lemma 3.7. *For every $\theta \in (0,1)$ under the law P_θ, $\{m(x)\}_{x \in \mathbb{N}_0}$ is a Markov chain on state space $\mathbb{N}$ with transition kernel*

$$P_\theta(i,j) = \binom{i+j-2}{i-1}\left(\frac{1+\theta}{2}\right)^i\left(\frac{1-\theta}{2}\right)^{j-1}, \qquad i,j \in \mathbb{N}. \qquad (3.34)$$

Proof. Fix x. If $m(x) = i$, then the edge $(x, x+1)$ receives i upcrossings and $i-1$ downcrossings. For $s = 1, \ldots, i-1$, let Z_s denote the number of upcrossings of $(x+1, x+2)$ in between the s-th upcrossing and the s-th downcrossing of $(x, x+1)$. Let Z denote the number of upcrossings of $(x+1, x+2)$ after the i-th upcrossing of $(x, x+1)$, which is different from the others because no further downcrossing of $(x, x+1)$ is allowed. Since the random walk has drift θ, the probability that it makes a loop excursion to the right of $x+1$ is $\frac{1-\theta}{2}$. Hence, we have

$$P_\theta(Z_s = k \mid m(x) = i) = \left(\frac{1+\theta}{1-\theta}\right)\left(\frac{1-\theta}{2}\right)^{k+1}, \qquad k \in \mathbb{N}_0, \ s = 1, \ldots, i-1,$$

$$P_\theta(Z = k \mid m(x) = i) = \frac{1+\theta}{1-\theta}\left(\frac{1-\theta}{2}\right)^k, \qquad k \in \mathbb{N}.$$
$$(3.35)$$

Since $m(x+1) = j$ means that $Z_1 + \cdots + Z_{i-1} + Z = j$, we see that our process is Markov: it is irrelevant for the outcome of $m(x+1)$ what the random walk does to the left of x, only the value of $m(x)$ matters. Moreover, $Z_s + 1$, $s = 1, \ldots, i-1$, have the same law as Z. Since $(Z_1 + 1) + \cdots + (Z_{i-1} + 1) + Z = i+j-1$ and since the number of ways $i+j-1$ can be divided into i pieces of length ≥ 1 equals the binomial factor in (3.34), we obtain the formula for $P_\theta(i,j)$ in (3.34). $\square$

The proof of Lemma 3.7 shows that $\{m(x)\}_{x \in \mathbb{N}_0}$ is a branching process with one immigrant and with an offspring distribution that has mean smaller than 1. Therefore it is a positive recurrent Markov chain.

3.4.3 Step 3: Key Variational Problem

We have now completed our rewrite of $K_n(\theta)$ in (3.18) and are ready to apply large deviation theory. In this section we derive the key variational problem underlying the LDP for bridges in Theorem 3.5. This can be done along fairly standard lines. However, in order not to get lost in too many technicalities, the reader is asked to make a few small "leaps of faith".

The nice fact about the representation in (3.33) is that $\widetilde{K}_n(\theta)$ can be expressed in terms of the *pair empirical measure* associated with $\{m(x)\}_{x \in \mathbb{N}_0}$. To that end, define

$$L_N^2 = \frac{1}{N}\sum_{x=0}^{N-1}\delta_{(m(x-1),m(x))}, \qquad N \in \mathbb{N}, \qquad (3.36)$$

with periodic boundary conditions $(m(-1) = m(N))$, and let

$$F_\beta(\nu) = -\beta \sum_{i,j \in \mathbb{N}} (i + j - 1)^2 \nu(i, j),$$

(3.37)

$$A_\theta = \left\{ \nu \in \widetilde{\mathcal{M}}_1(\mathbb{N} \times \mathbb{N}) \colon \sum_{i,j \in \mathbb{N}} (i + j - 1)\nu(i, j) = \frac{1}{\theta} \right\},$$

where

$$\widetilde{\mathcal{M}}_1(\mathbb{N} \times \mathbb{N}) = \text{the set of probability measures}$$
$$\text{on } \mathbb{N} \times \mathbb{N} \text{ with identical marginals.}$$

(3.38)

Then (3.33) becomes

$$\widetilde{K}_n(\theta) = e^{o(n)} E_\theta \left(e^{N F_\beta(L_N^2)} 1_{\{L_N^2 \in A_\theta\}} \right) \text{ with } N = \lceil \theta n \rceil + 1.$$

(3.39)

Indeed,

1. The exponential factor in (3.33) equals the one in (3.39), with a negligible error arising from forcing the periodic boundary condition in the definition of L_N^2.
2. The first constraint in (3.33) is asymptotically the same as the constraint in (3.39), because we replaced $(n+1)/(\lceil \theta n \rceil + 1)$ by $1/\theta$, which will *a posteriori* be justified by the continuity of the function $\theta \mapsto \widetilde{J}_\beta(\theta)$ appearing in Lemma 3.8 below (see Sections 3.4.4–3.4.5).
3. The second constraint in (3.33) is negligible as $n \to \infty$.

The reason for introducing the representation in (3.39) is that it allows us to use the LDP for the empirical pair measure L_N^2, based on the Markov property established in Lemma 3.7.

Lemma 3.8. *For every $\theta \in (0, 1)$,*

$$\lim_{n \to \infty} \frac{1}{n} \log \widetilde{K}_n(\theta) = -\widetilde{J}_\beta(\theta),$$

(3.40)

with

$$\widetilde{J}_\beta(\theta) = \theta \inf_{\nu \in A_\theta} \left[-F_\beta(\nu) + I_{P_\theta}^2(\nu) \right],$$

(3.41)

where

$$I_{P_\theta}^2(\nu) = \sum_{i,j \in \mathbb{N}} \nu(i, j) \log \left(\frac{\nu(i, j)}{\bar{\nu}(i) P_\theta(i, j)} \right).$$

(3.42)

Proof. The formula in (3.42) is the *weak* rate function in the *weak* LDP (see den Hollander [168], Section III.6) for $(L_N^2)_{N \in \mathbb{N}}$ under the law of the Markov chain $\{m(x)\}_{x \in \mathbb{N}_0}$ with transition kernel P_θ. In order to apply Varadhan's Lemma (see den Hollander [168], Section III.3) we need the LDP, i.e., we need to overcome the technical difficulty that the state space $\mathbb{N}$ is infinite. This can be handled via a truncation argument because, as was observed at the end of Section 3.4.2, the Markov chain $\{m(x)\}_{x\in\mathbb{N}_0}$ has strong recurrence properties (see Greven and den Hollander [130] for more details).

Next we apply Varadhan's Lemma to (3.39), which is an exponential integral restricted to the set A_θ. Here another technical difficulty arises: the weak LDP "needs to be transferred from $\widetilde{\mathcal{M}}_1(\mathbb{N}\times\mathbb{N})$ to A_θ" (see Dembo and Zeitouni [82], Lemma 4.1.5). The result of the usual manipulations reads, somewhat informally,

$$
\begin{aligned}
\widetilde{K}_n(\theta) &= e^{o(n)}\, E_\theta\Big(e^{NF_\beta(L_N^2)}1_{\{L_N^2\in A_\theta\}}\Big)\\
&= e^{o(n)}\int_{A_\theta} e^{NF_\beta(L_N^2)} P_\theta(L_N^2\in d\nu)\\
&= e^{o(n)}\, e^{N\sup_{\nu\in A_\theta}[F_\beta(\nu)-I_{P_\theta}^2(\nu)]},\qquad N\to\infty,
\end{aligned}
\tag{3.43}
$$

which proves the claim because $N=\lceil\theta n\rceil+1$. $\quad\square$

At this point we recall (3.22) and (3.23), and rewrite Lemma 3.8 as follows:

Lemma 3.9. *For every $\theta\in(0,1)$,*

$$
\lim_{n\to\infty}\frac{1}{n}\log\widehat{K}_n(\theta)=-\widehat{J}_\beta(\theta),
\tag{3.44}
$$

with

$$
\widehat{J}_\beta(\theta)=\theta\inf_{\nu\in A_\theta}\big[-F_\beta(\nu)+I_{P_0}^2(\nu)\big],
\tag{3.45}
$$

i.e., the same variational formula as in Lemma 3.8 but with P_θ replaced by P_0, given by (recall Lemma 3.7)

$$
P_0(i,j)=\binom{i+j-2}{i-1}\Big(\frac{1}{2}\Big)^{i+j-1},\qquad i,j\in\mathbb{N}.
\tag{3.46}
$$

Proof. Simply note that $\nu\in A_\theta$ implies $\sum_{i\in\mathbb{N}} i\bar{\nu}(i)=\frac{1+\theta}{2\theta}$, so that

$$
\begin{aligned}
I_{P_0}^2(\nu)-I_{P_\theta}^2(\nu) &= \sum_{i,j\in\mathbb{N}}\nu(i,j)\log\big[(1+\theta)^i(1-\theta)^{j-1}\big]\\
&= \frac{1+\theta}{2\theta}\log(1+\theta)+\frac{1-\theta}{2\theta}\log(1-\theta),\qquad \nu\in A_\theta,
\end{aligned}
\tag{3.47}
$$

which makes the prefactor in (3.22) cancel out. $\quad\square$

Thus we have identified $\widehat{J}_\beta(\theta)$ for $\theta\in(0,1)$, which is the function we were after in Section 3.3. The same formulas as in (3.44–3.45) apply for the degenerate case $\theta=1$, as is easily checked by direct computation. Finally, (3.21) gives us J_β, the rate function in the LDP for bridge soft polymers in Theorem 3.5.

3.4.4 Step 4: Solution of the Variational Problem in Terms of an Eigenvalue Problem

We next proceed to give the solution of the variational problem in (3.45), leading to the qualitative shape of the function $\theta \mapsto J_\beta(\theta)$ anticipated in Fig. 3.2. The variational problem requires us to minimize a non-linear functional under a linear constraint. It is possible to find the solution in terms of a certain eigenvalue problem that is well-behaved, and we will see that the outcome is relatively simple.

Fix $\beta \in (0, \infty)$ and $r \in \mathbb{R}$, and let $A_{r,\beta}$ be the $\mathbb{N} \times \mathbb{N}$ matrix with components

$$A_{r,\beta}(i,j) = e^{r(i+j-1)-\beta(i+j-1)^2} P_0(i,j), \qquad i,j \in \mathbb{N}. \tag{3.48}$$

The parameter r will be seen to play the role of a Lagrange multiplier needed to handle the constraint in (3.45).

Lemma 3.10. *Fix $\beta \in (0, \infty)$. For every $r \in \mathbb{R}$, $A_{r,\beta}$ is a self-adjoint operator on $l^2(\mathbb{N})$ having a unique largest eigenvalue $\lambda_{r,\beta}$ and corresponding eigenvector $\tau_{r,\beta}$ (normalized as $\|\tau_{r,\beta}\|_2 = 1$).*

Proof. Since $A_{r,\beta}$ is strictly positive and has rapidly decaying tails, the assertion follows from standard Perron-Frobenius theory. In fact, $A_{r,\beta}$ has the so-called Hilbert-Schmidt property $\sum_{i,j \in \mathbb{N}} A_{r,\beta}(i,j)^2 < \infty$ and, consequently, is a compact operator (see Dunford and Schwartz [94], Section XI.6). □

The eigenvalue $\lambda_{r,\beta}$ has the following properties:

Lemma 3.11. (i) $(r, \beta) \mapsto \lambda_{r,\beta}$ *is analytic on $\mathbb{R} \times (0, \infty)$.*
(ii) $\lim_{r \to -\infty} \frac{\partial}{\partial r} \log \lambda_{r,\beta} = 1$ *and* $\lim_{r \to \infty} \frac{\partial}{\partial r} \log \lambda_{r,\beta} = \infty$ *for all $\beta \in (0, \infty)$.*
(iii) $r \mapsto \log \lambda_{r,\beta}$ *is strictly convex for all $\beta \in (0, \infty)$.*

Proof. Here is a quick sketch. For details we refer to Greven and den Hollander [130].
(i) Analyticity holds because $\lambda_{r,\beta}$ has multiplicity 1 and all elements of the matrix $A_{r,\beta}$ are analytic.
(ii) This follows from straightforward estimates on the eigenvector $\tau_{r,\beta}$ for $r \to -\infty$ and $r \to \infty$, respectively.
(iii) Convexity follows from the observations:

1. $\lambda_{r,\beta} = \sup_{x \in l^2(\mathbb{N}): \ x > 0, \|x\|_2 = 1} \sum_{i,j \in \mathbb{N}} x(i) A_{r,\beta}(i,j) x(j)$;
2. $r \mapsto \log A_{r,\beta}(i,j)$ is linear for all $i,j \in \mathbb{N}$ and $\beta \in (0, \infty)$;
3. log-convexity is preserved under taking sums and suprema.

Strict convexity follows from convexity in combination with (i) and (ii). □

With the help of Lemma 3.11 we can express $\widehat{J}_\beta(\theta)$ in terms of the eigenvalue $\lambda_{r,\beta}$ for some r depending on θ.

Lemma 3.12. *Fix $\beta \in (0, \infty)$. Then, for every $\theta \in (0, 1)$,*

$$\widehat{J}_\beta(\theta) = r - \theta \log \lambda_{r,\beta} \Big|_{r=r_\beta(\theta)}, \tag{3.49}$$

where $r_\beta(\theta) \in \mathbb{R}$ is the unique solution of the equation

$$\frac{1}{\theta} = \frac{\partial}{\partial r} \log \lambda_{r,\beta}. \tag{3.50}$$

Proof. The fact that (3.50) has a solution for all $\theta \in (0, 1)$ and that this solution is unique follows from Lemma 3.11 (see Fig. 3.3).

Consider the following family of pair probability measures:

$$\nu_{r,\beta}(i, j) = \frac{1}{\lambda_{r,\beta}} \tau_{r,\beta}(i) A_{r,\beta}(i, j) \tau_{r,\beta}(j), \qquad i, j \in \mathbb{N}. \tag{3.51}$$

One easily checks that $\nu_{r,\beta} \in \widetilde{\mathcal{M}}_1(\mathbb{N} \times \mathbb{N})$. Compute

$$
\begin{aligned}
I^2_{P_0}(\nu_{r,\beta}) &= \sum_{i,j} \nu_{r,\beta}(i, j) \log \left(\frac{\nu_{r,\beta}(i, j)}{\bar{\nu}_{r,\beta}(i) P_0(i, j)} \right) \\
&= \sum_{i,j} \nu_{r,\beta}(i, j) \left[r(i+j-1) - \beta(i+j-1)^2 - \log \lambda_{r,\beta} + \log \frac{\tau_{r,\beta}(j)}{\tau_{r,\beta}(i)} \right],
\end{aligned}
\tag{3.52}
$$

where we use that $\bar{\nu}_{r,\beta}(i) = \tau^2_{r,\beta}(i)$. Since $\nu_{r,\beta}$ has identical marginals, the last term vanishes and we end up with the simple expression

$$I^2_{P_0}(\nu_{r,\beta}) = \frac{r}{\theta} + F_\beta(\nu_{r,\beta}) - \log \lambda_{r,\beta}, \tag{3.53}$$

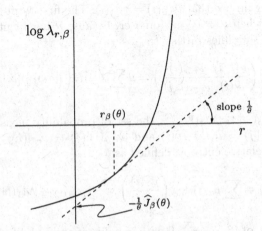

Fig. 3.3. Identification of $\widehat{J}_\beta(\theta)$.

provided (!) $\nu_{r,\beta} \in A_\theta$. This expression says that

$$-\theta\big[F_\beta(\nu_{r,\beta}) - I^2_{P_0}(\nu_{r,\beta})\big] = r - \theta \log \lambda_{r,\beta}. \tag{3.54}$$

We thus see from Lemma 3.9 that the claim in Lemma 3.12 is correct provided (!) we can prove the following two properties:

(i) $r = r_\beta(\theta)$ implies $\nu_{r,\beta} \in A_\theta$;
(ii) $\nu_{r_\beta(\theta),\beta}$ is a minimizer of the variational problem in (3.45).

Property (i): Compute

$$\sum_{i,j}(i + j - 1)\nu_{r,\beta}(i,j) = \frac{1}{\lambda_{r,\beta}} \sum_{i,j} \tau_{r,\beta}(i)\Big[\frac{\partial}{\partial r} A_{r,\beta}(i,j)\Big]\tau_{r,\beta}(j)$$

$$= \frac{1}{\lambda_{r,\beta}} \frac{\partial}{\partial r}\Big[\sum_{i,j}\tau_{r,\beta}(i)A_{r,\beta}(i,j)\tau_{r,\beta}(j)\Big] = \frac{1}{\lambda_{r,\beta}} \frac{\partial}{\partial r}\lambda_{r,\beta} = \frac{\partial}{\partial r}\log\lambda_{r,\beta}. \tag{3.55}$$

The second equality uses that $A_{r,\beta}\tau_{r,\beta} = \lambda_{r,\beta}\tau_{r,\beta}$ and $\|\tau_{r,\beta}\|_2 = 1$ for all r, β. Hence (3.50) indeed guarantees (i).

Property (ii): If $\nu \in A_\theta$, then we can write

$$- \theta\big[F_\beta(\nu) - I^2_{P_0}(\nu)\big]$$

$$= [r - \theta\log\lambda_{r,\beta}] + \theta\sum_{i,j}\nu(i,j)\log\left(\frac{\nu(i,j)}{\bar{\nu}(i)\frac{1}{\lambda_{r,\beta}}A_{r,\beta}(i,j)}\right) \tag{3.56}$$

$$= [r - \theta\log\lambda_{r,\beta}] + \theta\sum_{i,j}\nu(i,j)\log\left(\frac{\nu(i,j)}{\bar{\nu}(i)}\frac{\sqrt{\bar{\nu}_{r,\beta}(i)\bar{\nu}_{r,\beta}(j)}}{\nu_{r,\beta}(i,j)}\right),$$

where we once again use that $\bar{\nu}_{r,\beta}(i) = \tau^2_{r,\beta}(i)$. The first term is precisely the value we found when $\nu = \nu_{r,\beta}$. Moreover, because ν has identical marginals the second term simplifies further to

$$\theta\sum_{i,j}\nu(i,j)\log\left(\frac{\nu(i,j)}{\bar{\nu}(i)}\frac{\bar{\nu}_{r,\beta}(i)}{\nu_{r,\beta}(i,j)}\right) = \theta\sum_i\bar{\nu}(i)H\Big([\nu(i)] \,\big|\, [\nu_{r,\beta}(i)]\Big), \tag{3.57}$$

where $[\nu(i)], [\nu_{r,\beta}(i)] \in \mathcal{M}_1(\mathbb{N})$ (= the set of probability measures of $\mathbb{N}$) are defined by $[\nu(i)](j) = \nu(i,j)/\nu(i)$ and $[\nu_{r,\beta}(i)](j) = \nu_{r,\beta}(i,j)/\nu_{r,\beta}(i)$, while $H(\cdot \,|\, \cdot)$ denotes relative entropy, defined as

$$H(\mu_1 \,|\, \mu_2) = \sum_i \mu_1(i)\log\left(\frac{\mu_1(i)}{\mu_2(i)}\right), \qquad \mu_1, \mu_2 \in \mathcal{M}_1(\mathbb{N}). \tag{3.58}$$

Clearly, the r.h.s. of (3.57) is ≥ 0 with equality if and only if $\nu = \nu_{r,\beta}$. $\qquad\square$

3.4.5 Step 5: Identification of the Speed

Lemma 3.12 gives us a nice representation of $\widehat{J}_\beta(\theta)$, $\theta \in (0,1)$, in terms of the family of eigenvalues $\lambda_{r,\beta}$, $r \in \mathbb{R}$. As we saw in Fig. 3.2, $\theta^*(\beta)$ is to be identified as the unique minimum of $\theta \mapsto \widehat{J}_\beta(\theta)$.

Lemma 3.13. *Fix $\beta \in (0,\infty)$. Then*

$$\frac{1}{\theta^*(\beta)} = \frac{\partial}{\partial r} \log \lambda_{r,\beta}\Big|_{r=r^*(\beta)}, \tag{3.59}$$

with $r^*(\beta) \in (0,\infty)$ the unique solution of the equation

$$\lambda_{r,\beta} = 1. \tag{3.60}$$

Proof. The fact that (3.60) has a solution for all $\beta \in (0,\infty)$ and that this solution is unique follows from Lemma 3.11 (see Fig. 3.3).

Differentiate $\widehat{J}_\beta(\theta)$ with respect to θ to obtain

$$\frac{\partial}{\partial \theta} \widehat{J}_\beta(\theta) = \frac{\partial}{\partial \theta} r_\beta(\theta) - \log \lambda_{r_\beta(\theta),\beta} - \theta\Big[\frac{\partial}{\partial \theta} r_\beta(\theta)\Big]\Big[\frac{\partial}{\partial r} \log \lambda_{r,\beta}\Big]_{r=r_\beta(\theta)}. \tag{3.61}$$

However, the first and the third term cancel out because of (3.50), so we get

$$\frac{\partial}{\partial \theta} \widehat{J}_\beta(\theta) = -\log \lambda_{r_\beta(\theta),\beta}. \tag{3.62}$$

This is zero if and only if θ is such that $\lambda_{r_\beta(\theta),\beta} = 1$, i.e., the minimum $\theta^*(\beta)$ of $\theta \mapsto \widehat{J}_\beta(\theta)$ is found by solving (3.60). After that we put

$$r_\beta(\theta^*(\beta)) = r^*(\beta) \tag{3.63}$$

and use (3.50). Note that, by Lemma 3.12,

$$\frac{\partial^2}{\partial \theta^2} \widehat{J}_\beta(\theta) = -\frac{1}{\theta}\frac{\partial}{\partial \theta} r_\beta(\theta) > 0 \tag{3.64}$$

(see Fig. 3.3 and note that $\theta \mapsto r_\beta(\theta)$ has a negative slope), so that $r_\beta(\theta^*(\beta))$ is indeed the unique minimizer of $\theta \mapsto \widehat{J}_\beta(\theta)$. $\square$

Lemmas 3.11–3.13 yield Fig. 3.3. This finishes our analysis of the rate function J_β for bridge soft polymers, and the proof of Theorem 3.5 is now complete.

Note that $\theta^*(\beta)$ is the unique minimum of $\widehat{J}_\beta$ and the unique minimum and zero of J_β (recall (3.21) and Fig. 3.2).

3.5 The Large Deviation Principle without the Bridge Condition

In Sections 3.4.1–3.4.5 we have proved Theorem 3.5, the LDP for bridge soft polymers. In this section we give a quick sketch of how to obtain the LDP without the bridge condition, i.e., Theorem 3.3, which implies Theorems 3.1 and 3.2. Remarkably, it turns out that the rate function in this LDP has a linear piece between 0 and a *critical speed* $\theta^{**}(\beta)$ that is strictly smaller than $\theta^*(\beta)$ (see Fig. 3.4). What is written below developed out of discussions with W. König.

Theorem 3.14. *For every* $\beta \in (0, \infty)$ *the family* $(P_n^{+,\beta})_{n \in \mathbb{N}}$ *defined by*

$$P_n^{+,\beta}(\cdot) = P_n^{\beta}\left(\frac{1}{n}S_n \in \cdot \ \middle|\ S_n \geq 0\right) \tag{3.65}$$

satisfies the LDP *on* $[0, 1]$ *with rate* n *and with rate function* I_β *given by*

$$I_\beta(\theta) = \begin{cases} J_\beta(\theta), & \text{if } \theta \geq \theta^{**}(\beta), \\ I_\beta(0) + \frac{\theta}{\theta^{**}(\beta)}[J_\beta(\theta^{**}(\beta)) - I_\beta(0)], & \text{if } \theta \leq \theta^{**}(\beta), \end{cases} \tag{3.66}$$

where $\theta^{**}(\beta)$ *is the unique solution of the equation*

$$J_\beta(\theta) - I_\beta(0) = \theta \frac{\partial}{\partial \theta} J_\beta(\theta), \tag{3.67}$$

and $I_\beta(0)$ *is identified in Lemma* 3.15 *below. Moreover,* $\theta^{**}(\beta) \in (0, \theta^*(\beta))$.

The linear piece in (3.66) can be understood as follows. If the soft polymer is required to move at a speed $\theta < \theta^{**}(\beta)$, then it prefers to violate the bridge condition by moving at speed $\theta^{**}(\beta)$ between 0 and $\lceil \theta n \rceil$, and making two loops, one below 0 and one above $\lceil \theta n \rceil$. The penalty for making these loops

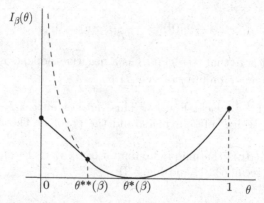

Fig. 3.4. The rate function I_β (compare with Fig. 3.2).

is less than the penalty for staying locked up like a bridge. The total length
of these loops is proportional to $\theta^{**}(\beta) - \theta$, i.e., the penalty for not behaving
like a bridge grows linearly with $\theta^{**}(\beta) - \theta$.

The analogue of Lemmas 3.12 and 3.13 reads as follows (compare (3.46)
and (3.48) with (3.69) and (3.68)):

Lemma 3.15. *Fix $\beta \in (0, \infty)$. For $r \in \mathbb{R}$, let $A_{r,\beta}^{\circlearrowleft}$ be the $\mathbb{N} \times \mathbb{N}$–matrix with
components*

$$A_{r,\beta}^{\circlearrowleft}(i,j) = e^{r(i+j)-\beta(i+j)^2} P_0^{\circlearrowleft}(i,j), \qquad i,j \in \mathbb{N}, \tag{3.68}$$

with

$$P_0^{\circlearrowleft}(i,j) = 1_{\{i \neq 0\}} P_0(i,j+1) + 1_{\{i=j=0\}}. \tag{3.69}$$

*Let $\lambda_{r,\beta}^{\circlearrowleft}$ be the unique largest eigenvalue of $A_{r,\beta}^{\circlearrowleft}$ acting as an operator on
$l^2(\mathbb{N})$. Then*

$$I_\beta(0) = r^{**}(\beta) - r^*(\beta), \tag{3.70}$$

with $r^{**}(\beta) \in (0, \infty)$ the unique solution of the equation

$$\lambda_{r,\beta}^{\circlearrowleft} = 1. \tag{3.71}$$

Proof. See den Hollander [168], Chapter IX. □

It is easy to show that $r^{**}(\beta) > r^*(\beta)$ for all $\beta \in (0, \infty)$. Since (3.67) says
that $r^{**}(\beta) = r_\beta(\theta^{**}(\beta))$, with $\theta \mapsto r_\beta(\theta)$ defined in Lemma 3.12, it follows
that $\theta^{**}(\beta) < \theta^*(\beta)$ for all $\beta \in (0, \infty)$.

In conclusion, we have proved Theorem 3.3 and identified the rate function
I_β in terms of the families of principal eigenvalues $(r, \beta) \mapsto \lambda_{r,\beta}$ and $(r, \beta) \mapsto
\lambda_{r,\beta}^{\circlearrowleft}$ of the operators $A_{r,\beta}$ and $A_{r,\beta}^{\circlearrowleft}$ defined in (3.48) and (3.68). These families
are analytically well-behaved and can also be easily computed numerically (see
Greven and den Hollander [130]).

3.6 Extensions

(1) Theorem 3.1 has been extended to a central limit theorem by König [218].
The standard deviation, denoted by $\sigma^*(\beta)$, turns out to be given by the
formula

$$\frac{1}{\sigma^{*2}(\beta)} = \frac{\partial^2}{\partial\theta^2} J_\beta(\theta)\Big|_{\theta=\theta^*(\beta)} = \frac{\partial^2}{\partial\theta^2} I_\beta(\theta)\Big|_{\theta=\theta^*(\beta)}. \tag{3.72}$$

Numerical computation gives the picture in Fig. 3.5.

(2) Van der Hofstad and den Hollander [157] prove that

$$\lim_{\beta \downarrow 0} \beta^{-\frac{1}{3}} \theta^*(\beta) = C \text{ for some } C > 0 \tag{3.73}$$

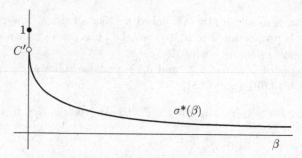

Fig. 3.5. The spread of the soft polymer.

(recall Fig. 3.1), while van der Hofstad, den Hollander and König [158] prove that

$$\lim_{\beta \downarrow 0} \sigma^*(\beta) = C' \text{ for some } C' \neq 1. \tag{3.74}$$

These asymptotic formulas show that, even in the limit of weak self-repellence, the behavior of the soft polymer cannot be understood via a perturbation argument around the non-repellent SRW. Van der Hofstad [153] derives rigorous bounds on C and C'. Numerically, $C \approx 1.1$ and $C' \approx 0.63$. These constants are related to a Brownian version of the polymer model – called the Edwards model – defined in (3.78) below. We refer to van der Hofstad, den Hollander and König [159] for the analogous law of large numbers (first proved by Westwater [312]) and central limit theorem.

The heuristics behind the scaling in (3.73) is as follows. Suppose that $S_n \approx \theta n$ and that $0 \leq S_i \leq S_n$ for all $0 \leq i \leq n$. The Hamiltonian $H_n = \beta \sum_{x \in \mathbb{Z}} \ell_n(x)^2$ is minimal when the local times are constant, i.e., when $\ell_n(x) \approx n/S_n \approx 1/\theta$ for $0 \leq x \leq S_n$, in which case $H_n \approx \beta(\theta n)(1/\theta)^2 = (\beta/\theta)n$. The probability under P, the law of SRW, that $S_n \approx \theta n$ is roughly $\exp[-(\theta n)^2/2n]$. Consequently, the contribution to the partition sum coming from paths with $S_n \approx \theta n$ is roughly $\exp[-\{(\beta/\theta) + \frac{1}{2}\theta^2\}n]$. The term between braces in the exponent is minimal when $\theta = \beta^{\frac{1}{3}}$.

(3) Van der Hofstad, den Hollander and König [158] prove that if β is replaced by β_n satisfying

$$\lim_{n \to \infty} \beta_n = 0 \quad \text{and} \quad \lim_{n \to \infty} n^{\frac{3}{2}} \beta_n = \infty, \tag{3.75}$$

then the law of large numbers and central limit theorem apply with $\theta^*(\beta)$ and $\sigma^*(\beta)$ replaced by

$$\theta^*(\beta_n) \sim C \, (\beta_n)^{\frac{1}{3}} \quad \text{and} \quad \sigma^*(\beta_n) \sim C'. \tag{3.76}$$

In comparison with (3.73–3.74), this shows that the weak interaction limit has a degree of *universality*. Van der Hofstad, den Hollander and König [161] offer a coarse-graining argument showing that, in one dimension, self-repellent random walks scale to self-repellent Brownian motions. The proof is based on cutting the path into pieces, controlling the interaction between the different pieces, and applying the invariance principle to the single pieces. The scaling properties are shown to be stable against adding self-attraction, provided the self-repellence remains dominant. We will return to polymers with self-repellence and self-attraction in Chapter 6.

(4) König [216], [217] (extending earlier work by Alm and Janson [9]) considers the case where the random walk is "spread out", e.g., it draws its step uniformly from the set $\{-L, \ldots, L\}$ for some $L \in \mathbb{N}$. For this case, the SAW problem is interesting. It is shown that for every $\beta \in (0, \infty]$ and $L \in \mathbb{N}$ the self-avoiding polymer has a speed $\theta^*(\beta, L) \in (0, L)$. Aldous [3] – assuming that the speed existed – had earlier conjectured that

$$\lim_{L \to \infty} L^{-\frac{2}{3}} \theta^*(\infty, L) = C'' \text{ for some } C'' \in (0, \infty). \tag{3.77}$$

This conjecture was based on a scaling result for the self-intersection local time of the spread-out random walk in the limit as $L \to \infty$. The conjecture in (3.77) was subsequently proved in van der Hofstad, den Hollander and König [161], where it was shown that $C'' = C3^{-\frac{1}{3}}$.

(5) A continuum version of the Domb-Joyce model – called the Edwards model – is analyzed in Kusuoka [224] and van der Hofstad, den Hollander and König [159]. The Hamiltonian for this model is

$$H\big((B_t)_{t \in [0,T]}\big)$$
$$= -\beta \int_0^T ds \int_0^T dt \, \delta(B_s - B_t) = -\beta \int_{\mathbb{R}} L(T, x)^2 \, dx, \qquad T \geq 0, \tag{3.78}$$

where $\delta(\cdot)$ is the Dirac delta-function, $(B_t)_{t \geq 0}$ is standard Brownian motion and $L(T, x)$ is its local time at position x up to time T. The behavior is ballistic, and both a law of large numbers and a central limit theorem apply, with speed $C\beta^{\frac{1}{3}}$ and spread C' (which provide the link with the weak interaction limit of the Domb-Joyce model). The corresponding LDP is proved in van der Hofstad, den Hollander and König [160].

3.7 Challenges

(1) Prove that $\beta \mapsto \theta^*(\beta)$ is non-decreasing (see Fig. 3.1). Even though this property seems intuitively plausible, it is actually deep (see Greven and den Hollander [130]) and remains open. Similarly, prove that $\beta \mapsto \sigma^*(\beta)$ is non-increasing (see Fig. 3.5). It is not hard to compute $\lambda_{r,\beta}$ numerically and get support for the monotonicity of both quantities. Coupling arguments do not work: P_n^β's for different values of β are hard to compare, because the normalizing partition sum depends on β (recall (3.3)).

(2) Prove the functional central limit theorem, i.e., show that under the law $P_n^{+,\beta}$ we have

$$\left(\frac{1}{\sigma^*(\beta)\sqrt{n}}\left[S_{\lfloor tn \rfloor} - \theta^*(\beta)\lfloor tn \rfloor\right]\right)_{0 \leq t \leq 1} \Longrightarrow (B_t)_{0 \leq t \leq 1} \quad \text{as } n \to \infty, \quad (3.79)$$

with standard Brownian motion as limit. The proof of (3.79) should not be hard: the method of local times employed in Sections 3.4.1–3.4.3 is very powerful and should allow us to get the multivariate analogues of the law of large numbers and the central limit theorem, together with the appropriate form of tightness.

(3) Derive a functional LDP extending Theorem 3.5.

(4) Van der Hofstad, den Hollander and König [161] have extended the LDP for the speed of the endpoint to the weak interaction limit in (3.75), but only for speeds that are not too small. Extend the proof to all speeds.

(5) Try to put some rigor into the following heuristic observation (put forward in van der Hofstad, den Hollander and König [161]), which argues in favor of the critical exponent $\nu = \frac{3}{4}$ for the two-dimensional soft polymer (recall (2.26)) based on the result in Section 3.6, Extension (3). Consider simple random walk on the slit $\{-L, \ldots, L\} \times \mathbb{Z}$ with periodic boundary conditions (see Fig. 3.6). Write

$$S = (S_i)_{i=0}^n = (S^{(1)}, S^{(2)}) = (S_i^{(1)}, S_i^{(2)})_{i=0}^n \quad (3.80)$$

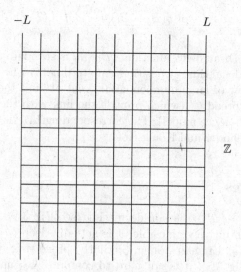

Fig. 3.6. The slit $\{-L, \ldots, L\} \times \mathbb{Z}$ with periodic boundary conditions.

for its two components, and note that S makes a self-intersection if and only if both $S^{(1)}$ and $S^{(2)}$ make a self-intersection. Under the soft polymer measure P_n^{β}, we have

$$|S_n^{(1)}| \asymp L \quad \text{and} \quad |S_n^{(2)}| \asymp L^{-\frac{1}{3}} n \quad \text{as } n \to \infty. \tag{3.81}$$

The first claim is trivial. The second claim comes from the fact that $S^{(1)}$ self-intersects one out of $(2L + 1)$ times. Hence, the motion of $S^{(2)}$ is comparable to that of the one-dimensional soft polymer with self-repellence parameter $\beta_n = \beta/(2L + 1)$. Therefore, according to (3.76), $S^{(2)}$ moves a distance of order

$$(\beta_n)^{\frac{1}{3}} n \asymp L^{-\frac{1}{3}} n \tag{3.82}$$

in time n. Now, the two scales in (3.81) coincide when $L = n^{\frac{3}{4}}$. Then, the two components run on the same scale, and consequently the slit is wide enough for the motion of S to be comparable to that of the two-dimensional soft polymer. For $L = n^{\frac{3}{4}}$, both $S_n^{(1)}$ and $S_n^{(2)}$ run on scale $n^{\frac{3}{4}}$, and hence so does S_n. (Note that $\beta_n \asymp n^{-\frac{3}{4}}$ when $L = n^{\frac{3}{4}}$, which indeed satisfies (3.75).)

4

Soft Polymers in High Dimension

In this chapter we consider the same model as in Chapter 3, but for $d \geq 5$ instead of $d = 1$. Our goal is to prove *diffusive behavior*. The tool to achieve this is the so-called *lace expansion*, a combinatorial technique that hinges on the idea that in high dimensions the soft polymer can be viewed as a "perturbation" of SRW. For the exposition below, we borrow from van der Hofstad [156], Section 2, and Slade [280], Chapter 3. Figs. 4.2–4.4 are borrowed from Gordon Slade, Fig. 4.5 from Bill Casselman and Gordon Slade.

4.1 Weakly Self-avoiding Walk in Dimension Five or Higher

The lace expansion was introduced by Brydges and Spencer [44] to prove diffusive behavior of weakly self-avoiding walk in $d \geq 5$. To obtain convergence of the lace expansion, a "small parameter" is needed. For that purpose, in [44] the parameter β was taken to be small. It is more interesting, however, to replace SRW by a random walk that is "spread out", i.e., to allow the walk to choose its steps randomly from the set

$$\Omega = \{x \in \mathbb{Z}^d \colon 0 < \|x\|_\infty \leq L\} \tag{4.1}$$

with $\| \cdot \|_\infty$ the supremum norm and $L \in \mathbb{N}$ (see Fig. 4.1). If L is large, then self-intersections are so few that the spread-out soft polymer can be viewed as a "perturbation" of the spread-out random walk with no interaction. The small parameter is $1/|\Omega|$.

Instead of (3.1–3.2), we pick as our set of n-step paths

$$\mathcal{W}_n = \{w = (w_i)_{i=0}^n \in (\mathbb{Z}^d)^{n+1} \colon w_0 = 0, \, w_{i+1} - w_i \in \Omega \; \forall \, 0 \leq i < n\} \tag{4.2}$$

and we keep as our Hamiltonian

$$H_n(w) = \beta I_n(w). \tag{4.3}$$

F. den Hollander, *Random Polymers*,
Lecture Notes in Mathematics 1974, DOI: 10.1007/978-3-642-00333-2_4,
© Springer-Verlag Berlin Heidelberg 2009

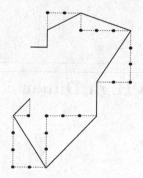

Fig. 4.1. A path drawing steps from Ω with $L = 3$.

Our soft polymer measure is

$$P_n^\beta(w) = \frac{1}{Z_n^\beta} e^{-\beta I_n(w)}, \qquad w \in \mathcal{W}_n, \tag{4.4}$$

which differs from (3.3) in that we drop the reference measure P_n. This is harmless because P_n is the uniform distribution. We do this because most of the present chapter is about combinatorics rather than about probability.

For SRW (corresponding to $L = 1$) we may take $1/2d$ as the small parameter for large d. However, choosing L large has the advantage that we will be able to prove diffusive behavior for *all* $d \geq 5$ and *all* $\beta \in (0, \infty)$. The threshold value for L will depend on d and β, but will be finite.

The goal of the present chapter is to derive Theorem 4.1 below, which is the analogue of (2.20–2.21), (2.25–2.26) and (2.29) for SAW in $d \geq 5$ stated in Section 2.2. To formulate this theorem, we introduce some notation. Let $\mathcal{W}_n(x) = \{w \in \mathcal{W}_n \colon w_n = x\}$, and

$$Z_n(x) = \sum_{w \in \mathcal{W}_n(x)} e^{-\beta I_n(w)}, \; x \in \mathbb{Z}^d,$$

$$Z_n = \sum_{x \in \mathbb{Z}^d} Z_n(x), \tag{4.5}$$

$$\widehat{Z}_n(k) = \sum_{x \in \mathbb{Z}^d} e^{i(k \cdot x)} Z_n(x), \; k \in [-\pi, \pi)^d.$$

where we suppress the dependence on β. In Section 2.2, where we were dealing with $\beta = \infty$ and $L = 1$, these objects were denoted by $c_n(x)$, c_n and $\widehat{c}_n(k)$, respectively. Note that

$$Z_n^{-1} Z_n(x) = P_n^\beta(S_n = x),$$

$$-Z_n^{-1} \nabla^2 \widehat{Z}_n(0) = E_n^\beta(\|S_n\|^2),$$

$$Z_n^{-1} \widehat{Z}_n(k) = E_n^\beta\left(e^{i(k \cdot S_n)}\right),$$

with $\nabla^2 \widehat{Z}_n = \nabla \cdot (\nabla \widehat{Z}_n)$. Let

$$D(x) = |\Omega|^{-1} \, 1\{x \in \Omega\} \tag{4.7}$$

denote the uniform distribution on Ω, and $\sigma^2 = \sum_{x \in \mathbb{Z}^d} \|x\|^2 D(x)$.

Theorem 4.1. *Fix $d \geq 5$ and $\beta \in (0, \infty)$. Then there exists an L_0 (depending on d, β) such that for all $L \geq L_0$ and as $n \to \infty$,*

$$Z_n = A\mu^n [1 + o(1)],$$
$$-Z_n^{-1} \nabla^2 \widehat{Z}_n(0) = \sigma^2 vn [1 + o(1)], \tag{4.8}$$
$$Z_n^{-1} \widehat{Z}_n \left(k/\sqrt{\sigma^2 vn} \right) = e^{-\frac{1}{2d}\|k\|^2} [1 + o(1)], \qquad k \in \mathbb{R}^d,$$

where $\mu, A, v > 0$ are constants (depending on d, L, β) and the error term in the last line is uniform in k provided $\|k\|^2 / \log n$ is sufficiently small.

In view of (4.6), the first line of (4.8) gives the scaling of the partition sum, with μ the analogue of the connective constant for SAW, the second line gives the scaling of the mean-square displacement, with $\sigma^2 v$ playing the role of the "renormalized" diffusion constant, while the third line is the central limit theorem in Fourier language.

Note that the self-repellence "renormalizes" the diffusion constant, which equals $\sigma^2 v$ instead of σ^2. We will see in Section 4.6 that $v > 1$, with v tending to 1 as $L \to \infty$.

The proof of Theorem 4.1 is given in Sections 4.2–4.6. In Sections 4.2–4.3 we derive the *lace expansion* for $Z_n(\cdot)$, which leads to a *recursion relation* for $Z_n(\cdot)$ in n. In Section 4.4 we describe *diagrammatic estimates* that bound the coefficients in this recursion relation. In Section 4.5 we state *induction hypotheses* that exploit these bounds to control the asymptotics of $Z_n(\cdot)$ as $n \to \infty$. Finally, in Section 4.6 the results are collected to complete the proof.

What follows is an *elegant but difficult computation*. The core idea is due to Brydges and Spencer [44]. Over the years, this idea has been developed and refined into a powerful tool capable of describing critical behavior in a variety of probabilistic models. For an overview we refer the reader to the Saint-Flour lectures by Slade [280].

4.2 Expansion

4.2.1 Graphs and Connected Graphs

For $w \in \mathcal{W}_n$ and $s, t \in \mathbb{N}_0$ with $s < t$, define

$$U_{st}(w) = \begin{cases} -\lambda(\beta), & \text{if } w_s = w_t, \\ 0, & \text{if } w_s \neq w_t, \end{cases} \tag{4.9}$$

with
$$\lambda(\beta) = 1 - e^{-\beta} \in (0,1). \tag{4.10}$$

Then the normalizing partition sum in (1.2) equals

$$Z_n = \sum_{x \in \mathbb{Z}^d} Z_n(x) \quad \text{with} \quad Z_n(x) = \sum_{w \in \mathcal{W}_n(x)} \prod_{0 \le s < t \le n} (1 + U_{st}(w)). \tag{4.11}$$

We take a closer look at the product.

For $w \in \mathcal{W}_n$ and $a, b \in \mathbb{N}_0$ with $a < b$, define

$$K[a,a](w) = 1 \quad \text{and} \quad K[a,b](w) = \prod_{a \le s < t \le b} (1 + U_{st}(w)), \tag{4.12}$$

and abbreviate $(a,b) \subset \mathbb{N}_0$ to denote the interval of integers strictly between a and b, and $[a,b] \subset \mathbb{N}_0$ to denote the interval including a and b.

Definition 4.2. (i) *Given an interval I, a pair $\{s,t\}$ of elements of I with $s < t$ is called an edge; st is short-hand notation for $\{s,t\}$. A set of edges is called a graph. The set of all graphs on $[a,b]$ is denoted by $\mathcal{G}[a,b]$.*
(ii) *A graph $\Gamma \in \mathcal{G}[a,b]$ is said to be connected if both a and b are endpoints of edges in Γ and, in addition, for any $c \in (a,b)$ there are $s,t \in [a,b]$ such that $s < c < t$ and $st \in \Gamma$, i.e., $\cup_{st \in \Gamma}(s,t) = (a,b)$ as intervals. The set of all connected graphs on $[a,b]$ is denoted by $\mathcal{C}[a,b]$.*

In words, a graph Γ on $[a,b]$ is connected if all the elements of (a,b) lie strictly under the "arc" of some edge in Γ (see Fig. 4.2). The edges symbolize the *self-intersections* of the path. Indeed, with the help of Definition 4.2(i) we may write

$$K[a,b](w) = \sum_{\Gamma \in \mathcal{G}[a,b]} \prod_{st \in \Gamma} U_{st}(w). \tag{4.13}$$

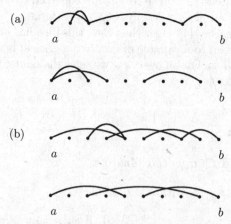

(a)

a　　　　　b

a　　　　　b

(b)

a　　　　　b

a　　　　　b

Fig. 4.2. Graphs in which an edge st is represented by an arc joining s and t. The graphs in (a) are not connected, whereas the graphs in (b) are connected.

Next, with the help of Definition 4.2(ii), define

$$J[a,a](w) = 1 \quad \text{or} \quad J[a,b](w) = \sum_{\Gamma \in \mathcal{C}[a,b]} \prod_{st \in \Gamma} U_{st}(w). \tag{4.14}$$

The following lemma, which is a kind of renewal equation, links $K[a,b](w)$ to $J[a,b](w)$.

Lemma 4.3. *For all $a, b \in \mathbb{N}_0$ with $a < b$,*

$$K[a,b](w) = K[a+1,b](w) + \sum_{m=a+1}^{b} J[a,m](w)K[m,b](w). \tag{4.15}$$

Proof. Note that $K[a+1,b](w)$ is the contribution to the sum in the right-hand side of (4.13) coming from all graphs Γ for which a is not in an edge of Γ. To resum the contribution due to the remaining graphs, we argue as follows. For Γ containing an edge that contains a, let $m = m(\Gamma)$ be the largest value of m such that the set of edges in Γ with both ends in the interval $[a, m]$ forms a connected graph on $[a, m]$. Then the sum over Γ factorizes into sums over connected graphs on $[a, m]$ and arbitrary graphs on $[m, b]$. Hence, we have

$$K[a,b](w) = K[a+1,b](w) + \sum_{m=a+1}^{b} \left(\sum_{\Gamma \in \mathcal{C}[a,m]} \prod_{st \in \Gamma} U_{st}(w) \right) K[m,b](w), \tag{4.16}$$

which proves the claim. □

4.2.2 Recursion Relation

The importance of Lemma 4.3 lies in the following. For $m \in \mathbb{N}_0$ and $x \in \mathbb{Z}^d$, define

$$\pi_m(x) = \sum_{w \in \mathcal{W}_n(x)} J[0,m](w), \tag{4.17}$$

and note that $\pi_1 \equiv 0$ because $0 \notin \Omega$ defined in (4.1).

Lemma 4.4. *For $n \in \mathbb{N}$ and $x \in \mathbb{Z}^d$,*

$$Z_n(x) = (|\Omega| D * Z_{n-1})(x) + \sum_{m=2}^{n} (\pi_m * Z_{n-m})(x), \tag{4.18}$$

with D defined in (4.7), $*$ denoting convolution in space, and $Z_0 \equiv 1$.

Proof. Recall from (4.11–4.12) that $Z_n(x) = \sum_{w \in \mathcal{W}_n(x)} K[0,n](w)$. Therefore, when we sum (4.15) with $a = 0$ and $b = n$ over $w \in \mathcal{W}_n$, note that the sums over disjoint intervals factorize, and use (4.17), we get the claim. □

As we will see, (4.17) will be the central object of our analysis. The key feature of (4.18) is that it is a *recursion relation*: $Z_n(\cdot)$ is expressed in terms of $Z_m(\cdot)$, $0 \leq m \leq n - 1$, and certain *unknown* $\pi_m(\cdot)$, $1 \leq m \leq n$. The idea is that the latter are small when L is large and fall off rapidly with m. Hence we may view (4.18) as a *perturbation* of the trivial recursion where the second term is absent, whose solution is

$$Z_n(x) = |\Omega|^n D^{*n}(x), \qquad n \in \mathbb{N}_0. \tag{4.19}$$

The second factor in the right-hand side of (4.19) is the distribution at time n of the random walk whose steps are drawn from D (the uniform distribution on Ω).

In order to make the perturbation argument work, we need to find a way to *bound* $\pi_m(\cdot)$, $1 \leq m \leq n$, in terms of $Z_m(\cdot)$, $0 \leq m \leq n - 1$. Indeed, this will open up the possibility of *induction* on n. Deriving this bound requires finding a more tractable representation of $\pi_m(\cdot)$, obtained by performing a resummation of (4.17). It is here that the notion of *lace* enters the stage.

4.3 Laces

4.3.1 Laces and Compatible Edges

Definition 4.5. *A lace is a minimally connected graph, i.e., a connected graph for which the removal of any edge results in a disconnected graph. The set of laces on $[a, b]$ is denoted by $\mathcal{L}[a, b]$. The set of laces on $[a, b]$ consisting of exactly N edges is denoted by $\mathcal{L}^{(N)}[a, b]$.*

We write $L \in \mathcal{L}^{(N)}[a, b]$ as $L = \{s_1 t_1, \ldots, s_N t_N\}$, with $s_l < t_l$ for each $l = 1, \ldots, N$. The fact that L is a lace is equivalent to a certain ordering of the s_l and t_l. For $N = 1$ we simply have $a = s_1 < t_1 = b$. For $N \geq 2$, we have

$$L \in \mathcal{L}^{(N)}[a, b] \Longleftrightarrow$$
$$a = s_1 < s_2, \ s_{l+1} < t_l \leq s_{l+2} \, (l = 1, \ldots, N - 2), \ s_N < t_{N-1} < t_N = b. \tag{4.20}$$

Thus, L divides $[a, b]$ into $2N - 1$ subintervals

$$[s_1, s_2], [s_2, t_1], [t_1, s_3], [s_3, t_2], \ldots, [s_N, t_{N-1}], [t_{N-1}, t_N]. \tag{4.21}$$

The intervals numbered $3, 5, \ldots, 2N - 3$ may have zero length when $N \geq 3$, while all other intervals have length at least 1 (see Fig. 4.3).

Given a connected graph $\Gamma \in \mathcal{C}[a, b]$, the following prescription *associates* to Γ a *unique* lace $L_\Gamma \subset \Gamma$: The lace L_Γ consists of edges $s_1 t_1, s_2 t_2, \ldots$ with $t_1, s_1, t_2, s_2, \ldots$ determined (in this order) by

$$t_1 = \max\{t: \, at \in \Gamma\},$$
$$s_1 = a, \tag{4.22}$$

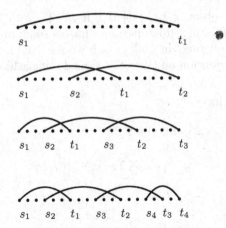

Fig. 4.3. Laces in $\mathcal{L}^{(N)}[a, b]$ for $N = 1, 2, 3, 4$, with $s_1 = a$ and $t_N = b$.

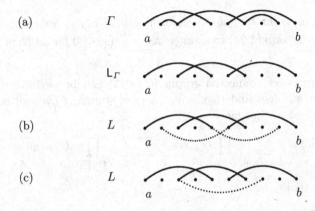

Fig. 4.4. (a) A connected graph Γ and its associated lace $L = \mathsf{L}_\Gamma$. (b) The dotted edges are compatible with the lace L. (c) The dotted edge is not compatible with the lace L.

and for $i = 1, 2, \ldots$ by

$$t_{i+1} = \max\{t : \exists s < t_i \text{ such that } st \in \Gamma\},$$
$$s_{i+1} = \min\{s : st_{i+1} \in \Gamma\}.$$

$$(4.23)$$

The prescription terminates when $t_{i+1} = b$.

Definition 4.6. *Given a lace L, the set of all edges $st \notin L$ such that $\mathsf{L}_{L\cup\{st\}} = L$ is denoted by $\mathcal{C}(L)$. Edges in $\mathcal{C}(L)$ are said to be compatible with L (see Fig. 4.4).*

Note that the construction in (4.22–4.23) looks for the longest edges possible. What Definition 4.6 says is that $st \in \mathcal{C}(L)$ if and only if the largest lace that can be found as a subset of $L \cup \{st\}$ is L itself. This will be important later on, because long loops in self-intersections have a small probability under the law of the spread-out random walk (which we removed from (4.4) because it is the uniform distribution on the set of spread-out paths defined in (4.2)).

4.3.2 Resummation

Lemma 4.7. *For $x \in \mathbb{Z}^d$ and $m \in \mathbb{N}$,*

$$\pi_m(x) = \sum_{N=1}^{\infty} (-1)^N \pi_m^{(N)}(x) \tag{4.24}$$

with

$$\pi_m^{(N)}(x) = \sum_{w \in \mathcal{W}_m(x)} \sum_{L \in \mathcal{L}^{(N)}[0,m]} \prod_{st \in L} (-U_{st}(w)) \prod_{s't' \in \mathcal{C}(L)} (1+U_{s't'}(w)), \tag{4.25}$$

where $\mathcal{L}^{(N)}[0,m]$ is the set of laces on $[0,m]$ with N edges. Note that the factor $(-1)^N$ is inserted into (4.24) to arrange that $\pi_m^{(N)}(x) \geq 0$ for all N, m, x (recall (4.9)).

Proof. The sum over connected graphs in (4.14) can be performed by first summing over all laces and then, given a lace, summing over all connected graphs associated to that lace:

$$J[a,b](w) = \sum_{L \in \mathcal{L}[a,b]} \prod_{st \in L} U_{st}(w) \sum_{\Gamma: \, L_\Gamma = L} \prod_{s't' \in \Gamma \backslash L} U_{s't'}(w). \tag{4.26}$$

Next we note that

$$L_\Gamma = L \text{ if and only if } L \text{ is a lace, } L \subset \Gamma, \quad \Gamma \backslash L \subset \mathcal{C}(L). \tag{4.27}$$

This is due to the fact that the lace L_Γ is obtained from Γ by looking for maxima and minima. Indeed, $L_\Gamma = L$ is equivalent to the property that an edge not in L is never chosen in (4.22–4.23). Using (4.27), we get that the second sum in (4.26) equals $\prod_{s't' \in \mathcal{C}(L)} (1 + U_{s't'}(w))$, which gives (recall (4.17))

$$\pi_m(x) = \sum_{w \in \mathcal{W}_m(x)} \sum_{L \in \mathcal{L}[0,m]} \prod_{st \in L} U_{st}(w) \prod_{s't' \in \mathcal{C}(L)} (1 + U_{s't'}(w)). \tag{4.28}$$

Splitting this out according to the number of edges in the laces, we get the claim. □

The factor $\prod_{st \in L} (-U_{st}(w))$ in (4.25) forces w to self-intersect at the times that are the endpoints of the edges in L (recall the second line of (4.9)). Hence, walks contributing to (4.25) are constrained to have the topology indicated

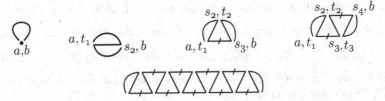

Fig. 4.5. Self-intersections required for a walk w with $\prod_{st} U_{st}(w) \neq 0$, for the laces with $N = 1, 2, 3, 4$ edges depicted in Fig. 4.3. The picture for $N = 11$ is added too.

by the diagrams in Fig. 4.5. The N-th diagram has $2N - 1$ subwalks. The factor $\prod_{s't' \in \mathcal{C}(L)}(1 + U_{s't'}(w))$ carries the self-repellence along the compatible edges. Hence, each subwalk is self-repellent (because all edges under a lace are compatible with that lace), while different subwalks are only partially (!) mutually self-repellent (because some of the compatible edges link the subwalks while others do not). The number of loops in a diagram is equal to the number of edges in the corresponding lace. The lines that are slashed correspond to subwalks that may have zero steps; the other lines correspond to subwalks that have at least one step. Note that the sum in (4.24) terminates at $N = m$.

A key feature of (4.25) is that $\mathcal{L}^{(N)}[0, m]$ is a *highly restrictive set of graphs*. Indeed, the N-th diagram requires the path to make N returns to sites visited earlier. If L is large, then such returns are "unlikely" under the law of the spread-out random walk, and so $\pi_m^{(N)}$ is expected to fall off rapidly with m and N.

Note that $|\mathcal{L}^{(N)}[0, n]|$ is bounded from above by the number of possible choices of $0 = s_0 < \cdots < s_N < n$ times the number of possible choices of $0 < t_1 < \cdots < t_N = n$, which is $[n^{N-1}/(N-1)!]^2$. Hence, summing on N, we see that

$$|\mathcal{L}[0, n]| \leq e^{2n} \ll 2^{\binom{n}{2}} = |\mathcal{G}[0, n]|. \tag{4.29}$$

4.4 Diagrammatic Estimates

The key idea to exploit Lemmas 4.4 and 4.7 is the following:

- Because the $2N - 1$ subwalks in the N-th diagram have a *repulsive* interaction, we get an upper bound on $\pi_m^{(N)}(x) \geq 0$ by ignoring the self-repellence *between* the different subwalks, retaining only the self-repellence *within* each subwalk.

After having done so, each subwalk becomes an independent self-repellent walk of a certain length, say l, starting and ending at certain sites, say u and v. The weight of this subwalk is precisely $Z_l(v - u)$. Thus, what we need to do to control the expansion is to keep track of the various summations encoded in the diagrams.

The following diagrammatic estimates make the above idea explicit.

Lemma 4.8. *Fix $d \geq 5$ and $\beta \in (0, \infty)$. Suppose that, for some $z \in (0, \infty)$,*

$$\|Z_m\|_1 z^m \leq K, \qquad \|Z_m\|_\infty z^m \leq K|\Omega|^{-1}(m+1)^{-d/2} \tag{4.30}$$

for all $1 \leq m \leq n$ and some K. Then there exist L_0 and C (both depending on d, K) such that, for all $L \geq L_0$,

$$z^m \sum_{x \in \mathbb{Z}^d} \|x\|^q \, |\pi_m(x)| \leq C|\Omega|^{-1}(m+1)^{-(d-q)/2} \tag{4.31}$$

for all $2 \leq m \leq n+1$ and $q = 0, 2, 4$.

Proof. We consider the case $q = 0$, and show how to estimate the terms with $N = 1$ and $N = 2$ (depicted by the first two figures in Fig. 4.5) for $2 \leq m \leq n+1$. A similar argument applies for $q = 2, 4$ and $N \geq 3$.

$\underline{N = 1}$: Return to (4.25). There is only one lace, namely, $L = \{0, m\}$, and all other edges are compatible. Hence we get (recall the first diagram in Fig. 4.5)

$$\sum_{x \in \mathbb{Z}^d} \pi_m^{(1)}(x) = \pi_m^{(1)}(0) \leq \sum_{y \in \mathbb{Z}^d} D(y) Z_{m-1}(y) \leq \|Z_{m-1}\|_\infty. \tag{4.32}$$

Inserting (4.30) into (4.32), we get

$$z^m \sum_{x \in \mathbb{Z}^d} \pi_m^{(1)}(x) \leq CK|\Omega|^{-1}(m+1)^{-d/2}. \tag{4.33}$$

$\underline{N = 2}$: Neglecting the self-repellence between the three subwalks in the second diagram in Fig. 4.5, we get

$$\sum_{x \in \mathbb{Z}^d} \pi_m^{(2)}(x) \leq \sum_{x \in \mathbb{Z}^d} \sum_{\substack{m_1, m_2, m_3 \geq 1 \\ m_1 + m_2 + m_3 = m}} Z_{m_1}(x) Z_{m_2}(x) Z_{m_3}(x)$$

$$\leq 3! \sum_{\substack{m_1 \geq m_2 \geq m_3 \geq 1 \\ m_1 + m_2 + m_3 = m}} \|Z_{m_1}\|_\infty \|Z_{m_2}\|_\infty \|Z_{m_3}\|_1. \tag{4.34}$$

Inserting (4.30) into (4.34), we get

$$z^m \sum_{x \in \mathbb{Z}^d} \pi_m^{(2)}(x) \leq 3! \sum_{\substack{m_1 \geq m_2 \geq m_3 \geq 1 \\ m_1 + m_2 + m_3 = m}} \|Z_{m_1}\|_\infty z^{m_1} \|Z_{m_2}\|_\infty z^{m_2} \|Z_{m_3}\|_1 z^{m_3}$$

$$\leq 3! K^3 |\Omega|^{-2} \sum_{\substack{m_1 \geq m_2 \geq m_3 \geq 1 \\ m_1 + m_2 + m_3 = m}} (m_1 + 1)^{-d/2} (m_2 + 1)^{-d/2}$$

$$\leq 3! K^3 |\Omega|^{-2} \left(\frac{m}{3} + 1\right)^{-d/2} \sum_{m_2 \geq m_3 \geq 1} (m_2 + 1)^{-d/2}$$

$$\leq CK^3 |\Omega|^{-2}(m+1)^{-d/2},$$

$$\tag{4.35}$$

where the last inequality uses $d > 4$ to make the sum in the third line finite.

For $N \geq 3$ similar estimates hold, which can be derived by induction on N. Each time N increases by 1, an extra factor $K^2 |\Omega|^{-1}$ appears, coming from the two extra subwalks in the diagram making an extra loop. The key point is that $K^2 |\Omega|^{-1} < 1$ when L is large enough, so that the prefactors are summable (!) in N. For details we refer to van der Hofstad and Slade [166]. □

The estimate in (4.31) for $q = 0, 2$ is needed to obtain the scaling in (4.8), the estimate for $q = 4$ to control error terms.

4.5 Induction

The key fact about Lemma 4.8 is that the bounds assumed in (4.30) for $1 \leq m \leq n$ are turned into a bound in (4.31) valid for $2 \leq m \leq n + 1$. This opens up the possibility of *induction* on n. Indeed, with (4.31) we have control over the coefficients in the recursion relation in Lemma 4.4.

4.5.1 Notation

The induction will be done in Fourier language. For $k \in [-\pi, \pi)^d$, let $\widehat{D}(k)$, $\widehat{\pi}_m(k)$ and $\widehat{\pi}_m^{(N)}(k)$ denote the Fourier transforms of $D(x)$, $\pi_m(x)$ and $\pi_m^{(N)}(x)$. It terms of the latter, (4.18) reads

$$\widehat{Z}_n(k) = |\Omega| \widehat{D}(k) \widehat{Z}_{n-1}(k) + \sum_{m=1}^{n} \widehat{\pi}_m(k) \widehat{Z}_{n-m}(k). \tag{4.36}$$

In what follows we will need four *induction hypotheses*. We will state these in Section 4.5.3, but first we introduce some notation and in Section 4.5.2 provide its background.

First, because

$$\|Z_n\|_1 = Z_n = \widehat{Z}_n(0), \qquad \|Z_n\|_\infty \leq \|\widehat{Z}_n\|_1, \tag{4.37}$$

Lemma 4.8 can be reformulated as saying that

- bounds on $\widehat{Z}_m(0)$ and $\|\widehat{Z}_m\|_1$ for $1 \leq m \leq n$ imply bounds on $\widehat{\pi}_m(0)$, $\nabla^2 \widehat{\pi}_m(0)$ for $2 \leq m \leq n + 1$.

Second, to formulate the induction hypotheses we need some more notation. Put $z_0 = z_1 = 1$, and define z_n recursively by

$$z_{n+1} = \frac{1}{|\Omega|} \left[1 - \sum_{m=2}^{n+1} \widehat{\pi}_m(0) [z_n]^m \right], \qquad n \in \mathbb{N}. \tag{4.38}$$

Also put

$$v_n(z) = \frac{B_n(z)}{1 + C_n(z)}, \qquad n \in \mathbb{N}, \tag{4.39}$$

with

$$B_n(z) = |\Omega|z - \sigma^{-2} \sum_{m=2}^{n} \nabla^2 \widehat{\pi}_m(0)\, z^m, \quad C_n(z) = \sum_{m=2}^{n} (m-1)\, \widehat{\pi}_m(0)\, z^m.$$

$$(4.40)$$

Fix $\delta_1, \delta_2, \delta_3 > 0$ such that

$$0 < \frac{d-4}{2} - \delta_1 < \delta_2 < \delta_2 + \delta_3 < 1 \wedge \frac{d-4}{2}. \quad (4.41)$$

We further need five constants satisfying

$$K_3 \gg K_1 \gg K_4 \gg 1, \quad K_2 \gg K_1, K_4, \quad K_5 \gg K_4. \quad (4.42)$$

Define the intervals

$$I_n = \left[z_n - K_1 |\Omega|^{-1} n^{-(d-2)/2}, z_n + K_1 |\Omega|^{-1} n^{-(d-2)/2} \right], \qquad n \in \mathbb{N}, \quad (4.43)$$

and abbreviate

$$\widehat{Z}_m(k; z) = \widehat{Z}_m(k)\, z^m, \qquad \widehat{\pi}_m(k; z) = \widehat{\pi}_m(k)\, z^m. \quad (4.44)$$

4.5.2 Heuristics

Here is the heuristics behind the above notation. In anticipation of the first line of (4.8), substitute $\widehat{Z}_n(0; \mu^{-1}) \to A$ as $n \to \infty$ into the recursion relation (4.36), with A and μ yet to be determined. This gives the relation

$$1 = |\Omega|\mu^{-1} + \sum_{m=2}^{\infty} \widehat{\pi}_m(0; \mu^{-1}). \quad (4.45)$$

The sum has not (!) yet been proved to converge. The recursion for $(z_n)_{n\in\mathbb{N}}$ defined in (4.38) approximates the relation in (4.45) by discarding the terms in the sum with $m > n$, which cannot be handled at the n-th stage of the induction argument. This is done in anticipation of $z_n \to \mu^{-1}$ as $n \to \infty$.

Differentiate the recursion relation (4.36) twice w.r.t. k, set $k = 0$, and use that odd derivatives vanish because of symmetry, to get

$$\nabla^2 \widehat{Z}_n(0; z) = |\Omega|\, z \left[\nabla^2 \widehat{Z}_{n-1}(0; z) - \sigma^2 \widehat{Z}_{n-1}(0; z) \right]$$

$$+ \sum_{m=2}^{n} \left[\widehat{\pi}_m(0; z) \nabla^2 \widehat{Z}_{n-m}(0; z) + \nabla^2 \widehat{\pi}_m(0; z) \widehat{Z}_{n-m}(0; z) \right],$$

$$(4.46)$$

using that $\widehat{D}(0) = 1$ and $\nabla^2 \widehat{D}(0) = -\sigma^2$. Approximate, for z close to z_n,

$$|\Omega|\, z \approx 1 - \sum_{m=2}^{n} \widehat{\pi}_m(0; z), \qquad \widehat{Z}_{n-m}(0; z) \approx \widehat{Z}_{n-1}(0; z). \quad (4.47)$$

Substitute this approximation into (4.46), and use the first half of (4.40), to get

$$\nabla^2 \widehat{Z}_n(0;z) - \nabla^2 \widehat{Z}_{n-1}(0;z)$$

$$\approx -\sigma^2 B_n(z)\widehat{Z}_{n-1}(0;z) + \sum_{m=2}^{n} \widehat{\pi}_m(0;z)\left[\nabla^2 \widehat{Z}_{n-m}(0;z) - \nabla^2 \widehat{Z}_{n-1}(0;z)\right].$$

$$(4.48)$$

Next, in anticipation of the second line of (4.8) and $v_n(z_n) \to v$ as $n \to \infty$, approximate, for z close to z_n,

$$\nabla^2 \widehat{Z}_{n-m}(0;z) - \nabla^2 \widehat{Z}_{n-1}(0;z) \approx (m-1)\sigma^2 v_n(z)\widehat{Z}_{n-1}(0;z). \qquad (4.49)$$

Substitute this approximation into (4.48), and use the second half of (4.40), to get

$$\nabla^2 \widehat{Z}_n(0;z) - \nabla^2 \widehat{Z}_{n-1}(0;z) \approx -\sigma^2 B_n(z)\widehat{Z}_{n-1}(0;z) + \sigma^2 v_n(z)C_n(z)\widehat{Z}_{n-1}(0;z).$$

$$(4.50)$$

The left-hand side of (4.50) is expected to be close to $-\sigma^2 v_n(z)\widehat{Z}_{n-1}(0;z)$. Equate the latter with the right-hand side of (4.50), to arrive at the definition of $v_n(z)$ in (4.39).

4.5.3 Induction Hypotheses

Our induction hypotheses read as follows.

- For all $z \in I_n$ and all $1 \le m \le n$:
  (IH1) $|z_m - z_{m-1}| \le K_1|\Omega|^{-1}m^{-d/2}$.
  (IH2) $|v_m(z) - v_{m-1}(z)| \le K_2|\Omega|^{-1}m^{-(d-2)/2}$.
  (IH3) For k such that $1 - \widehat{D}(k) \le \delta_2 m^{-1}\log m$,

$$\widehat{Z}_m(k;z) = \prod_{l=1}^{m}\left\{1 - v_l[1 - \widehat{D}(k)] + r_l(k;z)\right\} \qquad (4.51)$$

with

$$|r_l(0;z)| \le K_3|\Omega|^{-1}l^{-(d-2)/2},$$
$$|r_l(k;z) - r_l(0;z)| \le K_3|\Omega|^{-1}[1 - \widehat{D}(k)]l^{-\delta_3}. \qquad (4.52)$$

(IH4) For k such that $1 - \widehat{D}(k) > \delta_2 m^{-1}\log m$,

$$|\widehat{Z}_m(k;z)| \le K_4[1 - \widehat{D}(k)]^{-2-\delta_1}m^{-d/2},$$
$$|\widehat{Z}_m(k;z) - \widehat{Z}_{m-1}(k;z)| \le K_5[1 - \widehat{D}(k)]^{-1-\delta_1}m^{-d/2}. \qquad (4.53)$$

The role of (IH1–IH2) is to control the rate at which z_m and $v_m(z_m)$ tend to μ^{-1} and v, respectively, as $m \to \infty$. The role of (IH3–IH4) is to control the behavior of $\widehat{Z}_n(k;z)$ for small k and large k, respectively, the first coming from

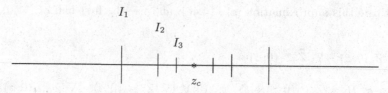

Fig. 4.6. Nested sequence of intervals $(I_n)_{n\in\mathbb{N}}$ converging to $z_c = 1/\mu$.

an expansion of the recursion relation in (4.36) around $k = 0$, the second being an error term that is needed to bound $\|\widehat{Z}_m\|_1$ (recall (4.37) and the remark following it). The factor $[1 - \widehat{D}(k)]^{-2-\delta_1}$ in (4.53) diverges as $k \to 0$, but is integrable for $d > 4$ when $\delta_1 > 0$ is taken small enough. It is *vital* that the induction hypotheses are taken for z on the *nested sequence* $(I_n)_{n\in\mathbb{N}}$ (see Fig. 4.6).

The key fact now is the following.

Lemma 4.9. (IH1)–(IH4) *can be advanced subject to* (4.41–4.42).

Proof. The advancement of (IH1)–(IH4) is technically involved. The proof starts from (4.36),

$$\widehat{Z}_{n+1}(k; z) = z\widehat{D}(k)\,\widehat{Z}_n(k; z) + \sum_{m=2}^{n+1} \widehat{\pi}_m(k; z)\,\widehat{Z}_{n+1-m}(k; z), \qquad (4.54)$$

and uses the bounds in the Fourier version of Lemma 4.8 (recall (4.37) and the remark following it). We refer to van der Hofstad [156], Section 2.4, for details. Note that *initialization* of the induction is trivial: the inequalities in (IH1–IH4) are easily seen to be true for $m = 1$, when the walk consists of a single step only. $\square$

4.6 Proof of Diffusive Behavior

With Lemma 4.9, we know that the inequalities in (IH1–IH4) are valid for *all* $n \in \mathbb{N}$. Finally, we show how this implies Theorem 4.1.

Proof. For $z \in I_n$, (IH1) gives

$$\begin{aligned}
|z_{n-1} - z| &\le |z_n - z| + |z_n - z_{n-1}| \\
&\le K_1|\Omega|^{-1}n^{-(d-2)/2} + K_1|\Omega|^{-1}n^{-d/2} \\
&\le K_1|\Omega|^{-1}(n-1)^{-(d-2)/2},
\end{aligned} \qquad (4.55)$$

and so $I_n \subset I_{n-1}$. This monotonicity is crucial, because it allows us to control z and shows that $\lim_{n\to\infty} z_n = z_c$ with $\{z_c\} = \cap_{n\in\mathbb{N}} I_n$ (see Fig. 4.6). Thus (IH1–IH4) can be applied at z_c.

By passing to the limit $n \to \infty$ in (4.38), we see that z_c solves the equation

$$z_c = \frac{1}{|\Omega|}\left[1 - \sum_{m=2}^{\infty} \widehat{\pi}_m(0; z_c)\right], \qquad (4.56)$$

where the sum converges because after the induction is completed we know that $|\widehat{\pi}_m(0; z_c)|$ falls off like $|\Omega|^{-1} m^{-d/2}$ for large m. Thus, μ in the first line of (4.8) equals (recall (4.45))

$$\mu = 1/z_c. \qquad (4.57)$$

For $k = 0$, (IH3) reduces to $\widehat{Z}_m(0; z) = \prod_{l=1}^{m}[1 + r_l(0; z)]$. Therefore the bound on $r_l(0)$ in (4.52) implies that

$$\lim_{n \to \infty} \widehat{Z}_n(0)\mu^{-n} = A = \prod_{l=1}^{\infty}[1 + r_l(0; \mu^{-1})]. \qquad (4.58)$$

This proves the first line in (4.8).

Picking $k/\sqrt{\sigma^2 vn}$ instead of k in (4.51), we get from (IH3) that

$$\widehat{Z}_n\left(\frac{k}{\sqrt{\sigma^2 vn}}\right)\mu^{-n}$$

$$= \prod_{l=1}^{n}\left\{1 - v_l(\mu^{-1})\left[1 - \widehat{D}\left(\frac{k}{\sqrt{\sigma^2 vn}}\right)\right] + r_l\left(\frac{k}{\sqrt{\sigma^2 vn}}; \mu^{-1}\right)\right\}$$

$$= [1 + o(1)] A \prod_{l=1}^{n}\left\{1 - v_l(\mu^{-1})\left[1 - \widehat{D}\left(\frac{k}{\sqrt{\sigma^2 vn}}\right)\right]\right\} \qquad (4.59)$$

$$= [1 + o(1)] A\, e^{-\frac{1}{2d}\|k\|^2},$$

where we use that for $l \to \infty$ (recall (4.7))

$$1 - v_l(\mu^{-1})\left[1 - \widehat{D}\left(\frac{k}{\sqrt{\sigma^2 vn}}\right)\right] = e^{-\frac{1}{vn} v_l(\mu^{-1})\frac{1}{2d}\|k\|^2[1+o(1)]},$$

$$v_l(\mu^{-1}) = v[1 + o(1)], \qquad (4.60)$$

with v given by (recall (4.40))

$$v = \frac{B_\infty(\mu^{-1})}{1 + C_\infty(\mu^{-1})}. \qquad (4.61)$$

This proves the third line in (4.8). The second part of (4.60) follows from (4.39) and (IH2), and needs $d > 4$ to guarantee that $B_\infty(\mu^{-1})$ and $C_\infty(\mu^{-1})$ are finite.

The second line in (4.8) follows by a Taylor expansion of $\widehat{Z}_n(k; \mu^{-1})$ around $k = 0$ together with the bounds in (IH3).

Note that (IH4) is not used in the above. However, this inequality is needed to control the error terms. $\quad\square$

Thus, we have finally completed the proof of Theorem 4.1. The constants in Theorem 4.1 are expressed in terms of $\widehat{\pi}_m$. The bounds on $\widehat{\pi}_m$, together with exact enumeration for small and moderate m, allow for an estimate of these constants (see e.g. Clisby, Liang and Slade [68] for SAW with $L = 1$). For instance, it can be deduced from (4.61) that $v > 1$ for L large enough, tending to 1 as $L \to \infty$.

4.7 Extensions

(1) The induction method in Section 4.5 was introduced in van der Hofstad, den Hollander and Slade [162] and was subsequently refined in van der Hofstad and Slade [166]. In the latter paper it is shown that, in some sense, (IH1–IH4) are universal induction hypotheses for diffusive scaling. The advantage of the induction method is that it avoids taking the Laplace transform in time, which is non-trivial to invert. Theorem 4.1 extends to spread-out random walk whose steps are not drawn uniformly from the set Ω defined in (4.1). See [166] for details.

(2) There are other ways to manipulate the recursion relation in Lemma 4.4 and the diagrammatic estimates in Lemmas 4.8. One is via generating functions (Hara and Slade [147]), an approach explained in detail in Madras and Slade [230]. An alternative approach, due to Bolthausen and Ritzmann [32], is via Banach fixed-point theorems. Here, (4.54) with $k = 0$ and $z = z_c$ is viewed as a fixed-point equation for the sequence $(\widehat{Z}_m[z_c]^m)_{m\in\mathbb{N}}$. The operator acting on this sequence is shown to be a contraction when β is sufficiently small. Control of the π_m is done via an induction scheme, reminiscent of that described in Section 4.5. The advantage of this approach is that it avoids taking the Fourier transform altogether, working in real space and time.

(3) Van der Hofstad, den Hollander and Slade [162] and van der Hofstad and Slade [166] extend the third line of (4.8) to a local central limit theorem. Rather than considering the probability that the right endpoint of the soft polymer is at a single site at a distance $O(\sqrt{n})$ from the origin, it is necessary to consider the probability that it lies in a box whose size tends to infinity as $n \to \infty$. To appreciate why, note that for SAW ($\beta = \infty$) the path cannot return to the origin.

(4) The third line of (4.8) is a statement about the asymptotic behavior of the two-point function $Z_n(x)$ as $n, \|x\| \to \infty$. Similar scaling results can be derived for the higher-point functions, defined by

$$Z_{n_1,\ldots,n_r}(x_1,\ldots,x_r) = \sum_{w\in\mathcal{W}_n} e^{-\beta I_n(w)} 1_{\{w_{n_1}=x_1,\ldots,w_{n_r}=x_r\}}, \qquad (4.62)$$

for $r = 2, 3, 4 \ldots$. Indeed, even the functional central limit can be proved, with Brownian motion as the scaling limit. We refer to van der Hofstad and Slade [167] for details.

(5) The estimates in Section 4.4–4.5 are valid uniformly in $\beta \in (0, \infty)$ for L sufficiently large. Indeed, the estimates only use that $\lambda(\beta)$ in (4.10) satisfies $\lambda(\beta) \leq 1$. Therefore, Theorem 4.1 equally well applies to the spread-out SAW ($\beta = \infty$). For $L = 1$, the first and second line of (4.8) were proved for $\beta = \infty$ in Hara and Slade [147], [148]. The proof is computer-assisted: for small and moderate m the diagrams are computed via exact enumeration, while for large m bounds on the diagrams are used. This is needed to get convergence without a small parameter.

(6) Heydenreich, van der Hofstad and Sakai [152] perform the lace expansion for SAW with long-range steps, i.e., the reference measure P_n is not that of SRW (which we actually dropped from (4.4) because it is uniform) but of a random walk whose step distribution has a polynomial tail. The critical dimension above which diffusive behavior occurs is computed as a function of the tail exponent, and turns out to drop from 4 and 0 as the tail becomes thicker. This is because self-intersections are less likely for long-range random walk. (A similar drop will be encountered for the elastic polymer described in Chapter 5, for a related but different reason.) Heydenreich [151] shows that the path measure converges weakly as $n \to \infty$ to that of an α-stable process, with α depending on the tail exponent.

(7) Ueltschi [299] looks at a spread-out SAW with self-attraction, i.e., (4.9) is replaced by

$$U_{st}(w) = \begin{cases} 1, & \text{if } \|w_s - w_t\| = 0, \\ -(1 - e^\gamma), & \text{if } \|w_s - w_t\| = 1, \\ 0, & \text{otherwise,} \end{cases} \tag{4.63}$$

where $\gamma \in (0, \infty)$ is a binding energy between neighboring monomers. It is shown that Theorem 4.1 carries over in the following sense: the first line of (4.8) is true when $d \geq 5$ and $\gamma \leq \gamma_0$, while the second line is true when $d \geq 5$, $\gamma \leq \gamma_0$ and $L \geq L_0$ (with γ_0 and L_0 depending on d). The step distribution of the SAW is assumed to have a particular scaling form and to be sufficiently "smooth", which makes the model somewhat restrictive. The proof relies on the lace expansion. The convergence of the lace expansion is the hard issue, because of the presence of the self-attraction (recall the remarks made at the beginning of Section 4.4). Convergence is achieved by playing out the self-attraction against the self-avoidance, showing that the effective interaction is repulsive. In particular, Z_n is shown to be submultiplicative ($Z_{m+n} \leq Z_m Z_n$ for all $m, n \in \mathbb{N}$), implying that the free energy exists. We will return to polymers with self-repellence and self-attraction in Chapter 6.

(8) Networks of mutually-avoiding self-avoiding walks have been considered in van der Hofstad and Slade [167] and in Holmes, Járai, Sakai and Slade [174]. Gaussian scaling behavior is proved for the spread-out model in $d > 4$. The diffusion constant depends on the topology of the network: each node of the network contributes to this constant in a way that depends on how the self-avoiding walks are tied together, i.e., on the topology of the network (for an

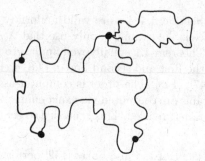

Fig. 4.7. An example of a network of SAWs.

example, see Fig. 4.7). This problem belongs to the topic of branched polymers, which is not addressed in the present monograph.

4.8 Challenges

(1) Prove that the constant v in Theorem 4.1 is a non-decreasing function of β. In view of the expression found in (4.61), this requires a strong control on the coefficients in the lace expansion.

(2) Extend Theorem 4.1 to SAW ($\beta = \infty$) in $d \geq 5$ with $L = 1$ without the help of the computer. This seems a very hard challenge.

(3) Derive an LDP for the endpoint of the soft polymer in $d \geq 5$, either with L large or β small. Determine how the rate function in this LDP depends on β. For $d = 1$, the LDP was derived in Chapter 3. The results in Bolthausen and Ritzmann [32] yield an upper bound on the rate function for $d \geq 5$ and β small.

(4) For the continuum version of the Domb-Joyce model – the Edwards model based on Brownian mentioned in Section 3.6, Extension (5) – the existence of the soft polymer measure is non-trivial when $d \geq 2$, unlike the situation in $d = 1$ where the Hamiltonian is given by (3.78). Indeed, for $d \geq 2$ the self-repellence via the Dirac delta-function is not properly defined and some regularization is needed. Varadhan [301] gives a construction of the soft polymer measure for $d = 2$, Bolthausen [26] for $d = 3$ (simplifying and improving earlier work by Westwater [310], [311] and Kusuoka [223]). The construction uses a renormalization procedure that is necessary to deal with the accumulation of short self-intersections in Brownian motion. For $d = 2$ the soft polymer measure is absolutely continuous w.r.t. the law of Brownian motion, for $d = 3$ it is not. For further details we refer to the Saint-Flour lectures by Bolthausen [28], Chapter 1. The challenge is to understand the behavior of this measure as a function of time and of the self-repellence parameter.

(5) What can be done for the spread-out random walk with self-attraction beyond the restrictive model studied in Ueltschi [299]?

5

Elastic Polymers

In this chapter we take a brief look at a version of the soft polymer where the penalty of the self-intersections decays with the loop length, i.e., the difference between the times at which the self-intersection occurs. This model is called the *elastic polymer*. Interestingly, it will turn out that this model has diffusive behavior in any $d \geq 1$ as soon as the decay is sufficiently fast, namely, the critical dimension for diffusive behavior is *lower* than $d = 4$ and depends on the parameter controlling the *rate of decay* of the penalty as a function of the loop length. We will see that the scaling behavior of the elastic polymer is in fact highly sensitive to the value of this parameter. The lace expansion that was used in Chapter 4 can be carried over with minor modifications to arrive at the main scaling result.

Section 5.1 defines the model, while Section 5.2 describes the main result, which is taken from van der Hofstad, den Hollander and Slade [162].

5.1 A Polymer with Decaying Self-repellence

We consider the model in which the set of paths and the Hamiltonian in (3.1–3.2) are replaced by

$$
\mathcal{W}_n = \{ w = (w_i)_{i=0}^n \in (\mathbb{Z}^d)^{n+1} \colon w_0 = 0, \ \|w_{i+1} - w_i\| = 1 \ \forall 0 \leq i < n \},
$$
$$
H_n(w) = \beta I_n^\kappa(w),
$$
(5.1)

where $\beta, \kappa \in (0, \infty)$ and

$$
I_n^\kappa(w) = \sum_{\substack{i,j=0 \\ i<j}}^n (j - i)^{-\kappa} 1_{\{w_i = w_j\}},
$$
(5.2)

and (3.3) is replaced by

$$
P_n^{\beta,\kappa}(w) = \frac{1}{Z_n^{\beta,\kappa}} \, e^{-\beta I_n^\kappa(w)}, \qquad w \in \mathcal{W}_n,
$$
(5.3)

F. den Hollander, *Random Polymers*,
Lecture Notes in Mathematics 1974, DOI: 10.1007/978-3-642-00333-2_5,
© Springer-Verlag Berlin Heidelberg 2009

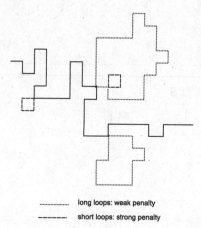

································ long loops: weak penalty

─ ─ ─ ─ ─ ─   short loops: strong penalty

Fig. 5.1. A path with long loops (thin dash) and short loops (thick dash). Long loops are less penalized than short loops.

where, as in (4.4), we drop the uniform reference measure P_n. Thus, self-intersections occurring after a short time lapse receive a larger penalty than self-intersections occurring after a long time lapse. In other words, short loops are more costly than long loops, and so we may think of the extra parameter κ as measuring the degree of *elasticity* of the polymer (see Fig. 5.1).

Our main theorem is the following analogue of Theorem 4.1, which holds under the restriction

$$\kappa + \frac{d-4}{2} > 0. \tag{5.4}$$

In what follows, β is the small parameter, rather than $|\Omega|^{-1}$ in Chapter 4.

Theorem 5.1. *Fix $d \geq 1$. If (5.4) holds, then there exists a constant $\beta_0 > 0$ (depending on d, κ) such that for all $\beta \in (0, \beta_0)$ and as $n \to \infty$,*

$$Z_n = A\mu^n \left[1 + o(1)\right],$$

$$-Z_n^{-1} \nabla^2 \widehat{Z}_n(0) = Dn \left[1 + o(1)\right], \tag{5.5}$$

$$Z_n^{-1} \widehat{Z}_n \left(k/\sqrt{Dn}\right) = e^{-\frac{1}{2d}\|k\|^2} \left[1 + o(1)\right], \qquad k \in \mathbb{R}^d,$$

where $\mu, A, D > 0$ are constants (depending on d, β, κ) and the error term in the last line is uniform in k provided $\|k\|^2 / \log n$ is sufficiently small.

Thus, the critical dimension is lowered from $d = 4$ to $d = 0 \vee (4 - 2\kappa)$. For instance, in $d = 1$ there is diffusive behavior as soon as $\kappa > \frac{3}{2}$.

5.2 The Lace Expansion Carries over

The proof of Theorem 5.1 can be built on the lace expansion approach in Chapter 4. Indeed, all we have to do is adapt the definition of $U_{st}(w)$ in (4.9) to

$$U_{st}(w) = \begin{cases} -\lambda_{t-s}(\beta, \kappa), & \text{if } w_s = w_t, \\ 0, & \text{if } w_s \neq w_t, \end{cases} \tag{5.6}$$

with

$$\lambda_m(\beta, \kappa) = 1 - e^{-\beta m^{-\kappa}}, \qquad m \in \mathbb{N}. \tag{5.7}$$

In particular, all the expansion formulas leading up to Lemma 4.4 and (4.24–4.25) carry over verbatim. The key difference is that the weights of the subwalks in the successive diagrams (recall Fig. 4.5) decay faster with their length as when $\kappa = 0$. This is why we get better convergence. More precisely, because

$$\lambda_m(\beta, \kappa) \leq \beta m^{-\kappa}, \tag{5.8}$$

Lemma 4.8 carries over with (4.31) replaced by

$$\sum_{x \in \mathbb{Z}^d} \|x\|^q |\pi_m(x)| \, z^m \leq C\beta(m+1)^{-(d+2\kappa-q)/2} \tag{5.9}$$

for $2 \leq m \leq n$ and $q = 0, 2, 4$, subject to (4.30) and β being small enough. Indeed, this is easily checked by repeating the estimates in (4.32–4.35), taking advantage of (5.8). In particular, for $q = 0, 2$ the bound in (5.9) yields (recall the notation introduced in (4.44))

$$|\widehat{\pi}_m(0; z_c)| \leq C\beta(m+1)^{-(\frac{d}{2}+\kappa)}, \quad |\nabla^2 \widehat{\pi}_m(0; z_c)| \leq C\beta(m+1)^{-(\frac{d}{2}+\kappa-1)}, \tag{5.10}$$

which shows that $\frac{d}{2} + \kappa > 2$ is needed to get convergence, for instance, for the quantity v_n in (4.61), based on the definitions of B_n and C_n in (4.39–4.40). This explains the restriction in (5.4).

For details of the full argument, we refer to van der Hofstad, den Hollander and Slade [162]. The induction hypotheses (IH1–IH4) formulated in Section 4.5 made their first appearance in that paper, dealing with the elastic polymer subject to (5.4). The soft polymer is the special case with $\kappa = 0$.

As in Sections 4.4–4.6, β may be replaced by $|\Omega|^{-1}$, in which case the lace expansion converges for L large enough.

It is important that the interaction in (5.2) is *repulsive*, resulting in $U_{st}(w) \leq 0$ in (5.6). Indeed, this played a key role in Section 4.4, where we estimated the quantity $\pi_m^{(N)}$ defined in (4.25). Removing the interaction *between* the subwalks in the diagrams in Fig. 4.3 results in an upper bound on $\pi_m^{(N)}$ for each N and m (recall the proof of Lemma 4.8)).

5.3 Extensions

(1) It is shown in van der Hofstad, den Hollander and Slade [162] that also the local central limit theorem holds for the elastic polymer subject to (5.4), provided a box of growing size is considered (recall Section 4.7, Extension (3)).

(2) Van der Hofstad [155], extending an earlier result by Kennedy [208], proves that in $d = 1$ the elastic polymer is ballistic for $\kappa \in [0, 1]$ and β sufficiently large. The proof uses the lace expansion, perturbing around the one-sided deterministic walk instead of SRW and employing induction hypotheses that are adapted to proving ballistic behavior. It has not yet been proved that the same result is true for all $\beta \in (0, \infty)$.

(3) The soft polymer with $\beta \in (-\infty, 0)$ (i.e., self-intersections are rewarded rather than penalized) is localized: with a probability tending to 1 as $n \to \infty$ the path eventually jumps back and forth between two sites (Oono [251], [252]). Brydges and Slade [42], [43] consider the soft polymer with β replaced by $\beta_n = \beta n^{-\kappa}$, $\kappa \in [0, \infty)$, allowing $\beta \in \mathbb{R}$. This model is attractive for $\beta < 0$ and repulsive for $\beta > 0$. They show the following:

(a) For $d \geq 1$ and $\kappa \in [0, 1)$ the polymer is *collapsed* for all $\beta \in (-\infty, 0)$, i.e., there is a finite box (whose size depends on d, κ and β) such that the polymer of length n lies entirely inside this box with a probability that tends to 1 exponentially fast as $n \to \infty$.

(b) For $d \geq 1$ and $\kappa = 1$ there is a critical β_c^* such that the polymer is *diffusive* for all $\beta \in (\beta_c^*, \infty)$, with $\beta_c^* = 0$ in $d = 1$ and $\beta_c^* < 0$ for $d \geq 2$. For $d \geq 3$ the diffusion constant equals 1 and the scaling limit is Brownian motion, while for $d = 2$ the diffusion constant is "renormalized" and the scaling limit is the two-dimensional (attractive) Edwards model constructed in Varadhan [301] (recall Challenge (3) in Section 4.8).

(c) For $d = 1$ and $\kappa = \frac{3}{2}$ the polymer is *diffusive* for all $\beta \in \mathbb{R}$. The diffusion constant is "renormalized" and the scaling limit is the one-dimensional (attractive) Edwards model with Hamiltonian (3.78).

Bolthausen and Schmock [33], [34] show that

(d) For $d \geq 1$ and $\kappa = 1$ there is a critical β_c such that the polymer is *collapsed* for all $\beta \in (-\infty, \beta_c)$, with $\beta_c = 0$ in $d = 1$ and $\beta_c < 0$ for $d \geq 2$.

Clearly, $\beta_c \leq \beta_c^*$. It is not known whether $\beta_c = \beta_c^*$ for $d \geq 2$ (see Fig. 5.2). Formulas for these two critical thresholds are

$$\beta_c = \sup \left\{ \beta \in \mathbb{R} : \limsup_{n \to \infty} Z_n^{1/n} > 1 \right\},$$

$$\beta_c^* = \inf \left\{ \beta \in \mathbb{R} : \sup_{n \in \mathbb{N}} Z_n^* < \infty \right\}, \tag{5.11}$$

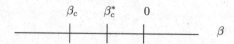

Fig. 5.2. Phase diagram for the weakly attracting soft polymer with $d \geq 2$ and $\kappa = 1$, with the collapsed phase for $\beta \in (-\infty, \beta_c)$ and the diffusive phase for $\beta \in (\beta_c^*, \infty)$.

where
$$Z_n = E\left(\exp\left[-\beta n^{-\kappa} I_n\right]\right),$$
$$Z_n^* = E\left(\exp\left[-\beta n^{-\kappa}(I_n - E(I_n))\right]\right),$$
(5.12)

with I_n the intersection local time in (3.2). Note that if $\beta_c = \beta_c^*$, then the phase transition between collapsed and diffusive is discontinuous in $d \geq 3$ (and possibly continuous in $d = 2$).

For further details we refer to the Saint-Flour lectures by Bolthausen [28], Chapter 2. Collapse transitions are well known in statistical physics, but mathematically they turn out to be hard. We return to this topic in Chapter 6.

5.4 Challenges

Very little information is available when $\kappa \in (0, \frac{4-d}{2}]$. Neither the large deviation approach in Chapter 3 nor the lace expansion approach in Chapter 4 are able to handle this regime.

(1) It was conjectured by Caracciolo, Parisi and Pelissetto [50] that for $1 \leq d \leq 4$ the critical exponent $\nu(\kappa)$ for the end-to-end distance, defined by

$$v_n = \sum_{w \in W_n} \|w_n\|^2 P_n^{\beta,\kappa}(w) = D n^{2\nu(\kappa)}[1 + o(1)] \qquad \text{as } n \to \infty,$$
(5.13)

equals (see Fig. 5.3)

$$\nu(\kappa) = \nu_{\text{SAW}} \wedge \nu_{\text{MF}}(\kappa)$$
(5.14)

with ν_{SAW} the critical exponent for the self-avoiding walk found in (2.26), and $\nu_{\text{MF}}(\kappa)$ the exponent for the "mean-field version" of the elastic polymer given by

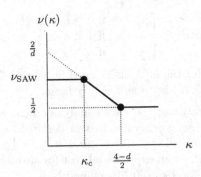

Fig. 5.3. Conjectured critical exponent of the elastic polymer for $1 \leq d \leq 4$. Here, $\kappa_c = \frac{4-d}{d+2}$ when $\nu_{\text{SAW}} = \frac{3}{d+2}$, the Flory value in (2.31). The slope of the linear piece between κ_c and $\frac{4-d}{2}$ is $-\frac{1}{4}$. Proofs are known only for $\kappa > \frac{4-d}{2}$ and for $\kappa \leq \kappa_c = 1$ when $d = 1$ and β is sufficiently large. Note that the behavior is superdiffusive and subballistic for $\kappa \in (\kappa_c, \frac{4-d}{2})$.

$$\nu_{\mathrm{MF}}(\kappa) = \begin{cases} \frac{1}{2}, & \text{if } \kappa \geq \frac{4-d}{2}, \\ \frac{2-\kappa}{d}, & \text{if } \kappa < \frac{4-d}{2}. \end{cases} \qquad (5.15)$$

The challenge is to provide a proof of this conjecture. In [50], the scaling in (5.13–5.15) is amply supported by simulations.

The formula for $\nu_{\mathrm{MF}}(\kappa)$ in (5.15) comes from a *Gaussian approximation* of the path measure in (5.3). Namely, (5.2) is replaced by

$$I_n^G(w) = \sum_{\substack{i,j=0 \\ i<j}}^{n} (G^{-1})(i,j)\,(w_i, w_j), \qquad w \in \mathcal{W}_n, \qquad (5.16)$$

for a given covariance matrix $G = (G(i,j))_{i,j \in \mathbb{N}_0}$, satisfying $G(i,j) = G(0, j-i)$, $i, j \in \mathbb{N}_0$, with $(\cdot, \cdot)$ the inner product on $\mathbb{Z}^d$. Subsequently, the functional

$$G \mapsto \lim_{n \to \infty} \frac{1}{n} \left\{ \sum_{w \in \mathcal{W}_n} P_n^{\beta,G}(w)\, \beta\, [I_n^\kappa(w) - I_n^G(w)] - \log Z_n^G \right\} \qquad (5.17)$$

is minimized w.r.t. to G, where $P_n^{\beta,G}$ is the path measure associated with I_n^G and Z_n^G is its normalizing partition sum. This variational computation leads to a minimizer $G_{\min}$ whose Fourier transform

$$\widehat{G}_{\min}(k) = \sum_{l \in \mathbb{N}_0} G_{\min}(0, l)\, e^{i\,kl}, \qquad k \in [0, 2\pi), \qquad (5.18)$$

scales like

$$\widehat{G}_{\min}(k) \asymp k^{-2\nu_{\mathrm{MF}}(\kappa)-1}, \qquad k \downarrow 0, \qquad (5.19)$$

with $\nu_{\mathrm{MF}}(\kappa)$ given by (5.15). The rationale behind (5.14) is that the Gaussian approximation overestimates the critical exponent $\nu(\kappa)$, because (5.16) carries long-range interactions whereas (5.2) carries only local interactions.

For $d = 1$, (5.14–5.15) amount to

$$\nu(\kappa) = \begin{cases} 1, & \text{if } 0 \leq \kappa \leq 1, \\ 2 - \kappa, & \text{if } 1 < \kappa < \frac{3}{2}, \\ \frac{1}{2}, & \text{if } \kappa \geq \frac{3}{2}. \end{cases} \qquad (5.20)$$

The result by van der Hofstad [155] mentioned in Section 5.3, Extension (3), proves the first line of (5.20) for β sufficiently large.

(2) The difficulty in analyzing the elastic polymer in $d = 1$ in the regime $\kappa \in (0, 1]$ is an underlying *degeneracy*. In van der Hofstad and König [165] it is conjectured that, for all $\beta \in (0, \infty)$ and $\kappa \in (0, 1)$, the speed of the polymer satisfies an LDP, at an exponential cost for speeds in $(\theta^*, 1]$ but at a subexponential cost for speeds in $[0, \theta^*]$, namely,

$$\lim_{n \to \infty} n^{-(1-\kappa)} \log P_n^{\beta,\kappa}(S_n = \lceil \theta n \rceil) = -\frac{\beta}{2(1-\kappa)}\, \frac{(\theta^* - \theta)^{1-\kappa}}{(\theta^*)^{2-\kappa}}, \qquad \theta \in [0, \theta^*],$$

$$(5.21)$$

with $\theta^* = \theta^*(\beta, \kappa) \in (0, 1)$ the speed. The challenge is to prove this conjecture.

A heuristic explanation of (5.21), similar in the spirit to what we found in Section 3.5 for the one-dimensional soft polymer, runs as follows. For $\theta \in [0, \theta^*)$, the best strategy for the polymer is to first move at speed θ^* for a time $\frac{(\theta^*+\theta)}{2\theta^*}n$ and afterwards move at speed $-\theta^*$ for a time $\frac{(\theta^*-\theta)}{2\theta^*}n$, so as to arrive at site θn at time n. In each of the two stretches, the polymer visits a site $1/\theta^*$ times (on average), so that the interaction it accumulates between the two stretches equals $n^{1-\kappa}$ times the right-hand side of (5.21). This is the cost compared to the strategy in which the polymer moves at speed θ^* for a time n.

(3) Find out whether for $d = 1$ and $\kappa = 1$ there is a critical value $\beta_c \in (0, \infty)$ such that the elastic polymer is ballistic for $\beta > \beta_c$ and subdiffusive for $\beta < \beta_c$. Or is $\beta_c = 0$? (Recall Fig. 5.3.)

(4) For $d = 1$, try to derive a functional central limit theorem and an LDP. These are the analogues of Challenges (2) and (3) in Section 3.7. The method of local times used in Chapter 3 is a powerful tool that can potentially be adapted to deal with the time lapses between the successive visits to a site.

(5) For the soft polymer with $d \geq 2$ and $\beta_n = \beta n^{-1}$, find out whether or not $\beta_c = \beta_c^*$. (See Extension (3) in Section 5.3.)

(6) For the soft polymer with $d = 1$ and $\beta_n = \beta n^{-\kappa}$, prove that the variance of the end-to-end distance of the polymer grows like $n^{\nu(\kappa)}$ as $n \to \infty$ with

$$\nu(\kappa) = \kappa - 1 \text{ for } \beta < 0 \text{ and } \kappa \in (1, \tfrac{3}{2}). \qquad (5.22)$$

If this were true, then as κ is varied (rather than β) the polymer makes a gradual crossover from collapsed to diffusive behavior. This is in contrast to $d \geq 2$, where we know from the observations made in Section 5.2, Extension (3), that the polymer makes a sharp crossover at $\kappa = 1$ when $\beta > \beta_c^*$. We already know from (3.75–3.76) that $\nu(\kappa) = 1 - \tfrac{1}{3}\kappa$ for $\beta > 0$ and $\kappa \in (0, \tfrac{3}{2})$.

6

Polymer Collapse

In this chapter we look at the soft polymer (studied in Chapters 3–4) and add to the penalty for self-intersections a reward for self-touchings, i.e., a negative energy is associated with contacts between any two monomers that are not connected to each other within the polymer chain. This is a model of a *polymer in a poor solvent*: when the polymer does not like to make contact with a solvent it is immersed in, it tries to make contact with itself in order to push out the solvent. An example is polystyrene dissolved in cyclohexane. At temperatures above 35 degrees Celsius the cyclohexane is a good solvent, at temperatures below 30 it is a poor solvent. When cooling down, the polystyrene undergoes a transition from a "random coil" to a "compact ball" (see e.g. S.-T. Sun, I. Nishio, G. Swislow and T. Tanaka [287] and S.F. Sun, C.-C. Chou and R.A. Nash [286]).

In Section 6.1 we consider an undirected model, in Section 6.2 a directed model. We will see that there are three phases: *extended*, *collapsed* and *localized*. The transition between the last two phases is relatively well understood, the transition between the first two phases is not. The localized phase is not present in the case of SAW (infinite self-repellence).

In Section 4.7, Extension (7), and Section 5.3, Extension (3), we already encountered a polymer with an attractive interaction. In the present chapter we are facing a model with mixed interactions, both repulsive and attractive. Due to the presence of the latter, the lace expansion method does not work without modification. We will use *local times* and *generating functions* instead.

6.1 An Undirected Polymer in a Poor Solvent

Our choice for the set of paths and for the Hamiltonian is

$$\mathcal{W}_n = \left\{ w = (w_i)_{i=0}^n \in (\mathbb{Z}^d)^{n+1} \colon w_0 = 0,\ \|w_{i+1} - w_i\| = 1\ \forall\, 0 \le i < n \right\},$$
$$H_n(w) = \beta I_n(w) - \gamma J_n(w),$$

$$(6.1)$$

F. den Hollander, *Random Polymers*,
Lecture Notes in Mathematics 1974, DOI: 10.1007/978-3-642-00333-2_6,
© Springer-Verlag Berlin Heidelberg 2009

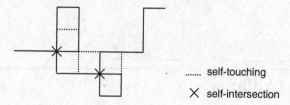

Fig. 6.1. A polymer with self-intersections and self-touchings.

where $\beta, \gamma \in (0, \infty)$ and

$$I_n(w) = \sum_{\substack{i,j=0 \\ i<j}}^{n} 1_{\{\|w_i - w_j\| = 0\}},$$

$$J_n(w) = \frac{1}{2d} \sum_{\substack{i,j=0 \\ i<j-1}}^{n} 1_{\{\|w_i - w_j\| = 1\}},$$

(6.2)

are the number of *self-intersections*, respectively, the number of *self-touchings* (divided by $2d$ to account for the $2d$ possible directions). In this model, the polymer suffers on-site repulsion and off-site attraction: each self-intersection contributes an energy β and is penalized by a factor $e^{-\beta}$, while each self-touching contributes an energy $-\gamma$ and is rewarded by a factor e^{γ} (see Fig. 6.1).

Our choice of the path measure in (1.2) is

$$P_n^{\beta,\gamma}(w) = \frac{1}{Z_n^{\beta,\gamma}} e^{-\beta I_n(w) + \gamma J_n(w)} P_n(w), \qquad w \in \mathcal{W}_n, \tag{6.3}$$

where P_n is the law of SRW. In what follows we will drop the orderings $i < j$ and $i < j - 1$ from (6.2), as in (3.8). This amounts to replacing β, γ by $2\beta, 2\gamma$ in the Hamiltonian and adding to it a constant term $\beta(n+1) - \gamma n$. The advantage of dropping the ordering is that we can write

$$H_n(w) = \beta \widehat{I}_n(w) - \gamma \widehat{J}_n(w),$$

$$\widehat{I}_n(w) = \sum_{x \in \mathbb{Z}^d} \ell_n(x)^2,$$

$$\widehat{J}_n(w) = \frac{1}{2d} \sum_{\substack{x \in \mathbb{Z}^d \\ \|e\|=1}} \ell_n(x)\ell_n(x+e),$$

(6.4)

which brings us back to the *local times* that played such an important role in Chapter 3. We henceforth write $I_n(w), J_n(w)$ again, dropping the overscript.

We will prove the following two theorems, both of which are due to van der Hofstad and Klenke [163]. For $L \in \mathbb{N}$, abbreviate $\Lambda_L = [-L, L]^d \cap \mathbb{Z}^d$.

Theorem 6.1. *If $\gamma < \beta$, then the polymer is minimally extended, i.e., there exists an $\epsilon_0 = \epsilon_0(\beta, \gamma) > 0$ such that for all $0 < \epsilon \leq \epsilon_0$ there exists a $c = c(\beta, \gamma, \epsilon) > 0$ such that*

$$P_n^{\beta,\gamma}\left(S_i \in \Lambda_{\lceil \epsilon n^{1/d}\rceil} \; \forall 0 \leq i \leq n\right) \leq e^{-cn} \qquad \forall n \in \mathbb{N}. \tag{6.5}$$

Theorem 6.2. *If $\gamma > \beta$, then the polymer is exponentially localized, i.e., there exist $c = c(\beta, \gamma) > 0$ and $L_0 = L_0(\beta, \gamma) \in \mathbb{N}$ such that*

$$P_n^{\beta,\gamma}\left(S_i \in \Lambda_L \; \forall 0 \leq i \leq n\right) \geq 1 - e^{-cLn} \qquad \forall n \in \mathbb{N}, \, L \geq L_0. \tag{6.6}$$

Thus, at $\gamma = \beta$ a transition takes place from a phase in which the polymer is spread out over a box of volume at least order n to a phase in which it is confined to a finite box.

The proofs of Theorems 6.1 and 6.2 are given in Sections 6.1.1 and 6.1.2. We will see that the key ingredients in the proofs are the following properties for the partition sum:

$$\gamma < \beta: \quad \limsup_{n \to \infty} \frac{1}{n} \log Z_n^{\beta,\gamma} \in (-\infty, 0),$$

$$\gamma > \beta: \quad \liminf_{n \to \infty} \frac{1}{n^2} \log Z_n^{\beta,\gamma} \in (0, \infty). \tag{6.7}$$

Note that existence of the limits does not follow from subadditivity arguments, because of the mixture of repellent and attractive interactions (recall Section 1.3).

6.1.1 The Minimally Extended Phase

Proof. Fix $n \in \mathbb{N}$ and $\gamma < \beta$. From (6.4), we have

$$H_n(w) = \beta I_n(w) - \gamma J_n(w)$$

$$= (\beta - \gamma) \sum_{x \in \mathbb{Z}^d} \ell_n(x)^2 + \frac{\gamma}{4d} \sum_{\substack{x \in \mathbb{Z}^d \\ \|e\|=1}} [\ell_n(x) - \ell_n(x+e)]^2 \tag{6.8}$$

$$\geq (\beta - \gamma) \sum_{x \in \mathbb{Z}^d} \ell_n(x)^2.$$

For $L \in \mathbb{N}$, let

$$C_{n,L} = \{w \in \mathcal{W}_n : w_i \in \Lambda_L \; \forall 0 \leq i \leq n\} \tag{6.9}$$

be the event that the path stays confined to Λ_L up to time n. Since $\sum_{x \in \mathbb{Z}^d} \ell_n(x) = n+1$, it follows from (6.8) that

$$H_n(w) \geq (\beta - \gamma)(n+1)^2 |\Lambda_L|^{-1} \qquad \forall w \in C_{n,L}. \tag{6.10}$$

On the other hand, writing w^* for the path that makes n steps in the 1-st direction, we have from (6.1) and (6.3–6.4) that

$$Z_n^{\beta,\gamma} \geq \frac{1}{(2d)^n} \, e^{-H_n(w^*)}$$

$$= \frac{1}{(2d)^n} \, e^{-\beta(n+1)+(\gamma/2d)[2+2(n-1)]} \tag{6.11}$$

$$= e^{-\beta} \left[\frac{1}{2d} \, e^{-\beta+(\gamma/d)} \right]^n.$$

Combining (6.3) and (6.10–6.11) with the bound $|\mathcal{C}_{n,L}| \leq (2d)^n$, we obtain

$$P_n^{\beta,\gamma}(\mathcal{C}_{n,L}) \leq e^{-(\beta-\gamma)(n+1)^2} |\Lambda_L|^{-1} \, e^{\beta} \left[2d \, e^{\beta-(\gamma/d)} \right]^n \leq e^{-cn}, \tag{6.12}$$

with the last inequality being true for all $L \geq L_n = \lceil \epsilon n^{1/d} \rceil$ and $0 < \epsilon \leq \epsilon_0$, for some $\epsilon_0 = \epsilon_0(\beta,\gamma) > 0$ and $c = c(\beta,\gamma,\epsilon) > 0$. This proves (6.5). $\quad\square$

Note that $Z_n^{\beta,\gamma} \leq Z_n^{\beta-\gamma,0}$ by (6.8), while subadditivity gives

$$\lim_{n\to\infty} \frac{1}{n} \log Z_n^{\beta-\gamma,0} = \inf_{n\in\mathbb{N}} \frac{1}{n} \log Z_n^{\beta-\gamma,0} \leq \frac{1}{2} \log Z_2^{\beta-\gamma,0}$$

$$= \frac{1}{2} \log \left[\left(1 - \frac{1}{2d} \right) + \frac{1}{2d} e^{-(\beta-\gamma)} \right] < 0. \tag{6.13}$$

This explains the first line of (6.7).

6.1.2 The Localized Phase

Proof. The proof uses two lemmas. The first is an estimate of the minimum of the Hamiltonian, the second says that an energetic penalty of order Ln is associated with leaving the box Λ_L prior to time n, provided L is large enough to contain a minimizing path for the Hamiltonian.

For $n, L \in \mathbb{N}$, define

$$M_n = M_n(\beta,\gamma) = \min_{w\in\mathcal{W}_n} H_n(w),$$

$$\mathcal{W}_n \setminus \mathcal{C}_{n,L} = \{w \in \mathcal{W}_n : w_i \notin \Lambda_L \text{ for some } 0 \leq i \leq n\}. \tag{6.14}$$

Lemma 6.3. *If $\gamma > \beta$, then there exist $a = a(\beta,\gamma) > 0$ and $b = b(\beta,\gamma) > 0$ such that $-bn \leq M_n + an^2 \leq bn$ for all $n \in \mathbb{N}$.*

Lemma 6.4. *If $\gamma > \beta$, then $H_n(w) \geq M_n + \frac{1}{6}aLn$ for all $w \in \mathcal{W}_n \setminus \mathcal{C}_{n,L}$ when L is large enough.*

Before giving the proof of these lemmas, we complete the proof of Theorem 6.2. Estimate

$$P_n^{\beta,\gamma}(\mathcal{W}_n \setminus \mathcal{C}_{n,L}) = \frac{1}{Z_n^{\beta,\gamma}} \sum_{w \in \mathcal{W}_n \setminus \mathcal{C}_{n,L}} e^{-H_n(w)} \frac{1}{(2d)^n}$$

$$\leq \left[\frac{1}{(2d)^n} e^{-M_n} \right]^{-1} e^{-M_n - \frac{1}{6}aLn} |\mathcal{W}_n \setminus \mathcal{C}_{n,L}| \frac{1}{(2d)^n} \quad (6.15)$$

$$\leq (2d)^{2n} e^{-\frac{1}{6}aLn}$$

$$\leq e^{-cLn},$$

where the first inequality uses Lemma 6.4 and the bound $Z_n^{\beta,\gamma} \geq (2d)^{-n} e^{-M_n}$, the second inequality uses the bound $|\mathcal{W}_n \setminus \mathcal{C}_{n,L}| \leq (2d)^n$, while the third inequality is true for some $c > 0$ provided $L \geq L_0$ with $L_0 = \frac{12}{a} \log(2d)$. $\quad \square$

Note that $(2d)^{-n} e^{-M_n} \leq Z_n^{\beta,\gamma} \leq e^{-M_n}$, and so Lemma 6.3 gives

$$\lim_{n \to \infty} \frac{1}{n^2} \log Z_n^{\beta,\gamma} = a. \quad (6.16)$$

This explains the second line of (6.7).

Proof of Lemma 6.3

Proof. The idea is that the minimum in (6.14) is achieved by a path that stays inside Λ_L for some $L \in \mathbb{N}$ large enough, making a minimal number of self-intersections and a maximal number of self-touchings. To get a crude idea, consider a path w^* that fills Λ_L approximately uniformly. Then $[\ell_n(x)](w^*) \approx n/|\Lambda_L|$ for $x \in L$ and hence, by (6.4),

$$H_n(w^*) \approx \beta |\Lambda_L| \left(\frac{n}{|\Lambda_L|} \right)^2 - \frac{\gamma}{2d} \left[|\Lambda_L| 2d + O(|\partial \Lambda_L|) \right] \left(\frac{n}{|\Lambda_L|} \right)^2 \quad (6.17)$$

$$= -(\gamma - \beta) \frac{n^2}{|\Lambda_L|} \left[1 + O(|\partial \Lambda_L|/|\Lambda_L|) \right].$$

By picking L large enough, we get $M_n \leq H_n(w^*) \leq -a_1 n^2$ for some $a_1 > 0$. Conversely, since $\sum_{x \in \mathbb{Z}^d} [\ell_n(x)](w) = n + 1$ for all $w \in \mathcal{W}_n$, we see from (6.8) that $H_n(w) \geq -(\gamma - \beta)(n+1)^2$ for all $w \in \mathcal{W}_n$, and so we also get $M_n \geq -a_2 n^2$ for some $a_2 > 0$. Thus, M_n indeed is of order $-n^2$.

It turns out that in a minimizing path the local times are not close to uniform on Λ_L, but rather have a certain profile that is given by a variational problem. To understand why, let $F = F^{\beta,\gamma}$ be the $\mathbb{Z}^d \times \mathbb{Z}^d$-matrix defined by

$$F(x,y) = \begin{cases} -\beta, & \text{if } \|x - y\| = 0, \\ \gamma/2d, & \text{if } \|x - y\| = 1, \\ 0 & \text{otherwise.} \end{cases} \quad (6.18)$$

Then the Hamiltonian in (6.4) can be written as the symmetric bilinear form $-H_n = \langle \ell_n, F\ell_n \rangle$ (i.e., the inner product of ℓ_n with $F\ell_n$). Define

$$[\tilde{\ell}_n](w) = [\ell_n](w) - \tfrac{1}{2}1_{w_0} - \tfrac{1}{2}1_{w_n}, \tag{6.19}$$

i.e., we take away $\tfrac{1}{2}$ from the local times at the endpoints. Since $|F(x,y)| \leq \gamma$ for all $x, y \in \mathbb{Z}^d$ and $\sum_{x \in \mathbb{Z}^d} \tilde{\ell}_n(x) = n$, we have

$$-\gamma(2n+1) \leq H_n(w) + \langle \tilde{\ell}_n, F\tilde{\ell}_n \rangle \leq \gamma(2n+1). \tag{6.20}$$

Now, let $\mathcal{F}$ denote the set of all probability measures on $\mathbb{Z}^d$ that may arise as limit points of the sequence $(\tfrac{1}{n}\tilde{\ell}_n)_{n \in \mathbb{N}}$. Then, clearly,

$$\frac{1}{n^2} \langle \tilde{\ell}_n, F\tilde{\ell}_n \rangle \leq a \quad \text{with } a = \max_{f \in \mathcal{F}} \langle f, Ff \rangle, \tag{6.21}$$

and so (6.20) gives $-H_n(w) \leq an^2 + \gamma(2n+1)$ for all $w \in \mathcal{W}_n$. This proves the lower bound on M_n with $b = 3\gamma$. The upper bound on M_n follows by noting that for every $f \in \mathcal{F}$ and $n \in \mathbb{N}$ there exists a $w \in \mathcal{W}_n$ such that $\sum_{x \in \mathbb{Z}^d} |\tilde{\ell}_n(x) - nf(x)| \leq c$ for some $c > 0$. For details we refer to van der Hofstad and Klenke [163]. □

It is easy to show that the variational formula for a in (6.21) has a maximizer with finite support. Uniqueness, however, is not obvious. In $d = 1$ the maximizer is known to be unique modulo translations and can be computed explicitly (van der Hofstad, Klenke and König [164]).

Proof of Lemma 6.4

Proof. The proof is based on a cutting argument. If $w \in \mathcal{W}_n \setminus \mathcal{C}_{n,L}$, then w leaves Λ_L. Cutting Λ_L by some hyperplane, we can look at the local times on the two sides of this hyperplane and estimate the loss associated with these local times not contributing to the self-touchings. Below we only give the main steps of the argument. For details we again refer to van der Hofstad and Klenke [163].

Fix $L \in 3\mathbb{N}$ and $w \in \mathcal{W}_n \setminus \mathcal{C}_{n,L}$. Without loss of generality we may assume that $w_i^1 > L$ for some $0 \leq i \leq n$. Consider the hyperplanes

$$\mathrm{HP}(y) = \{x \in \mathbb{Z}^d \colon x^1 = y\}, \qquad y \in \mathbb{Z}. \tag{6.22}$$

Since $\sum_{x \in \mathbb{Z}^d} \tilde{\ell}_n(x) = n$, there exists a $y_0 \in (\tfrac{1}{3}L, \tfrac{2}{3}L) \cap \mathbb{N}$ such that

$$\sum_{x \in \mathrm{HP}(y_0-1) \cup \mathrm{HP}(y_0) \cup \mathrm{HP}(y_0+1)} \tilde{\ell}_n(x) \leq 3\,\frac{n}{\tfrac{1}{3}L - 1} = \frac{9n}{L-3}. \tag{6.23}$$

Decompose the path into pieces left of $\mathrm{HP}(y_0)$ and right of $\mathrm{HP}(y_0)$. These pieces give rise to their own local times (with the endpoint correction), which are denoted by $\tilde{\ell}_n^+$ and $\tilde{\ell}_n^-$, satisfying $\tilde{\ell}_n = \tilde{\ell}_n^+ + \tilde{\ell}_n^-$. By (6.20), we have

$$-H_n - \gamma(2n+1) \leq \langle \tilde{\ell}_n, F\tilde{\ell}_n \rangle = \langle \tilde{\ell}_n^+, F\tilde{\ell}_n^+ \rangle + \langle \tilde{\ell}_n^-, F\tilde{\ell}_n^- \rangle + 2\langle \tilde{\ell}_n^-, F\tilde{\ell}_n^+ \rangle, \tag{6.24}$$

where, by (6.21),

$$\langle \tilde{\ell}_n^+, F\tilde{\ell}_n^+ \rangle \leq a(m^+)^2,$$
$$\langle \tilde{\ell}_n^-, F\tilde{\ell}_n^- \rangle \leq a(m^-)^2,$$

(6.25)

with $m^+ = \sum_{x \in \mathbb{Z}^d} \tilde{\ell}_n^+(x)$ and $m^- = \sum_{x \in \mathbb{Z}^d} \tilde{\ell}_n^-(x)$. Moreover, we may estimate

$$2\langle \tilde{\ell}_n^-, F\tilde{\ell}_n^+ \rangle$$
$$= \frac{\gamma}{d} \sum_{x \in \mathrm{HP}(y_0)} [\tilde{\ell}_n^-(x - e^1)\tilde{\ell}_n^+(x) + \tilde{\ell}_n^-(x)\tilde{\ell}_n^+(x + e^1)]$$

$$\leq \frac{\gamma}{d} \left[\sum_{x \in \mathrm{HP}(y_0-1)} \tilde{\ell}_n^-(x) \sum_{x \in \mathrm{HP}(y_0)} \tilde{\ell}_n^+(x) + \sum_{x \in \mathrm{HP}(y_0)} \tilde{\ell}_n^-(x) \sum_{x \in \mathrm{HP}(y_0+1)} \tilde{\ell}_n^+(x) \right]$$

$$\leq \frac{\gamma}{d} \frac{20n}{L} (m^- \wedge m^+),$$

(6.26)

where in the last line we use (6.23) and note that $9n/(L-3) \leq 10n/L$ when $L \geq 30$. Now, $m^- \wedge m^+ \geq \frac{1}{3}L$ by construction and, since $m^- + m^+ = n$, we therefore have $m^- m^+ \geq (m^- \wedge m^+)\frac{1}{2}n \geq \frac{1}{6}Ln$. Consequently, combining Lemma 6.3 and (6.24–6.26), we get

$$-H_n - \gamma(2n+1) \leq a(m^+)^2 + a(m^-)^2 + \frac{20\gamma n}{dL}(m^- \wedge m^+)$$

$$= a(m^+ + m^-)^2 + \left[\frac{20\gamma n}{dL}(m^- \wedge m^+) - 2am^-m^+ \right]$$

$$\leq -M_n + bn + \left(\frac{20\gamma}{dL} - a \right) n(m^- \wedge m^+).$$

(6.27)

The term between brackets in the last line is negative for L large enough, in which case we get

$$-H_n \leq -M_n + bn + \left(\frac{20\gamma}{dL} - a \right) \frac{1}{3}Ln + \gamma(2n+1).$$

(6.28)

The sum of the last three terms is $\leq -\frac{1}{6}aLn$ for L large enough, and so we get the claim. □

6.1.3 Conjectured Phase Diagram

It is conjectured in van der Hofstad and Klenke [163] that on the critical line $\gamma = \beta$ the polymer lives on scale $n^{1/(d+1)}$, more precisely,

$$\sum_{w \in \mathcal{W}_n} \|w_n\|^2 P_n^{\beta,\gamma}(w) \sim D n^{2/(d+1)} (\log n)^{-1/(d+1)}$$

(6.29)

for some $D > 0$. For $d = 1$, this conjecture is proved in van der Hofstad, Klenke and König [164].

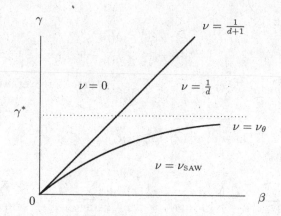

Fig. 6.2. Conjectured phase diagram and critical exponents for the polymer with self-repellence and self-attraction in $d \geq 2$. Only the existence of the critical line $\gamma = \beta$ and the critical exponent $\nu = 0$ in the region above the critical line have been proved. Below the critical line it is known that $\nu \geq 1/d$.

Theorem 6.1 says nothing about the scale of the polymer in the minimally extended phase other than that it is at least of order $n^{1/d}$. As explained in Brak, Owczarek and Prellberg [38], it is believed that there is a second phase transition at a critical value $\gamma_c = \gamma_c(\beta) < \beta$ at which the polymer moves from scale $n^{1/d}$ ("collapsed phase") to scale $n^{\nu_{\text{SAW}}}$ ("extended phase"), with ν_{SAW} the critical exponent for SAW in (2.26). This second critical curve is believed to have a finite asymptote $\lim_{\beta \to \infty} \gamma_c(\beta) = \gamma^*$, the latter being the critical value for the SAW-version of the model with $\beta = \infty$ and $\gamma \in (0, \infty)$. At this critical curve, the critical exponent is believed to be (see B. Duplantier and H. Saleur [95], F. Seno and A. Stella [273])

$$\nu_\theta = \begin{cases} \frac{4}{7}, & \text{if } d = 2, \\ \frac{1}{2}, & \text{if } d \geq 3. \end{cases} \tag{6.30}$$

(The index θ is used here because in the physics literature the phase transition point for the SAW-version of the model is called the θ-point.)

Summarizing the above remarks, we get that the phase diagram for $d \geq 2$ is believed to look like Fig. 6.2. For $d = 1$ there is no collapse transition. Indeed, we know from Theorem 6.5 that below the critical line $\gamma = \beta$ the polymer moves past the point ϵn with high probability for ϵ small enough, so that its motion is ballistic like SAW in $d = 1$.

6.2 A Directed Polymer in a Poor Solvent

In this section we turn to a *directed* version of the model in order to get our hands on the collapse transition, which for the undirected model is conjectured to occur at the second critical curve in Fig. 6.2. Rather than looking at the mean-square displacement, which for the directed model is of less interest, we will prove the existence of a critical value $\gamma_c > 0$ for the free energy at $\beta = \infty$. The results described below are taken from Brak, Guttmann and Whittington [36].

Our choice for the set of paths and the Hamiltonian in (1.1) is

$$\mathcal{W}_n = \{w = (w_i)_{i=0}^n \in (\mathbb{N}_0 \times \mathbb{Z})^{n+1} : w_0 = 0, \, w_1 - w_0 = \rightarrow,$$
$$w_{i+1} - w_i \in \{\uparrow, \downarrow, \rightarrow\} \, \forall 0 < i < n, \, w_i \neq w_j \, \forall 0 \leq i < j \leq n\},$$
$$H_n(w) = -\gamma J_n(w),$$

$$(6.31)$$

where $\uparrow, \downarrow$ and $\rightarrow$ denote steps between neighboring sites in the north, south and east direction, respectively, the first step is required to be east for later convenience, $\gamma \in \mathbb{R}$ and

$$J_n(w) = \sum_{\substack{i,j=0 \\ i<j-1}}^n 1_{\{\|w_i - w_j\| = 1\}}.$$ $$(6.32)$$

An example of a path in $\mathcal{W}_n$ is drawn in Fig. 6.3 (recall also Fig. 1.4). Our choice for the path measure in (1.2) is

$$P_n^\gamma(w) = \frac{1}{Z_n^\gamma} \, e^{\gamma J_n(w)}, \qquad w \in \mathcal{W}_n, \qquad (6.33)$$

where we again drop the uniform reference measure P_n. Thus, each self-touching is weighted by a factor e^γ, and we allow for both positive (=attractive) and negative (=repulsive) values of γ. The case $\gamma = 0$ corresponds to the non-interacting directed random walk. Note that because the path is self-avoiding the law in (6.33) corresponds to the law in (6.3) with $\beta = \infty$.

In Section 6.2.1 we prove that there is a collapse transition at a critical value $\gamma_c > 0$. In Section 6.2.2 we look at the limiting number of self-touchings per monomer in the collapsed phase.

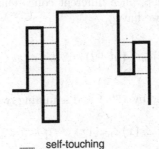

‑‑‑‑‑‑ self-touching

Fig. 6.3. A directed SAW with self-touchings.

6.2.1 The Collapse Transition

The free energy of the directed polymer defined in (6.31–6.33) is given by (recall (1.6))

$$f(\gamma) = \lim_{n \to \infty} \frac{1}{n} \log Z_n^\gamma, \qquad (6.34)$$

whenever the limit exists. The following theorem establishes existence and shows that there are two phases: a *collapsed* phase and an *extended* phase.

Theorem 6.5. *The free energy exists, is finite, and has a collapse transition at $\gamma_c = \log x_c$, with $x_c \approx 3.382975$ the unique positive solution of the cubic equation $x^3 - 3x^2 - x - 1 = 0$. The collapsed phase corresponds to $\gamma > \gamma_c$, the extended phase to $\gamma < \gamma_c$ (see Fig. 6.5).*

Proof. We begin by noting that the partition sum $Z_n^\gamma = \sum_{w \in \mathcal{W}_n} e^{-\gamma J_n(w)}$ can be written as the power series

$$Z_n(x) = \sum_{m=0}^{\infty} c_n(m) x^m, \qquad x \in [0, \infty), \ n \in \mathbb{N}_0, \qquad (6.35)$$

with $x = e^\gamma$ and

$$\begin{aligned} c_n(m) &= |\{w \in \mathcal{W}_n : J_n(w) = m\}| \\ &= \text{the number of } n\text{-step paths with } m \text{ self-touchings} \end{aligned} \qquad (6.36)$$

$(c_0(0) = 1)$. For each n the sum in (6.35) is finite, because $c_n(m) = 0$ when $m \geq n$.

1. The existence of the free energy can be proved with the help of a standard subadditivity argument, as explained in Section 1.3. If we *concatenate* two paths, one with k steps and l self-touchings and one with $n - k$ steps and $m - l$ self-touchings – putting a single step to the east in between to guarantee that no self-touchings arise between the two paths – then we get a path with $n + 1$ steps and m self-touchings. Since different concatenated paths are obtained for different choices of l, we have

$$c_{n+1}(m) \geq \sum_{l=0}^{m} c_k(l) c_{n-k}(m - l) \qquad \forall\, k, m, n \in \mathbb{N}_0, \ k \leq n. \qquad (6.37)$$

Multiplying this inequality by x^{m+1} and summing over m, we get

$$Z_{n+1}(x) \geq x Z_k(x) Z_{n-k}(x) \qquad \forall\, k, n \in \mathbb{N}_0, \ k \leq n. \qquad (6.38)$$

Putting $\widetilde{Z}_{n+1}(x) = x Z_n(x)$, $n \in \mathbb{N}_0$, we get the inequality

$$\widetilde{Z}_{n+2}(x) \geq \widetilde{Z}_{k+1}(x) \widetilde{Z}_{n-k+1}(x) \qquad \forall\, k, n \in \mathbb{N}_0, \ k \leq n, \qquad (6.39)$$

i.e., $n \mapsto \log \widetilde{Z}_n(x)$ is superadditive on $\mathbb{N}$. Consequently,

$$\widetilde{f}(x) = \lim_{n \to \infty} \frac{1}{n} \log \widetilde{Z}_n(x) \qquad (6.40)$$

exists and equals

$$\widetilde{f}(x) = \sup_{n \in \mathbb{N}} \frac{1}{n} \log \widetilde{Z}_n(x) \in (-\infty, \infty]. \qquad (6.41)$$

The free energy in (6.34) is $f(\gamma) = \widetilde{f}(e^\gamma)$.

2. The finiteness of the free energy follows from the observation that $c_n(m) = 0$ for $m \geq n$ and $\sum_{m=0}^{\infty} c_n(m) \leq 3^n$, which give $\widetilde{Z}_{n+1}(x) = xZ_n(x) \leq [3(x \vee 1)]^n$ and $\widetilde{f}(x) \leq \log[3(x \vee 1)]$.

3. Lemma 6.6 below gives a closed form expression for the generating function

$$G = G(x, y) = \sum_{n=0}^{\infty} Z_n^\gamma y^n = \sum_{n=0}^{\infty} \sum_{m=0}^{n} c_n(m) x^m y^n, \quad x, y \in [0, \infty). \qquad (6.42)$$

By analyzing the singularity structure of $G(x, y)$ we will be able to compute $f(\gamma)$. Indeed, our task will be to identify the critical curve $x \mapsto y_c(x)$ in the (x, y)-plane below which $G(x, y)$ has no singularities and on or above which it does, because this identifies the free energy as

$$f(\gamma) = -\log y_c(e^\gamma), \quad \gamma \in \mathbb{R}. \qquad (6.43)$$

We will see that this curve has the shape given in Fig. 6.4.

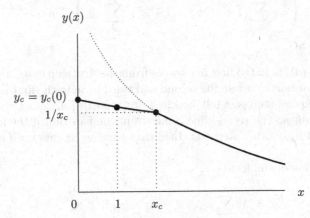

Fig. 6.4. The domain of convergence of the generating function $G(x, y)$ for the directed SAW with self-touchings lies below the solid curve. The dotted line and the piece of the solid curve on $[x_c, \infty)$ are the hyperbola $xy = 1$. The point x_c is identified with the collapse transition, because this is where the free energy $f(\gamma) = f(\log x)$ is non-analytic.

Lemma 6.6. *For $x, y \in [0, \infty)$ the generating function is given by the formal power series*

$$G(x,y) = -\frac{aH(x,y) - 2y^2}{bH(x,y) - 2y^2}, \tag{6.44}$$

where

$$a = y^2(2 + y - xy), \ b = y^2(1 + x + y - xy), \ H(x,y) = y\frac{g_{0,1}(x,y)}{g_{1,1}(x,y)}, \tag{6.45}$$

with

$$g_{r,1}(x,y) = y^r\left(1 + \sum_{k=1}^{\infty} \frac{(y-q)^k y^{2k} q^{\frac{1}{2}k(k+1)}}{\prod_{l=1}^{k}(yq^l - y)(yq^l - q)} q^{kr}\right), \quad q = xy, \ r = 1, 2. \tag{6.46}$$

Proof. For $r \in \mathbb{N}_0$, let $c_{n,r}(m)$ denote the number of n-step paths with m self-touchings that make precisely r steps either north or south immediately after the first step east. Let

$$g_r = g_r(x,y) = \sum_{n=0}^{\infty} \sum_{m=0}^{\infty} c_{n,r}(m) \, x^m y^n. \tag{6.47}$$

Then

$$G = G(x,y) = \sum_{r=0}^{\infty} g_r(x,y). \tag{6.48}$$

The generating functions g_r satisfy the recursion relation

$$g_r = y^{r+1}\left(2 + \sum_{s=0}^{r}(1 + x^s)g_s + (1 + x^r)\sum_{s=r+1}^{\infty} g_s\right), \quad r \in \mathbb{N},$$
$$g_0 = y + yG, \qquad\qquad\qquad\qquad\qquad\qquad\qquad\qquad r = 0. \tag{6.49}$$

Indeed, the first term in the first line comes from the first step east followed by r steps south (or north), while the second and third term in the first line come from the subsequent step east followed by k steps north (or south), leading to $k \wedge r$ self-touchings. The second line comes from the fact that if the first step east is followed by another step east, then the counting repeats itself after the first step.

Try a solution of the form

$$g_r = \lambda^r \sum_{l=0}^{\infty} p_l(q) \, q^{lr}, \quad q = xy, \ r \in \mathbb{N}, \tag{6.50}$$

with $p_0(q) = 1$. Substitution of (6.50) into the first line of (6.49) gives

$$p_l(q) = \frac{\lambda\,(y-q)\,yq^l}{(\lambda q^l - y)(\lambda q^l - q)} p_{l-1}(q), \quad l \in \mathbb{N}, \tag{6.51}$$

provided λ solves $\lambda^2 - (y + q)\lambda + yq = 0$, i.e.,

$$\lambda \in \{\lambda_1, \lambda_2\} = \{y, q\}. \tag{6.52}$$

Thus, the general solution of (6.49) is of the form

$$g_r = C_1 g_{r,1} + C_2 g_{r,2}, \quad r \in \mathbb{N}, \tag{6.53}$$

with C_1, C_2 arbitrary functions of y, q, to be determined by the boundary conditions, and

$$g_{r,i} = (\lambda_i)^r \left(1 + \sum_{k=1}^{\infty} \frac{(\lambda_i)^k (y - q)^k y^k q^{\frac{1}{2}k(k+1)}}{\prod_{l=1}^{k} (\lambda_i q^l - y)(\lambda_i q^l - q)} q^{kr} \right), \quad i = 1, 2, r \in \mathbb{N}_0. \tag{6.54}$$

Next, for $x > 1$ and $0 < y < 1$ such that $xy < 1/(1 + \sqrt{2})$, we have

$$\lim_{r \to \infty} q^{-r} g_{r,1} = 0, \quad \lim_{r \to \infty} q^{-r} g_{r,2} = 1, \quad \lim_{r \to \infty} q^{-r} g_r = 0. \tag{6.55}$$

Indeed, the first two limits follow straight from (6.54), while the last limit follows from monotonicity in x, y and the observation that $f(0) = \log(1 + \sqrt{2})$. Hence $C_2 = 0$, and we are left with

$$g_r = C_1 g_{r,1}, \quad r \in \mathbb{N}. \tag{6.56}$$

To determine C_1, we note that

$$g_0 = y + yG = \tfrac{1}{2} C_1 g_{0,1}, \quad g_1 = a + bG = C_1 g_{1,1}, \tag{6.57}$$

with a and b given by (6.45). The first identity follows from (6.49), because for $r = 0$ the right-hand side of the first line of (6.49) reduces to $2(y + yG)$ and equals $C_1 g_{0,1}$ by the above argument (note that (6.54) includes $r = 0$). The second identity follows from the first line of (6.49) for $r = 1$, giving $g_1 = y^2(2 + 2g_0 + (1 + x)(G - g_0))$. Solving (6.57) for G, we get the expression in (6.44). $\square$

4. The function $G(x, y)$ in (6.44) can be analyzed via a closer study of the function $H(x, y)$ in (6.45–6.46). The latter is a quotient of two q-hypergeometric functions that – as shown in Brak, Guttmann and Whittington [36] – can be expressed in the form of a *continued fraction*. The advantage is that this continued fraction is easy to treat numerically. The analysis of the continued fraction representation was subsequently refined in Owczarek, Prellberg and Brak [259]. It turns out that $G(x, y)$ has no zeroes in the denominator when $x > x_c$ and $0 \leq y \leq 1/x$, with x_c the unique positive solution of the cubic equation $x^3 - 3x^2 - x - 1 = 0$, i.e., $x_c \approx 3.382975$. The denominator has three zeroes:

- $(x, y) = (0, y_c)$, with y_c the solution of the cubic equation $1 - 2y - y^3 = 0$, i.e., $y_c \approx 0.453397$.
- $(x, y) = (1, -1 \pm \sqrt{2})$.

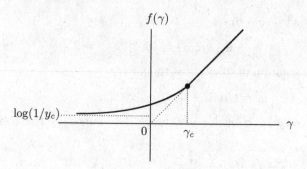

Fig. 6.5. Plot of the free energy per monomer.

Two of these zeroes are indicated in Fig. 6.4. In this way, three points are identified on the piece of the curve in Fig. 6.4 on $[0, x_c]$. The rest of this piece follows from a numerical analysis of the continued fraction representation of $H(x, y)$. We refer to [36] and [259] for details.

Fig. 6.4 yields the free energy via (6.43). This completes the proof of Theorem 6.5. □

By the convexity of the free energy as a function of $\log x$, it follows that the curve in Fig. 6.4 is continuous on $(0, \infty)$. This curve has not been shown to be analytic on $(0, x_c)$, although this is expected to be true.

The regime $x > 1$ corresponds to attraction, the regime $0 \leq x < 1$ to repulsion. The collapse transition occurs at x_c, corresponding to $\gamma_c = \log x_c$. Note that (see Fig. 6.5)

$$f(\gamma) \begin{cases} = \gamma, & \text{if } \gamma \geq \gamma_c, \\ > \gamma, & \text{if } \gamma < \gamma_c. \end{cases} \tag{6.58}$$

6.2.2 Properties of the Two Phases

As in (1.14), the derivative of the free energy, whenever it exists, is the limiting number of self-touchings per monomer:

$$f'(\gamma) = \lim_{n \to \infty} \frac{1}{n} \sum_{w \in \mathcal{W}_n} J_n(w) \, P_n^\gamma(w). \tag{6.59}$$

Theorem 6.7. *The limiting number of self-touchings per monomer is 1 in the collapsed phase and < 1 in the extended phase. The phase transition is second order.*

Proof. A closer analysis of the curve in Fig. 6.4 based on Lemma 6.6, i.e., on the identification of $G(x, y)$ in (6.44) and the continued fraction representation of $H(x, y)$ in (6.45), reveals that

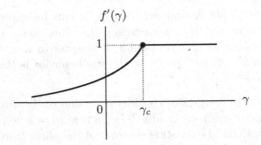

Fig. 6.6. Plot of the number of self-touchings per monomer.

$$f'(\gamma) \begin{cases} = 1, & \text{if } \gamma \geq \gamma_c, \\ < 1, & \text{if } \gamma < \gamma_c, \end{cases} \tag{6.60}$$

with $\gamma \mapsto f'(\gamma)$ continuous at γ_c (see Fig. 6.6). We again refer to [36] and [259] for details. $\square$

The case $\gamma = -\infty$ ($x = 0$) corresponds to the directed polymer avoiding self-touchings altogether, and $\log(1/y_c)$ is its entropy per step. In the collapsed phase the entropy per step is zero, in the extended phase it is strictly positive.

6.3 Extensions

(1) Van der Hofstad and Klenke [163] show that for the model in Section 6.1 a central limit holds when $d = 1$ and $\gamma \leq \beta - \frac{1}{2} \log 2$, i.e., under $P_n^{\beta,\gamma}$ as $n \to \infty$,

$$\frac{|S_n| - \theta^* n}{\sigma^* \sqrt{n}} \Longrightarrow Z, \tag{6.61}$$

where Z is a standard normal random variable, and $\theta^* = \theta^*(\beta, \gamma) \in (0, 1]$ and $\sigma^* = \sigma^*(\beta, \gamma) \in (0, \infty)$ are the speed, respectively, the spread of the polymer. This is the analogue of Extension (1) in Section 3.6 for the soft polymer.

(2) Van der Hofstad, den Hollander and König [161] study the weak interaction limit in $d = 1$ of the model in Section 6.1, i.e., β, γ in (6.1–6.3) are replaced by β_n, γ_n tending to zero as $n \to \infty$. Provided the self-repellence dominates, the behavior is shown to be similar as that of the soft polymer treated in Chapter 3. Compare with Section 3.6, Extension (3).

(3) Owczarek, Prellberg and Brak [259] consider a version of the directed model in Section 6.2 in which the horizontal steps take discrete values, as before, but the vertical steps take continuous values. The number of self-touchings is replaced by the total length where vertical pieces lie next to each other (compare with Fig. 6.3). It turns out that the corresponding generating

function replacing (6.44) is simpler, namely, it can be expressed in terms of Bessel functions rather than q-hypergeometric functions. As a result, the singularity analysis is easier. In particular, it is easier to show that the phase transition is of second order. Qualitatively, the behavior is the same as that of the original model.

(4) Owczarek and Prellberg [258] consider a version of the directed model in Section 6.2 to which "stiffness" is added, i.e., a reward or a penalty is given to two consecutive horizontal steps. It is shown that the phase transition becomes first order in the case of positive stiffness (=reward), but remains second order in the case of negative stiffness (=penalty).

(5) Tesi, Janse van Rensburg, Orlandini and Whittington [289], [290] perform a Monte Carlo study of SAWs in $d = 3$ with self-attraction. They find that $\gamma_c \in [0.274, 0.282]$ and $\nu_\theta \in [0.48, 0.50]$. In [290] it is shown that for SAWs the free energy exists for $\gamma \leq 0$, while this remains open for $\gamma > 0$. On the other hand, it is shown that for self-avoiding polygons the free energy exists for all $\gamma \in \mathbb{R}$. For $\gamma \leq 0$ the two free energies are shown to coincide. This is expected to be the case also for $\gamma > 0$, as the numerical estimates indicate.

Janse van Rensburg, Orlandini, Tesi and Whittington [193] consider self-avoiding polygons consisting of a random concatenation of different types of monomers, allowing for different interaction strengths between different pairs of monomers. The free energy is shown to exist and to be self-averaging, i.e., to be sample-independent. The proof requires a delicate concatenation technique, in which the randomness of the polymer is changed along. Polymers with a random interaction will be treated in Part B of the present monograph.

(6) Kumar, Jensen, Jacobsen and Guttmann [222] look at a two-dimensional version of the polymer with self-attraction in which the path is a SAW constrained to be a bridge, i.e., constrained to lie between the vertical lines through its endpoints, and a force $\phi \in (0, \infty)$ is applied to the right endpoint of the path in the positive vertical direction. This force is modelled by adding a term $-\phi w_n$ to the Hamiltonian: ϕw_n is the work exerted by the force to move the right endpoint a distance w_n away from the left endpoint. With the help of exact enumeration (up to $n = 55$), combined with extrapolation techniques (Guttmann [137]), the "force-temperature diagram" is computed, showing for what values of ϕ and γ the polymer is collapsed, extended or stretched, with the mean-square displacement being of order $n^{1/d}$, n^ν and n, respectively, with ν the critical exponent of SAW. The model is a caricature for folding-unfolding transitions in proteins. We will return to polymers and forces in Section 7.3: see, in particular, (7.62) and Figs. 7.8–7.11. See Haupt, Senden and Sevick [150], Gunari, Balasz and Walker [135], and Gunari and Walker [136] for experiments on a force applied to a polymer in a poor solvent.

(7) Ioffe and Velenik [184] consider a version of the model in Section 6.1 in which the Hamiltonian takes the form (compare with (6.4))

$$H_n^{\psi,\phi}(w) = \sum_{x \in \mathbb{Z}^d} \psi(\ell_n(x)) - (\phi, w_n), \qquad (6.62)$$

where $\psi \colon \mathbb{N}_0 \to [0, \infty)$ is non-decreasing with $\psi(0) = 0$, and $\phi \in \mathbb{R}^d$ is a force acting on the endpoint of the polymer. The reference measure P_n in the path measure $P_n^{\psi,\phi} = (Z_n^{\psi,\phi})^{-1} \exp[-H_n^{\psi,\phi}] P_n$ can be either SRW or a random walk with drift. Two cases are considered: (1) ψ is superlinear (repulsive interaction); (2) ψ is sublinear (attractive interaction), with the additional assumption that $\lim_{\ell \to \infty} \psi(\ell)/\ell = 0$. It is shown that in the attractive case there is a convex set $K = K(\psi) \subset \mathbb{R}^d$, with $\mathrm{int}(K) \ni 0$, such that the polymer is in a collapsed phase when $\phi \in \mathrm{int}(K)$ and in an extended phase when $\phi \notin K$. It is further shown that in the repulsive case the polymer is in an extended phase for all $\phi \in \mathbb{R}^d$. With the help of a coarse-graining technique it is shown that in the extended phase the transversal fluctuations of the polymer (in the direction perpendicular to ϕ or to the drift of the reference random walk) are diffusive.

6.4 Challenges

(1) Extend the central limit theorem in (6.61) for $d = 1$ to the full minimally extended phase, i.e., $\gamma < \beta$. Subsequently, extend the central limit theorem to an LDP, analogous to the LDP derived in Chapter 3 for the soft polymer.

(2) Show that $\theta^*(\beta, \gamma)$ in (6.61) is increasing in β and deceasing in γ, with $\lim_{\gamma \uparrow \beta} \theta^*(\beta, \gamma) = 0$ for all $\beta \in (0, \infty)$. This is the analogue of Challenge (1) in Section 3.7 for the soft polymer.

(3) As indicated in Fig. 6.2, show that $\nu = \frac{1}{d+1}$ at the critical curve $\gamma = \beta$ for $d \geq 2$.

(4) Prove that the second critical curve in Fig. 6.2 exists and has the qualitative shape that is drawn, and show that the critical exponents on either side take the values as indicated. For $2 \leq d \leq 4$ this is a very serious challenge indeed, because little is known rigorously in these dimensions even for the soft polymer. But for $d \geq 5$ there is some hope, because the critical exponents are known for the soft polymer (see Chapter 4). The lace expansion does not work without modification because of the presence of attractive interactions. Recall the remark made at the beginning of Section 4.4. Recall also Section 4.7, Extension (7).

(5) Write down a continuum version of the model defined in (6.1–6.3) and investigate its behavior. For $d = 1$ this is in the spirit of Extension (5) in Section 3.6. What about $d = 2, 3$, in the spirit of Challenge (3) in Section 4.8?

(6) Study the version of the directed model in Section 6.2 in which the vertical steps take values in $\mathbb{Z}^d$ rather than in $\mathbb{Z}$, i.e., a $(1+d)$-dimensional directed SAW. Does the model have the same qualitative behavior in $d \geq 2$ as in $d = 1$?

(7) What is the order of the phase transition between the collapsed phase and the extended phase in the model mentioned in Section 6.3, Extension (7)? Does the order depend on d?

7

Polymer Adsorption

In this chapter we study a polymer in the vicinity of a *linear substrate*. Each monomer on the substrate feels a *binding energy*, resulting in an attractive interaction between the polymer and the substrate. We will consider the two situations where the substrate is:

(1) *penetrable* ('pinning at an interface"),
(2) *impenetrable* ("wetting of a surface").

Our focus will be on the free energy and on the occurrence of a phase transition between a *localized phase*, where the polymer stays close to the substrate, and a *delocalized phase*, where it wanders away from the substrate (see Fig. 7.1). The main technical tools that we will use are *generating functions* and the *method of excursions*. In Sections 7.1–7.4 we focus on directed SAW's, in Section 7.5 on undirected SAW's.

Early references on pinning and wetting of polymers (both for directed and undirected random walk models) are Rubin [272], Burkhardt [47], van Leeuwen and Hilhorst [228], Fisher [98], Privman, Forgacs and Frisch [269], Carvalho and Privman [63], Forgacs, Privman and Frisch [105], and an overview by De'Bell and Lookman [81]. Early results on scaling properties were obtained by Brak, Essam and Owczarek [35] (for ballot paths; recall Fig. 1.4), Whittington [315], Isozaki and Yoshida [186] (for generalized ballot paths), and Roynette, Vallois and Yor [271] (for Brownian paths). More recent references will be mentioned as we go along.

Pinning and wetting may occur not only when the polymer is attracted by the substrate, but also when the polymer is repelled by the solvent(s) around the substrate. Alternatively, the polymer may sit at an interface between two incompatible solvents to reduce the interfacial tension. The monograph by Fleer, Cohen Stuart, Cosgrove, Scheutjens and Vincent [99] describes a wealth of experiments with polymers at interfaces.

In Section 7.1 we look at a directed model for the "pinned polymer", compute the free energy, derive path properties in the two phases, and identify the order of the phase transition. In Section 7.2 we do the same for the "wetted

F. den Hollander, *Random Polymers*,
Lecture Notes in Mathematics 1974, DOI: 10.1007/978-3-642-00333-2_7,
© Springer-Verlag Berlin Heidelberg 2009

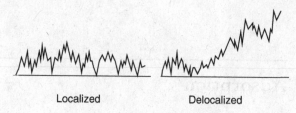

Localized   Delocalized

Fig. 7.1. The localized and delocalized phase for the wetted polymer.

polymer", exploiting the fact that the mathematics is essentially the same. In Section 7.3 we look at the pinned and the wetted polymer when a force is applied to its right endpoint perpendicular to and away from the substrate (e.g. with the help of optical tweezers or atomic force microscopy), we compute the free energy and compute the force that is needed to pull the polymer off the substrate when the interaction parameters are in the localized phase. In Section 7.4 we look at a polymer in a slit between two impenetrable substrates. We show that such a polymer exerts an effective force on the two substrates, which is repulsive or attractive depending on the relative strengths of the adsorption energies at the two walls. In Section 7.5, finally, we describe what is known about the pinned and the wetted polymer when it is undirected, for which the analysis is much harder.

We refer to Giacomin [116], Chapter 2, for an account of the main mathematical ingredients of directed polymer pinning and wetting via the *method of excursions*. Much of what is written in Sections 7.1–7.3 follows the arguments put forward there.

7.1 A Polymer Near a Linear Penetrable Substrate: Pinning

In this section we consider the model where the set of paths and the Hamiltonian are

$$\mathcal{W}_n = \left\{ w = (i, w_i)_{i=0}^n : w_0 = 0, \ w_i \in \mathbb{Z} \ \forall \, 0 \le i \le n \right\},$$
$$H_n(w) = -\zeta K_n(w),$$
(7.1)

with $\zeta \in \mathbb{R}$ and

$$K_n(w) = \sum_{i=1}^n 1_{\{w_i = 0\}}, \qquad w \in \mathcal{W}_n.$$
(7.2)

In words, $\mathcal{W}_n$ denotes the set of all n-step directed paths in $\mathbb{N}_0 \times \mathbb{Z}$ starting from the origin (see Fig. 1.4 for the special cases of ballot paths and generalized ballot paths), and $H_n(w)$ assigns an energy $-\zeta$ to each visit of w to $\mathbb{N} \times \{0\}$. The path measure is

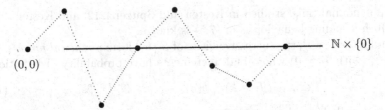

Fig. 7.2. A 7-step path in $\mathbb{N}_0 \times \mathbb{Z}$ that makes 2 visits to $\mathbb{N} \times \{0\}$.

$$P_n^\zeta(w) = \frac{1}{Z_n^\zeta}\, e^{\zeta K_n(w)}\, P_n(w), \qquad w \in \mathcal{W}_n, \tag{7.3}$$

where

- P_n is the projection onto $\mathcal{W}_n$ of the path measure P of a directed irreducible *random walk*, i.e., under P the increments $w_{i+1} - w_i$, $i \in \mathbb{N}_0$, are i.i.d. $\mathbb{Z}$-valued with a common law whose support generates $\mathbb{Z}$.

The choice in (7.1–7.3) models a two-dimensional directed polymer in $\mathbb{N}_0 \times \mathbb{Z}$ where each visit to the substrate $\mathbb{N} \times \{0\}$ contributes a weight factor e^ζ, which is a reward when $\zeta > 0$ and a penalty when $\zeta < 0$ (see Fig. 7.2).

Let $S = (S_i)_{i \in \mathbb{N}_0}$ denote the random walk with law P starting from $S_0 = 0$. For $k \in \mathbb{N}$, define

$$\begin{aligned} a(k) &= P(S_i \neq 0 \ \forall\, 1 \leq i \leq k), \\ b(k) &= P(S_i \neq 0 \ \forall\, 1 \leq i < k,\ S_k = 0). \end{aligned} \tag{7.4}$$

Throughout the sequel we will assume that $\sum_{k \in \mathbb{N}} b(k) = 1$, i.e., S is a recurrent random walk under the law P. In addition, we will assume that $b(\cdot)$ is *regularly varying at infinity*, i.e.,

$$b(k) \sim k^{-1-a}\, L(k) \quad \text{as } k \to \infty \text{ through } \Lambda \tag{7.5}$$

for some $a \in (0, \infty)$ and some slowly varying function L, where $\Lambda = \{k \in \mathbb{N}: b(k) > 0\}$. Since $a(k) = \sum_{l=k+1}^{\infty} b(l)$, $k \in \mathbb{N}$, it follows from (7.5) that

$$a(k) \sim \frac{1}{a}\, k^{-a}\, L(k) \quad \text{as } k \to \infty. \tag{7.6}$$

Note that (7.5–7.6) imply that $a(k)/b(k) \leq Ck$, $k \in \Lambda$, for some $C < \infty$.

For SRW we have $\Lambda = 2\mathbb{N}$, $a = \frac{1}{2}$ and L constant (Spitzer [284], Section 1). To exhibit another example, take the random walk whose increments have distribution $P(S_1 = x) = |x|^{-1-\chi}/2\zeta(1+\chi)$, $x \in \mathbb{Z}\backslash\{0\}$, with ζ the Riemann zeta-function and $\chi \in (0, \infty)$. This random walk has $\Lambda = \mathbb{N}\backslash\{1\}$ and is recurrent when $\chi \in [1, \infty)$ (Spitzer [284], Section 8). Moreover, if $\chi \in (1, \infty)$, then $a = \frac{1}{2} \wedge [1 - (1/\chi)]$ and L is constant, except when $\chi = 2$, in which case L is logarithmic. This is an example of a much larger class of random walks with polynomial tails, studied in Kesten and Spitzer [212] and Kesten [210], for which a scaling behavior as in (7.5) is known.

Under the law P, the successive visits of S to 0 form a *renewal process* $T = (T_l)_{l \in \mathbb{N}_0}$ (with $T_0 = 0$) whose i.i.d. increments have probability distribution

$$P(T_{l+1} - T_l = k) = b(k), \qquad k \in \mathbb{N}, l \in \mathbb{N}_0. \tag{7.7}$$

This fact will drive the computations. Indeed, we will see that the results below are valid for general renewals not necessarily arising from random walk, as long as (7.5) is in force. For convenience, in the proofs of the theorems to be stated below we will assume that $\Lambda = \mathbb{Z}$. The proofs are easily adapted to the situation where the support of $b(\cdot)$ is smaller.

7.1.1 Free Energy

Our starting point is the existence of the free energy.

Theorem 7.1. $f(\zeta) = \lim_{n \to \infty} \frac{1}{n} \log Z_n^\zeta$ *exist for all* $\zeta \in \mathbb{R}$ *and is non-negative.*

Proof. Let

$$Z_n^{\zeta,0} = E \left(\exp \left[\zeta \sum_{i=1}^n 1_{\{S_i = 0\}} \right] 1_{\{S_n = 0\}} \right) \tag{7.8}$$

be the partition sum for the polymer constrained to end at the substrate. We begin by showing that there exists a $C < \infty$ such that

$$Z_n^{\zeta,0} \leq Z_n^\zeta \leq (1 + Cn) Z_n^{\zeta,0} \qquad \forall n \in \mathbb{N}. \tag{7.9}$$

The lower bound is obvious. The upper bound is proved as follows. By conditioning on the last hitting time of 0 prior to time n, we may write

$$Z_n^\zeta = Z_n^{\zeta,0} + \sum_{k=1}^n Z_{n-k}^{\zeta,0} a(k) = Z_n^{\zeta,0} + \sum_{k=1}^n Z_{n-k}^{\zeta,0} b(k) \frac{a(k)}{b(k)}. \tag{7.10}$$

As noted below (7.6), we have $a(k)/b(k) \leq Ck$, $k \in \mathbb{N}$, for some $C < \infty$. Without the factor $a(k)/b(k)$, the last sum in (7.10) is precisely $Z_n^{\zeta,0}$, and so we get the upper bound in (7.9).

The proof of the existence of the free energy is completed by noting that

$$Z_n^{\zeta,0} \geq Z_m^{\zeta,0} Z_{n-m}^{\zeta,0} \qquad \forall m, n \in \mathbb{N}, m \leq n, \tag{7.11}$$

which follows by inserting an extra indicator $1_{\{S_m = 0\}}$ into (7.8) and using the Markov property of S at time m. This inequality says that $n \mapsto \log Z_n^{\zeta,0}$ is superadditive, which implies that (recall Section 1.3)

$$f(\zeta) = \lim_{n \to \infty} \frac{1}{n} \log Z_n^{\zeta,0} \quad \text{exists.} \tag{7.12}$$

The limit equals the free energy because of (7.9).

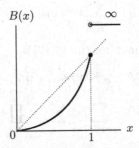

Fig. 7.3. Qualitative picture of $x \mapsto B(x)$.

Finally, the trivial bound $Z_n^\zeta \geq a(n)$, obtained by inserting into (7.8) the indicator of the event that the path never returns to the interface, implies that $f(\zeta) \geq 0$ because of (7.6). $\square$

Our next theorem relates the free energy to the generating function for the length of the *excursions away from the interface*. Let (see Fig. 7.3)

$$B(x) = \sum_{k=1}^{\infty} x^k \, b(k), \qquad x \in [0, \infty). \tag{7.13}$$

Theorem 7.2. *The free energy is given by*

$$f(\zeta) = \begin{cases} 0, & \text{if } \zeta \leq 0, \\ r(\zeta), & \text{if } \zeta > 0, \end{cases} \tag{7.14}$$

where $r(\zeta)$ is the unique solution of the equation

$$B(e^{-r}) = e^{-\zeta}, \qquad \zeta > 0. \tag{7.15}$$

Proof. For $\zeta \leq 0$, we have the trivial bounds

$$a(n) \leq Z_n^\zeta \leq 1 \qquad \forall \, n \in \mathbb{N}, \tag{7.16}$$

which by (7.6) imply that $f(\zeta) = 0$.

For $\zeta > 0$, we look at the constrained partition sum $Z_n^{\zeta,0}$ and write

$$Z_n^{\zeta,0} = \sum_{m=1}^{n} \sum_{\substack{j_1, \ldots, j_m \in \mathbb{N} \\ j_1 + \cdots + j_m = n}} \prod_{i=1}^{m} e^\zeta \, b(j_i). \tag{7.17}$$

Let

$$b^\zeta(k) = e^{\zeta - r(\zeta)k} \, b(k), \qquad k \in \mathbb{N}. \tag{7.18}$$

By (7.15), this is a probability distribution on $\mathbb{N}$. Moreover, because $r(\zeta) > 0$, we have

$$M_{b^\zeta} = \sum_{k \in \mathbb{N}} k b^\zeta(k) < \infty. \tag{7.19}$$

With the help of (7.18), we may rewrite (7.17) as

$$Z_n^{\zeta,0} = e^{r(\zeta)n} P^\zeta(n \in T),$$

$$P^\zeta(n \in T) = \sum_{m=1}^{n} \sum_{\substack{j_1,\ldots,j_m \in \mathbb{N} \\ j_1 + \cdots + j_m = n}} \prod_{i=1}^{m} b^\zeta(j_i), \tag{7.20}$$

where $T = (T_l)_{l \in \mathbb{N}_0}$ (with $T_0 = 0$) is the renewal process whose i.i.d. increments have probability distribution

$$P^\zeta(T_{l+1} - T_l = k) = b^\zeta(k), \qquad k \in \mathbb{N}, \, l \in \mathbb{N}_0. \tag{7.21}$$

By the renewal theorem (Asmussen [10], Theorem I.2.2), we have

$$\lim_{n \to \infty} P^\zeta(n \in T) = \frac{1}{M_{b^\zeta}}. \tag{7.22}$$

Combining (7.9), (7.20) and (7.22), we find that $f(\zeta) = \lim_{n\to\infty} \frac{1}{n} \log Z_n^{\zeta,0} = r(\zeta)$. $\square$

Theorem 7.2 shows that $\zeta \mapsto f(\zeta)$ is non-analytic at $\zeta_c = 0$. Since $x \mapsto B(x)$ is strictly increasing and analytic on $(0,1)$, it follows from (7.15) that $\zeta \mapsto f(\zeta)$ is strictly increasing and analytic on $(0,\infty)$. Consequently, $\zeta_c = 0$ is the only point of non-analyticity of f.

For SRW we have (Spitzer [284], Section 1)

$$B(x) = 1 - \sqrt{1 - x^2}, \qquad x \in [0,1], \tag{7.23}$$

and so (see Fig. 7.4)

$$r(\zeta) = \tfrac{1}{2}\zeta - \tfrac{1}{2}\log(2 - e^{-\zeta}), \qquad \zeta > 0. \tag{7.24}$$

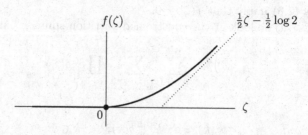

Fig. 7.4. Plot of the free energy for pinned SRW as found in (7.14) and (7.24).

7.1.2 Path Properties

Our next theorem identifies the path behavior in the two phases, and confirms the naive view of localized and delocalized behavior based on the free energy.

Note that the proof of Theorem 7.2 shows that P_∞^ζ, the weak limit as $n \to \infty$ of P_n^ζ defined in (7.3), is the path measure whose excursions are i.i.d. with length distribution

$$P^\zeta(T_1 - T_0 = k) = e^{\zeta - r(\zeta)k} b(k), \qquad k \in \mathbb{N}, \tag{7.25}$$

and with distribution at length k equal to that of the random walk conditioned on returning to the origin after k steps. For $\zeta > 0$ this process is positive recurrent $(r(\zeta) > 0)$, for $\zeta = 0$ it is positive recurrent when $M_b = \sum_{k \in \mathbb{N}} kb(k) < \infty$ and null recurrent $(r(\zeta) = 0)$ when $M_b = \infty$, while for $\zeta < 0$ it is transient $(r(\zeta) = 0)$.

Theorem 7.3. *Under P_n^ζ as $n \to \infty$:*
(a) *If $\zeta > 0$, then the path hits $\mathbb{N} \times \{0\}$ with a strictly positive density, while the length and the height of the largest excursion away from $\mathbb{N} \times \{0\}$ up to time n is of order $\log n$.*
(b) *If $\zeta < 0$, then the path hits $\mathbb{N} \times \{0\}$ finitely many times.*

Proof. (a) The first claim follows from the observation that P_∞^ζ is positive recurrent when $\zeta > 0$. Note also that, as argued in Section 1.3, the derivative of the free energy equals the limiting average fraction of monomers on the substrate, i.e.,

$$f'(\zeta) = \lim_{n \to \infty} \frac{1}{n} \sum_{w \in \mathcal{W}_n} K_n(w) P_n^\zeta(w), \qquad \zeta \neq 0, \tag{7.26}$$

where we recall (7.2–7.3). Theorem 7.2 shows that this derivative is strictly positive for all $\zeta > 0$.

To prove the second claim, we define, for $k \in \mathbb{N}_0$,

$$\text{gap}_k(T) = [T_{l(k)+1} - T_{l(k)}] 1_{\{k \notin T\}}, \qquad l(k) = \max\{l \in \mathbb{N} : T_l \leq k\}. \tag{7.27}$$

and, for $n \in \mathbb{N}$,

$$\text{maxgap}_n(T) = \max_{0 \leq k < n} \text{gap}_k(T). \tag{7.28}$$

Then

$$P_n^{\zeta,0}\big(\text{maxgap}_n(T) > C \log n\big) \leq \sum_{k=0}^{n-1} P_n^{\zeta,0}\big(\text{gap}_k(T) > C \log n\big). \tag{7.29}$$

To estimate the right-hand side we observe that, for every $0 \leq k_1 < k_2 \leq n$,

$$P_n^{\zeta,0}\big(k_1 \in T, k_2 \in T, (k_1, k_2) \cap T = \emptyset\big) = \frac{1}{Z_n^{\zeta,0}} Z_{k_1}^{\zeta,0} b(k_2 - k_1) e^\zeta Z_{n-k_2}^{\zeta,0}$$

$$\leq \frac{1}{Z_{k_2-k_1}^{\zeta,0}} b(k_2 - k_1) e^\zeta,$$

$$(7.30)$$

where in the last line we use the inequality $Z_n^{\zeta,0} \geq Z_{k_1}^{\zeta,0} Z_{k_2-k_1}^{\zeta,0} Z_{n-k_2}^{\zeta,0}$. It follows from (7.12) and (7.30) that there exists a $C_1 < \infty$ such that

$$P_n^{\zeta,0}\big(\mathrm{gap}_k(T) > N\big) \leq C_1 e^{-\frac{1}{2}Nf(\zeta)} \qquad \forall N, n \in \mathbb{N}, \ 0 \leq k < n. \qquad (7.31)$$

Substituting this bound into (7.29), we get

$$\lim_{n \to \infty} P_n^{\zeta,0}\big(\mathrm{maxgap}_n(T) > C_2 \log n\big) = 0 \qquad (7.32)$$

as soon as $C_2 > 2/f(\zeta)$, which is finite for all $\zeta > 0$. Hence, the maximal gap is of order $\log n$ as $n \to \infty$.

(b) The claim follows from the observation that P_∞^ζ is transient when $\zeta < 0$. Note also that Theorem 7.2 shows that $f'(\zeta) = 0$ for all $\zeta < 0$, i.e., the density at which the path hits $\mathbb{N} \times \{0\}$ is zero. $\square$

The fact that in the localized phase the largest excursion away from the substrate up to time n is of order $\log n$ has the following intuitive explanation. The free energy f is strictly positive, and the probability for the polymer to be away from the substrate during a time j decays like e^{-fj} as $j \to \infty$. This is because the polymer contributes an amount fj to the free energy when it stays near the substrate, but contributes 0 when it moves away from the substrate.

The average fraction of adsorbed monomers is linear near $\zeta_c = 0$ when the phase transition is second order (see Fig. 7.5), and discontinuous at $\zeta_c = 0$ when the phase transition is first order.

It follows from Theorem 7.2 that

$$\lim_{\zeta \to \infty} f'(\zeta) = \frac{1}{k^*} \quad \text{with } k^* = \min \Lambda = \min\{k \in \mathbb{N} : b(k) > 0\}. \qquad (7.33)$$

This is the maximal fraction of monomers that can be adsorbed when the reward for adsorption tends to infinity.

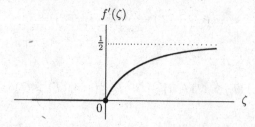

Fig. 7.5. Plot of the average fraction of adsorbed monomers for SRW as found in (7.14) and (7.24).

7.1.3 Order of the Phase Transition

Theorem 7.4. *Recall* (7.5). *There exists a slowly varying function L^* such that*

$$f(\zeta) \sim \zeta^{1/(1 \wedge a)} L^*(1/\zeta), \qquad \zeta \downarrow 0. \tag{7.34}$$

Moreover, L^* is constant in the following three cases: (1) $a \in (1, \infty)$; (2) $a = 1$ *and* $M_b = \sum_{k=1}^{\infty} k b(k) < \infty$; (3) $a \in (0,1)$ *and* L *is constant.*

Proof. It follows from (7.5) and a standard Abelian theorem (see Bingham, Goldie and Teugels [15], Corollary 8.1.7) that, in the limit as $r \downarrow 0$, the generating function defined in (7.13) scales like

$$1 - B(e^{-r}) \sim \begin{cases} \frac{1}{a}\, \Gamma(1-a)\, r^a\, L(1/r), & \text{if } a \in (0,1), \\[2mm] \left[\int_1^\infty \frac{1}{s} L(s)\, e^{-rs}\, ds \right] r, & \text{if } a = 1,\, M_b = \infty, \\[2mm] M_b\, r, & \text{if } a = 1,\, M_b < \infty \text{ or } a \in (1,\infty), \end{cases} \tag{7.35}$$

where Γ is the Gamma-function. Inserting this into (7.15) and using (7.14), we get the claim. For details, see Giacomin [116], Section 2.1. $\square$

Theorem 7.4 shows that the order of the phase transition depends on a and L. Indeed, the phase transition is first order when $a \in (1, \infty)$ or when $a = 1$ and $M_b < \infty$, while for $m \in \mathbb{N} \setminus \{1\}$ it is m-th order when $a \in [\frac{1}{m}, \frac{1}{m-1})$ and L is constant.

For SRW, we have $a = \frac{1}{2}$ and L constant, and so the phase transition is second order (see Fig. 7.4).

7.2 A Polymer Near a Linear Impenetrable Substrate: Wetting

In this section we investigate in what way the results in Section 7.1 are to be modified when the substrate is impenetrable, i.e., when the set of paths $\mathcal{W}_n$ in (7.1) is replaced by (see Fig. 7.6)

$$\mathcal{W}_n^+ = \big\{ w = (i, w_i)_{i=0}^n : w_0 = 0,\, w_i \in \mathbb{N}_0 \ \forall\, 0 \le i \le n \big\}. \tag{7.36}$$

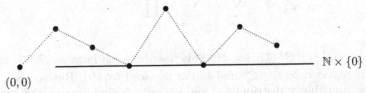

Fig. 7.6. A 7-step path in $\mathbb{N}_0 \times \mathbb{N}_0$ that makes 2 visits to $\mathbb{N} \times \{0\}$.

Accordingly, we write $P_n^{\zeta,+}(w)$, $Z_n^{\zeta,+}$ and $f^+(\zeta)$ for the path measure, partition sum and free energy in this version of the model.

The free energy can be computed using the same excursion approach as in Section 7.1. Theorem 7.1 carries over immediately. The analogue of Theorem 7.2 reads as follows. For $k \in \mathbb{N}$, let

$$b^+(k) = P(S_i > 0 \; \forall 1 \le i < k, \, S_k = 0). \qquad (7.37)$$

This is a *defective* probability distribution, i.e., $\sum_{k \in \mathbb{N}} b^+(k) < 1$. Define

$$B^+(x) = \sum_{k=1}^{\infty} x^k \, b^+(k), \qquad x \in [0, \infty), \qquad (7.38)$$

and put

$$\widetilde{B}(x) = \frac{B^+(x)}{B^+(1)}, \qquad \zeta_c^+ = \log \frac{1}{B^+(1)} > 0. \qquad (7.39)$$

Theorem 7.5. *The free energy is given by*

$$f^+(\zeta) = \begin{cases} 0, & \text{if } \zeta \le \zeta_c^+, \\ r^+(\zeta), & \text{if } \zeta > \zeta_c^+, \end{cases} \qquad (7.40)$$

where $r^+(\zeta)$ *is the unique solution of the equation*

$$\widetilde{B}(e^{-r}) = e^{-(\zeta - \zeta_c^+)}, \qquad \zeta > \zeta_c^+. \qquad (7.41)$$

Proof. The same bounds as in (7.9) hold, so that

$$f^+(\zeta) = \lim_{n \to \infty} \frac{1}{n} \log Z_n^{\zeta,0,+}, \qquad (7.42)$$

where, as before, the upper index 0 refers to the polymer being constrained to end at the substrate.

Write, similarly as in (7.17),

$$Z_n^{\zeta,0,+} = \sum_{m=1}^{n} \sum_{\substack{j_1,\dots,j_m \in \mathbb{N} \\ j_1+\cdots+j_m=n}} \prod_{i=1}^{m} e^{\zeta} \, b^+(j_i). \qquad (7.43)$$

Shifting ζ, we may rewrite this as

$$Z_n^{\zeta,0,+} = \sum_{m=1}^{n} \sum_{\substack{j_1,\dots,j_m \in \mathbb{N} \\ j_1+\cdots+j_m=n}} \prod_{i=1}^{m} e^{\zeta - \zeta_c^+} \, \widetilde{b}(j_i). \qquad (7.44)$$

with $\widetilde{b}(k) = b^+(k)/B^+(1)$, $k \in \mathbb{N}$. This has the same form as (7.17), except that ζ is replaced by $\zeta - \zeta_c^+$ and $b(\cdot)$ is replaced by $\widetilde{b}(\cdot)$. But $\widetilde{b}(\cdot)$ is a *nondefective* probability distribution, and so we can simply copy the proof of Theorem 7.2 to get the claim. $\square$

Localization on an impenetrable substrate is harder than on a penetrable substrate, because the polymer suffers a larger loss of entropy. This is the reason why $\zeta_c^+ > 0$ in Theorem 7.5.

To get the analogue of Theorem 7.4, we need to assume, as in (7.5), that

$$b^+(k) \sim k^{-1-a} L(k) \quad \text{as } k \to \infty \text{ through } \Lambda^+, \tag{7.45}$$

for some $a \in (0, \infty)$ and some slowly varying function L, where $\Lambda^+ = \{k \in \mathbb{N} : b^+(k) > 0\} \subseteq \Lambda$. Under this assumption, the theorem and its proof carry over verbatim, after we shift ζ by ζ_c^+. Note that $\widetilde{b}(\cdot)$ inherits the scaling of $b^+(\cdot)$ in (7.45).

For SRW we have $b^+(k) = \frac{1}{2}b(k)$, $k \in \mathbb{N}$. Hence

$$\zeta_c^+ = \log 2, \quad \widetilde{b}(\cdot) = b(\cdot), \quad \widetilde{B}(\cdot) = B(\cdot), \tag{7.46}$$

implying that $f^+(\zeta) = f(\zeta - \zeta_c^+)$, $\zeta \in \mathbb{R}$. Thus, for SRW the free energy *simply suffers a shift*. For more general random walks, however, it is *not* the case that $\widetilde{b}(\cdot) = b(\cdot)$. Even their tail exponents may be different. The latter may be seen from the relation

$$1 - \sum_{k=1}^{\infty} x^k b^+(k) = \exp\left[-\sum_{k=1}^{\infty} \frac{1}{k} x^k P(S_k = 0)\right], \quad x \in [0, \infty), \tag{7.47}$$

for which we refer the reader to Spitzer [284], Section 17. Indeed, this relation shows that $b^+(k) \asymp P(S_k = 0)/k$ as $k \to \infty$ (subject to a regulary varying tail condition on $P(S_k = 0)$ implied by (7.45)). For instance, for the Riemann random walk mentioned at the end of Section 7.1, if $\chi \in (1, 2)$, then $P(S_k = 0) \asymp k^{-1/\chi}$ as $k \to \infty$ (where $\asymp$ means that the ratio of the two sides is bounded above and below by strictly positive and finite constants), and so the tail exponent of $\widetilde{b}(\cdot)$ is $1 + (1/\chi)$, which is different from the tail exponent $2 - (1/\chi)$ we found for $b(\cdot)$. Only for $\chi \in [2, \infty)$ are both exponents the same, and equal to $\frac{3}{2}$ (with a logarithmic correction for $\chi = 2$).

7.3 Pulling a Polymer off a Substrate by a Force

A polymer can be pulled off a substrate by a force. This can be done, for instance, with the help of *optical tweezers*. Here, a focussed laser beam is used, containing a narrow region – called the beam waist – in which there is a strong electric field gradient. When a dielectric particle, a few nanometers in diameter, is placed in the waist, it feels a strong attraction towards the center of the waist. One can chemically attach such a particle to the end of the polymer and then pull on the particle with the laser beam, thereby effectively exerting a force on the polymer itself. Current experiments allow for forces in the range of 10^{-12} Newton. With such microscopically small forces the structural, mechanical and elastic properties of polymers can be probed.

In Section 7.3.1 we look at the pinned polymer, in Section 7.3.2 at the wetted polymer.

7.3.1 Force and Pinning

To model the effect of the force on a pinned polymer we return to (7.1–7.3) and replace the Hamiltonian by

$$H_n(w) = -\zeta K_n(w) - \phi w_n, \qquad (7.48)$$

where $\phi \in \mathbb{R}$ is a force in the upward direction acting on the right endpoint of the polymer (ϕw_n is the work exerted by the force to move the endpoint a distance w_n away from the interface). We write $Z_n^{\zeta,\phi}$ to denote the partition sum and

$$f(\zeta, \phi) = \lim_{n \to \infty} \frac{1}{n} \log Z_n^{\zeta,\phi} \qquad (7.49)$$

to denote the free energy. Without loss of generality, we will assume that $\phi > 0$.

For convenience of exposition, we restrict to the case where the reference random walk can only make steps of size ≤ 1, namely,

$$P(S_1 = -1) = P(S_1 = +1) = \tfrac{1}{2}p, \quad P(S_1 = 0) = 1 - p, \qquad p \in (0,1]. \quad (7.50)$$

Theorem 7.6. *For every $\zeta \in \mathbb{R}$ and $\phi > 0$, the free energy exists and is given by*

$$f(\zeta, \phi) = f(\zeta) \vee g(\phi), \qquad (7.51)$$

with $f(\zeta)$ the free energy of the pinned polymer without force given by Theorem 7.2, and

$$g(\phi) = \log\left[p\cosh(\phi) + (1 - p)\right]. \qquad (7.52)$$

Proof. We have

$$Z_n^{\zeta,\phi} = Z_n^{\zeta,0} + \sum_{k=1}^{n} Z_{n-k}^{\zeta,0} \, Z_k^{+,\phi} \qquad (7.53)$$

with $Z_n^{\zeta,0}$ the partition sum of the pinned polymer without force constrained to end at the substrate, and

$$Z_k^{+,\phi} = \sum_{x \in \mathbb{Z}} e^{\phi x} b^+(k,x), \qquad b^+(k,x) = P(S_i > 0 \ \forall 1 \leq i < k, \ S_k = x). \qquad (7.54)$$

We will show that

$$g(\phi) = \lim_{k \to \infty} \frac{1}{k} \log Z_k^{+,\phi} \qquad (7.55)$$

exists and is given by (7.52). Combining (7.12), (7.53) and (7.55), we get the claim.

The contribution to the sum in (7.54) coming from $x \in \mathbb{Z} \backslash \mathbb{N}$ is bounded above by $1/(1 - e^{-\phi}) < \infty$. For $x \in \mathbb{N}$, on the other hand, we can use the *reflection principle*, which gives

$$b^+(k, x) = \tfrac{p}{2}\left[P(S_{k-1} = x - 1) - P(S_{k-1} = x + 1)\right], \qquad k \in \mathbb{N}. \tag{7.56}$$

Hence, for $k \in \mathbb{N}$,

$$
\begin{aligned}
Z_k^{+,\phi} &= O(1) + \tfrac{p}{2}\sum_{x \in \mathbb{N}} e^{\phi x}\left[P(S_{k-1} = x - 1) - P(S_{k-1} = x + 1)\right] \\
&= O(1) + p\sinh(\phi)\sum_{x \in \mathbb{N}} e^{\phi x}\, P(S_{k-1} = x) \\
&= O(1) + p\sinh(\phi)\sum_{x \in \mathbb{Z}} e^{\phi x}\, P(S_{k-1} = x) \\
&= O(1) + p\sinh(\phi)\, E(e^{\phi S_{k-1}}) \\
&= O(1) + p\sinh(\phi)\left[p\cosh(\phi) + (1-p)\right]^{k-1},
\end{aligned}
\tag{7.57}
$$

where the error term changes from line to line. This proves that, indeed, the limit in (7.55) exists and is given by (7.52). $\square$

Formula (7.51) tells us that the force either leaves most of the polymer adsorbed (when $f(\zeta, \phi) = f(\zeta) > g(\phi)$) or pulls most of the polymer off (when $f(\zeta, \phi) = g(\phi) > f(\zeta)$). It further tells us that a phase transition occurs at those values of ζ and ϕ where $f(\zeta) = g(\phi)$, i.e., the critical value of the force is given by

$$\phi_c(\zeta) = g^{-1}\big(f(\zeta)\big), \qquad \zeta \in \mathbb{R}, \tag{7.58}$$

with g^{-1} the inverse of g. Note that $\phi_c(\zeta) = 0$ for $\zeta \le 0$ and $\phi_c(\zeta) > 0$ for $\zeta > 0$. In the latter regime, the phase transition is first order because $\partial f(\zeta, \phi)/\partial \phi$ is discontinuous at $\phi = \phi_c(\lambda)$. Also note that the phase transition is first order in ζ when $\phi > 0$ and second order in ζ when $\phi = 0$.

For our choice of random walk the analogue of (7.23) reads

$$B_p(x) = 1 - \sqrt{[1 - (1-p)x]^2 - [px]^2}, \qquad x \in [0, 1]. \tag{7.59}$$

Consequently, $f(\zeta) = -\log B_p^{-1}(e^{-\zeta})$, with B_p^{-1} the inverse of B_p, and so

$$\phi_c(\zeta) = \cosh^{-1}\left(\frac{1}{p}\left[\frac{1}{B_p^{-1}(e^{-\zeta})} - (1-p)\right]\right), \qquad \zeta > 0. \tag{7.60}$$

In what follows we look at $p = 1$ and $p \in (0, 1)$ separately.

● **Case $p = 1$**

For $p = 1$ (SRW), the formula in (7.60) reduces to

$$\phi_c(\zeta) = \tanh^{-1}(1 - e^{-\zeta}), \qquad \zeta > 0, \tag{7.61}$$

which we plot in Fig. 7.7. It is easily checked that $\zeta \mapsto \phi_c(\zeta)$ is strictly concave, with $\phi_c(\zeta) \sim \zeta$, $\zeta \downarrow 0$, and $\phi_c(\zeta) \sim \tfrac{1}{2}\zeta$, $\zeta \to \infty$.

An alternative way of exhibiting Fig. 7.7 is to put

$$\zeta = 1/T, \quad \phi = F/T, \quad F_c(T) = \phi_c(1/T)/T, \tag{7.62}$$

with $T > 0$ the physical temperature and F the physical force. This leads to the picture in Fig. 7.8, which shows that the *force-temperature diagram* is strictly increasing.

● **Case $p \in (0, 1)$**

From (7.59) we have

$$B_p(x) \sim \begin{cases} (1 - p)x, & \text{as } x \downarrow 0, \\ 1 - \sqrt{2p(1 - x)}, & \text{as } x \uparrow 1. \end{cases} \tag{7.63}$$

It therefore follows from (7.60) and (7.62) that

$$F_c(T) = \begin{cases} \frac{1}{p}, & \text{as } T \to \infty, \\ 1 + T[\log(1 - p) - \log(\frac{1}{2}p)] + o(T), & \text{as } T \downarrow 0, \end{cases} \tag{7.64}$$

which is plotted in Fig. 7.9. Note that $T \mapsto F_c(T)$ goes through a *minimum* when $p \in (\frac{2}{3}, 1)$. This is remarkable, since it says that at a fixed force F slightly below 1 the polymer is adsorbed both for small and for large temperatures, but is desorbed for moderate temperatures. This type of behavior is referred to as a *re-entrant phase transition*.

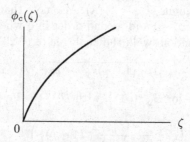

Fig. 7.7. Plot of $\zeta \mapsto \phi_c(\zeta)$ for $p = 1$.

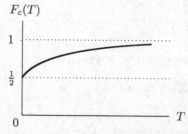

Fig. 7.8. Plot of $T \mapsto F_c(T)$ for $p = 1$.

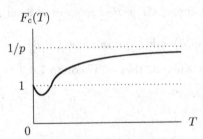

Fig. 7.9. Plot of $T \mapsto F_c(T)$ for $p \in (\frac{2}{3}, 1)$.

Since $\zeta_c = 0$, at zero force the polymer is adsorbed (pinned) for every $0 \leq T < \infty$, and becomes desorbed (depinned) only at $T = \infty$, implying that $F_c(T) > 0$ for all $0 \leq T < \infty$. This explains why the critical curves in Figs. 7.8–7.9 do not hit the horizontal axis.

Here is a heuristic explanation of Fig. 7.9. First consider the case $p = \frac{2}{3}$, where the steps in the east, north-east and south-east direction are equally probable, so that all paths have the same probability. For every $0 < T < \infty$, the polymer makes excursions away from the substrate and therefore has a strictly positive entropy. Part of this entropy is lost when a force is applied and part of the polymer (near its endpoint) is pulled away from the substrate and is forced to move upwards steeply. As T increases, the effect of this entropy loss on the free energy increases (because "free energy = energy − [temperature × entropy]"). This in turn must be counterbalanced by a larger force to achieve desorption and, consequently, $\partial F_c(T)/\partial T > 0$ for $0 < T < \infty$. The case $T = 0$ is special, because the polymer is fully pinned and has no entropy, implying that $\lim_{T \downarrow 0} \partial F_c(T)/\partial T = 0$. Next consider the case $p \in (\frac{2}{3}, 1)$. Then the steps in the east direction are disfavored over the steps in the north-east and south-east direction, and this tends to move the polymer farther away from the substrate for every $0 < T < \infty$. Consequently, $F_c(T) < F_c(0)$ for small T, i.e., $T \mapsto F_c(T)$ decreases for small T, an effect that becomes more pronouced as p increases. At larger T, entropy again dominates in the free energy and hence $T \mapsto F_c(T)$ increases when T is large enough. Finally, for the case $p \in (0, \frac{2}{3})$ the effect is precisely the opposite for small T.

7.3.2 Force and Wetting

A similar calculation can be done for the wetting version of the model. The main difference is that the critical curve shifts over ζ_c^+. The analogue of Theorem 7.6 reads as follows.

Theorem 7.7. *For every $\zeta \in \mathbb{R}$ and $\phi > 0$, the free energy exists and is given by*

$$f^+(\zeta, \phi) = f^+(\zeta) \vee g(\phi), \tag{7.65}$$

with $f^+(\zeta)$ the free energy of the wetted polymer without force given by Theorem 7.5, and

$$g(\phi) = \log\left[p\cosh(\phi) + 1 - p\right]. \qquad (7.66)$$

Proof. The proof is the same as that of Theorem 7.6. $\square$

- **Case $p = 1$**

The analogue of (7.61), with $\zeta_c^+ = \log 2$ for $p = 1$, reads

$$\phi_c^+(\zeta) = \tanh^{-1}\left(1 - e^{-(\zeta - \zeta_c^+)}\right), \qquad \zeta > \zeta_c^+, \qquad (7.67)$$

which is Fig. 7.7 shifted to the right by ζ_c^+, and gives a force-temperature diagram as plotted in Fig. 7.10.

- **Case $p \in (0, 1)$**

The analogue of (7.64) reads

$$F_c^+(T) = \begin{cases} 0, & \text{as } T \to \infty, \\ 1 + T[\log(1 - p) - \log(\tfrac{1}{2}p)] + o(T), & \text{as } T \downarrow 0, \end{cases} \qquad (7.68)$$

which is plotted in Fig. 7.11. Note that $T \mapsto F_c(T)$ goes through a *maximum* when $p \in (0, \tfrac{2}{3})$. At a fixed force F slightly above 1 the polymer is desorbed

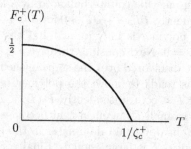

Fig. 7.10. Plot of $T \mapsto F_c^+(T)$ for $p = 1$.

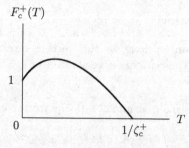

Fig. 7.11. Plot of $T \mapsto F_c^+(T)$ for $p \in (0, \tfrac{2}{3})$.

both for small and for large temperatures, but is adsorbed for moderate temperatures. This is precisely the *reverse* of what we had for pinning.

Since $\zeta_c^+ > 0$, at zero force the polymer is adsorbed (wetted) for small temperature and becomes desorbed (dewetted) at large temperature, which is why the critical curves in Figs. 7.10–7.11 hit the horizontal axis at $T = 1/\zeta_c^+$.

A heuristic explanation of Fig. 7.11 runs parallel to that of Fig. 7.9 given at the end of Section 7.3.1. When a force is applied, part of the polymer (near its endpoint) is pulled away from the substrate. For the case $p = \frac{2}{3}$, where all paths have equal probability, this comes with a gain of entropy (unlike the loss of entropy we had for pinning), because the polymer feels less of the hard-wall constraint. As T increases, the effect of this entropy gain on the free energy increases. This is counterbalanced by a smaller force to achieve desorption and, consequently, $\partial F_c(T)/\partial T < 0$. For the case $p \in (\frac{2}{3}, 1)$ (where steps in the east direction are disfavored), the polymer is farther away from the substrate, resulting in $F_c(T) < F_c(0)$, while for the case $p \in (0, \frac{2}{3})$ (where steps in the east direction are favored), the effect is precisely the opposite. For small T this explains the sign of the slope of the critical curve. At larger T, entropy dominates in the free energy and $T \mapsto F_c(T)$ decreases until it hits the horizontal axis.

7.4 A Polymer in a Slit between Two Impenetrable Substrates

In this section we consider a two-dimensional directed polymer in a slit between two walls. The lower wall is at height 0, the upper wall at height $N \in \mathbb{N}$. The polymer follows a ballot path (see Fig. 7.12). Each visit to the lower wall comes with an adsorption energy σ, each visit to the upper wall with an adsorption energy ξ, both of which are taken to be non-negative.

7.4.1 Model

We fix $N \in \mathbb{N}$ and replace the set of paths and the Hamiltonian in (7.1–7.2) by

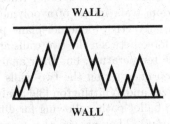

Fig. 7.12. A directed polymer in a slit.

$$\mathcal{W}_{n,N} = \big\{ w = (i, w_i)_{i=0}^n : w_0 = 0, w_{i+1} - w_i = \pm 1 \ \forall 0 \le i < n,$$
$$0 \le w_i \le N \ \forall 0 \le i \le n \big\}, \tag{7.69}$$
$$H_{n,N}(w) = -\sigma K_n(w) - \xi K_{n,N}(w),$$

with $\sigma, \xi \in [0, \infty)$ and

$$K_n(w) = \sum_{i=1}^n 1_{\{w_i = 0\}}, \quad K_{n,N}(w) = \sum_{i=1}^n 1_{\{w_i = N\}}, \qquad w \in \mathcal{W}_{n,N}. \tag{7.70}$$

The path measure replacing (7.3) becomes

$$P_{n,N}^{\sigma,\xi}(w) = \frac{1}{Z_{n,N}^{\sigma,\xi}} e^{\sigma K_n(w) + \xi K_{n,N}(w)}, \qquad w \in \mathcal{W}_{n,N}, \tag{7.71}$$

where we drop the reference measure P_n, which is harmless because P_n is the uniform distribution on ballot paths of length n. The main objects of interest are the finite width free energy

$$f_N(\sigma, \xi) = \lim_{n \to \infty} \frac{1}{n} \log Z_{n,N}^{\sigma,\xi} \tag{7.72}$$

and its infinite width counterpart

$$f(\sigma, \xi) = \lim_{N \to \infty} f_N(\sigma, \xi). \tag{7.73}$$

The motivation for this model is as follows. Consider a *colloidal dispersion* consisting of particles on which polymers are adsorbed. When two colloidal particles approach one another, the adsorbed polymers become confined to the space between the particles. The resulting loss of entropy induces a repulsive force between the particles, preventing them from clustering. This is called *steric stabilization*. On the other hand, when the adsorbed polymers bridge the space between the two particles and become adsorbed on both, the result is an additional attractive force, which may or may not exceed the repulsive force. In the former case the particles cluster. This is called *sensitized flocculation*. If the colloidal particles are large compared to the adsorbed polymers, then it is fair to approximate the space between the two particles by a slit between two parallel planes. Colloidal dispersions are used e.g. in paints and in pharmaceuticals. For more background, see Napper [247].

Pioneering work was done by DiMarzio and Rubin [91], who considered a three-dimensional random walk model of a polymer between two parallel walls with a short-range attractive monomer-plane potential. They showed that there is a repulsive force between the two walls at high temperature and an attractive force at low temperature, but their analysis is restricted to the case where the adsorption energies on the two walls are the same ($\sigma = \xi$). Brak, Owczarek, Rechnitzer and Whittington [39] analyzed a two-dimensional directed model, based on ballot paths, allowing for different adsorption energies on the two walls ($\sigma \neq \xi$). They arrived at a complete description of the phase diagram and the force diagram. We will describe their work below.

7.4.2 Generating Function

Let $\mathcal{W}_{n,N}^0$ be the subset of paths in $\mathcal{W}_{n,N}$ that end at the lower wall, i.e., $w_n = 0$ (see Fig. 7.12). Abbreviate $a = e^\sigma$ and $b = e^\xi$. Define the generating function

$$G_N(z; a, b) = \sum_{n=0}^\infty Z_{n,N}^0 \, z^n, \qquad z \in \mathbb{C}, \, N \in \mathbb{N}, \tag{7.74}$$

where

$$Z_{n,N}^0 = \sum_{w \in \mathcal{W}_{n,N}^0} a^{l(w)} b^{u(w)} \tag{7.75}$$

is the partition sum for paths ending at the lower wall, with $l(w)$ and $u(w)$ denoting the number of visits by w to the lower, respectively, upper wall. Our task is to identify the singularity $z_N(a, b)$ of $z \mapsto G_N(z, a, b)$ that lies closest to the origin on $(0, \infty)$, since

$$f_N(a, b) = \lim_{n \to \infty} \frac{1}{n} \log Z_{n,N}^0 = -\log z_N(a, b). \tag{7.76}$$

(It is trivial to show that $Z_{n,N}^0 / Z_{n,N} \le 1$ decays at most polynomially fast in n for fixed N, implying the first equality in (7.76).)

The following theorem gives an explicit formula for $G_N(z; a, b)$.

Theorem 7.8. *Let* $q = (1 - 2z^2 - \sqrt{1 - 4z^2})/2z^2$. *Then, for all* $z \in \mathbb{C}$ *and* $N \in \mathbb{N}$,

$$G_N(z; a, b) = \frac{(1+q)[(1+q-bq) + (1+q-b)q^N]}{(1+q-aq)(1+q-bq) - (1+q-a)(1+q-b)q^N}. \tag{7.77}$$

Proof. The argument given here is a bit sketchy. For details we refer to [39]. The computation of $G_N(z; a, b)$ proceeds via induction on N. A functional recursion relation is obtained by building up the paths in a slit of width $N+1$ from the paths in a slit of width N after replacing each hit of the top wall by a zig-zag path (with increments $\nearrow$ and $\searrow$ alternating) of an arbitrary length. The generating function of a zig-zag path equals $1/(1 - bz^2)$, and so we get the recursion

$$G_{N+1}(z; a, b) = G_N\big(z; a, 1/(1 - bz^2)\big), \qquad N \in \mathbb{N}, \tag{7.78}$$

which is to be solved subject to the initial condition $G_1(z; a, b) = 1/(1 - abz^2)$, the generating function for zig-zag paths in a slit of width 1.

By induction on N it can be shown that the solution of (7.78) is of the form

$$G_N(z; a, b) = \frac{P_N(z; 0, b)}{P_N(z; a, b)} \tag{7.79}$$

with $P_N(z; a, b)$ a polynomial in z. Inserting (7.79) into (7.78), we get the recursion

$$P_{N+1}(z;a,b) = (1 - bz^2) P_N(z;a, 1/(1 - bz^2)), \qquad N \in \mathbb{N}, \qquad (7.80)$$

which is to be solved subject to the initial condition $P_1(z;a,b) = 1/(1 - abz^2)$. In terms of the variable q defined by $z = \sqrt{q}/(1+q)$, the solution reads (with a little help from MAPLE)

$$P_N(z;a,b) = \frac{(1+q-aq)(1+q-bq) - (1+q-a)(1+q-b)q^N}{(1-q)(1+q)^{N+1}}. \qquad (7.81)$$

Substituting (7.81) into (7.79), we get the claim. $\square$

The dominant singularity of the r.h.s. of (7.77) is $q_N(a,b)$, defined to be the zero of the denominator with the smallest value of $|z|$, solving

$$q^N = \frac{(1+q-aq)(1+q-bq)}{(1+q-a)(1+q-b)}. \qquad (7.82)$$

A closed form solution is possible only for special values of (a,b), which we list next:

$$\begin{aligned}
(a,b) = (1,1) \qquad & q_N = e^{2\pi i/(N+1)} \qquad f_N(1,1) = \log(2\cos[\pi/(N+2)]), \\
(a,b) = (2,1) \qquad & q_N = e^{2\pi i/(2N+2)} \qquad f_N(2,1) = \log(2\cos[\pi/(2N+2)]), \\
ab = a+b, \, a > 2 \quad & q_N = 1/(a-1) \qquad f_N(a,b) = \log(a/\sqrt{a-1}),
\end{aligned}$$
$$(7.83)$$

and similar formulas when $(a,b) = (1,2)$ and $ab = a+b$, $b > 2$. For other values of (a,b) the solution can be found numerically (see [39]). Note that the curve $ab = a+b$ is special: along this curve the free energy is *independent* of N, and is equal to the free energy for wetting (one wall).

7.4.3 Effective Force

The *effective force* exerted by the polymer on the two walls at width N is

$$\phi_N(a,b) = f_{N+1}(a,b) - f_N(a,b). \qquad (7.84)$$

We thus find that for all $N \in \mathbb{N}$ the effective force $\phi_N(a,b)$ is zero on the curve $ab = a+b$, strictly negative above the curve (attraction) and strictly positive below the curve (repulsion). The force tends to zero as $N \to \infty$. The rate at which this happens depends on (a,b). As shown in [39], the decay can be *exponential* (short-range force) or *polynomial* (long-range force) in N. The phase diagram is drawn in Fig. 7.13, the force diagram in Fig. 7.14. Both these figures are valid in the limit $n \to \infty$ followed by $N \to \infty$. Note that these limits may in general not be interchanged.

In [39], the phase transitions in Fig. 7.13 are shown to be second order along the horizontal and vertical critical lines, and first order along the oblique critical line. Across the latter, the densities of visits by the polymer to the

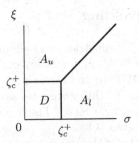

Fig. 7.13. Phase diagram for a directed polymer in a slit. There are three phases: D (= desorbed), A_l (= adsorbed on lower wall), A_u (= adsorbed on upper wall); $\zeta_c^+ = \log 2$ is the critical value for wetting in halfspace.

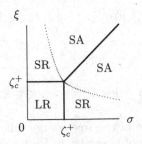

Fig. 7.14. Force diagram for a directed polymer in a slit. There are three types of forces: LR (= long-range repulsion), SR (= short-range repulsion), SA (= short-range attraction). On the dotted curve the force is zero.

walls make a jump. Remarkably, on this critical line the force turns out to be, twice as large (for large N) as near the line, so that there is also a discontinuity in the strength of the force.

In terms of colloidal dispersions, Fig. 7.14 shows that long-range repulsive forces promote steric stabilization (entropy dominates), while short-range attractive forces promote sensitized flocculation (energy dominates). The special curve $ab = a + b$ is not a phase boundary.

Owczarek, Prellberg and Rechnitzer [260] provide a complete finite-size scaling analysis, i.e., they show how the free energy and the force scale as the length n of the polymer and the width N of the slit tend to infinity simultaneously. Brak, Iliev, Rechnitzer and Whittington [37] carry out the same analysis as in Brak, Owczarek, Rechnitzer and Whittington [39], but for generalized ballot paths rather than ballot paths. The formulas are more complicated, but the results are similar. They use the so-called "wasp-waist"-method, which leads to a system of difference equations that is somewhat easier to analyze than the functional equations encountered in Section 7.4.2. The method is flexible enough to allow for generalization to partially directed self-avoiding walks in higher dimension.

7.5 Adsorption of Self-avoiding Walks

Early studies of SAWs in confined geometries can be found in Guttmann and Whittington [139], Whittington [314], Hammersley and Whittington [144], and Whittington and Soteros [317]. In the meantime, a vast combinatorial literature exists on such problems. In this section we ask the question: What happens when we consider wetting by a SAW? The following result is due to Hammersley, Torrie and Whittington [142].

Consider SAWs on $\mathbb{Z}^d$, $d \geq 2$, that start at the origin and are confined to the halfspace $\{x_1 \geq 0\}$. Let $c_n^+(v)$, $n, v \in \mathbb{N}$, be the number of halfspace SAWs with $n + 1$ vertices that have $v + 1$ vertices in the hyperplane $\{x_1 = 0\}$. Consider the partition sum

$$Z_n^+(\zeta) = \sum_{v \in \mathbb{N}} c_n^+(v) \, e^{\zeta v}, \qquad n \in \mathbb{N}, \zeta \in \mathbb{R}. \qquad (7.85)$$

The associated free energy is

$$f^+(\zeta) = \lim_{n \to \infty} \frac{1}{n} \log Z_n^+(\zeta), \qquad \zeta \in \mathbb{R}. \qquad (7.86)$$

As shown in [142], the existence of the free energy follows from a standard subadditivity argument. The proof is similar to Part 1 of the proof of Theorem 6.5, and therefore we skip the details.

Theorem 7.9. *For every $d \geq 2$, the function $\zeta \mapsto f^+(\zeta)$ is non-analytic at a critical value $\zeta_c^+ \in (0, \log\{\mu(d)/\mu(d-1)\})$, where $\mu(d)$ is the connective constant of* SAW *in $\mathbb{Z}^d$ defined in (2.19) (with $\mu(1) = 1$).*

Proof. First, for $\zeta \leq 0$ we may estimate

$$Z_n^+(\zeta) \leq Z_n^+(0) = \sum_{v \in \mathbb{N}} c_n^+(v) = c_n^+, \qquad (7.87)$$

with c_n^+ the number of halfspace SAWs. It is known (see Whittington [313]) that

$$\lim_{n \to \infty} \frac{1}{n} \log c_n^+ = \log \mu(d). \qquad (7.88)$$

Next, taking only the single term with $v = 0$ in (7.85), we have

$$Z_n^+(\zeta) \geq c_n^+(0). \qquad (7.89)$$

By translating all halfspace SAWs one unit in the positive x_1-direction and adding an edge from the origin to the new starting point, we see that $c_n^+(0) = c_{n-1}^+$. Together with (7.87–7.89), this yields

$$f^+(\zeta) = \log \mu(d) \qquad \forall \zeta \leq 0. \qquad (7.90)$$

To control $f^+(\zeta)$ for $\zeta \geq 0$, we note that a lower bound is obtained by restricting (7.85) to halfspace SAW's that lie entirely in the hyperplane $\{x_1 = 0\}$, i.e.,

$$Z_n^+(\zeta) \geq c_n^+(n)\, e^{\zeta n}, \qquad \zeta \in \mathbb{R}. \tag{7.91}$$

Since $c_n^+(n)$ is the number of SAWs in $\mathbb{Z}^{d-1}$ with $n+1$ vertices, we thus obtain

$$f^+(\zeta) \geq \log \mu(d-1) + \zeta, \qquad \zeta \in \mathbb{R}. \tag{7.92}$$

By combining (7.90) and (7.92), we see that $\zeta \mapsto f^+(\zeta)$ must be non-analytic at some critical value $\zeta_c^+ \in [0, \log\{\mu(d)/\mu(d-1)\}]$.

Considerable extra work is needed to show that ζ_c^+ does not lie on the edges of this interval, in particular, $\zeta_c^+ > 0$ as in the directed walk model studied in Section 7.2. For details we refer the reader to Hammersley, Torrie and Whittington [142]. Janse van Rensburg [187] gives an easier proof of $\zeta_c^+ > 0$ based on a comparison with wetting of self-avoiding polygons. This proof in fact yields the lower bound $\zeta_c^+ \geq [8(d-1)]^{-1} \log \sqrt{1 + [\mu(d)]^{-2}}$. $\qquad \square$

Janse van Rensburg and Rechnitzer [195] estimate ζ_c^+ for SAWs in $d = 2, 3$ with the help of Monte Carlo simulation. This yields $\zeta_c^+(2) \approx 0.56$ and $\zeta_c^+(3) \approx 0.29$. The phase transition clearly comes out as second order.

The same result as in Theorem 7.9 applies to the pinned SAW, i.e., when we drop the hard wall constraint. Also here $\zeta_c \in [0, \log\{\mu(d)/\mu(d-1)\}]$. It is believed that $\zeta_c = 0$, but a proof is not known. For both models the phase transition is expected to be second order. Proofs are missing, but simulations support the claim.

7.6 Extensions

(1) It is possible to include the boundary cases $a = 0$ and $a = \infty$ in (7.5). In the former case (where $b(\cdot)$ is marginally summable) the phase transition is infinite order, in the latter case (where the tail of $b(\cdot)$ is smaller than polynomial) the behavior is similar to that when $a \in (1, \infty)$. See Giacomin [116], Chapter 2, for details. If the random walk S is transient, i.e., $B(1) = \sum_{k \in \mathbb{N}} b(k) < 1$, then the same theory as in Section 7.1 can be used after normalization of $b(\cdot)$. In this case $\zeta_c = \log[1/B(1)] > 0$ (compare with (7.39)). Thus, the results equally well apply to a $(1 + d)$-dimensional pinning model, where the substrate is a d-dimensional hyperplane and the path measure P is that of a d-dimensional random walk, which for $d \geq 3$ is always transient (Spitzer [284], Section 8).

(2) Deuschel, Giacomin and Zambotti [89] and Caravenna, Giacomin and Zambotti [56] show that the path measure P_n^+, the analogue of P_n in (7.3) for the wetting model, in the limit as $n \to \infty$ converges weakly to a path measure P_∞^+ on the set of infinite directed non-negative paths. There are three relevant regimes (see Fig. 7.15): $\zeta < \zeta_c^+$, $\zeta = \zeta_c^+$ and $\zeta > \zeta_c^+$. In these regimes, P_∞^+ is the law of a transient, null-recurrent, respectively, positive recurrent Markov

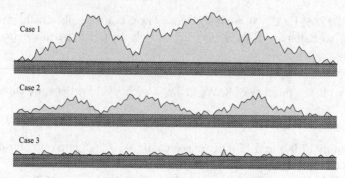

Fig. 7.15. A visual representation of the three regimes for the wetted polymer (courtesy of J.-D. Deuschel, G. Giacomin and L. Zambotti). Case 1: $\zeta < \zeta_c^+$; Case 2: $\zeta = \zeta_c^+$; Case 3: $\zeta > \zeta_c^+$.

process, whose transition probabilities can be written down explicitly. In the subcritical regime the scaling limit of P_∞^+ is Brownian meander, i.e., Brownian motion conditioned to stay positive. In the critical regime the intersection set under P_∞^+ turns out to be a regenerative set process with a *fractal structure*: it has "Lévy exponent" equal to $1 \wedge \zeta$. In the supercritical regime, finally, P_∞^+ has exponentially decaying correlations, and the distribution of the length of an excursion away from the wall has an exponential tail. (Similar results holds for the pinning model.)

In [89] continuous increments of the paths are considered, with an interaction Hamiltonian that sums a general potential at the values of the increments and sums the interactions with the wall. An explicit formula is derived for ζ_c^+ in terms of the partition sum of the model without interaction with the wall. In [56] the results are refined and also discrete increments of the path are allowed. In all cases, $\zeta_c^+ > 0$, i.e., the wetting transition occurs at a strictly positive value of the adsorption energy. Sohier [281] identifies the finite-size scaling limits of the path measure, i.e., when the polymer is close to criticality and has a length that is of the order of the correlation length.

(3) The results in Section 7.3 have been extended to SAWs (Orlandini, Tesi and Whittington [256]). For the wetting version the force-temperature diagram looks as drawn in Fig. 7.16, which is re-entrant. The reason is that, unlike a ballot path, a SAW has a positive entropy when it is adsorbed. This entropy gradually gets lost as more and more of the polymer desorbes when an increasing force is applied to it. Consequently, at low temperature the critical force needs to increase to counterbalance this loss. Simulations can be found in Krawczyk, Prellberg, Owczarek and Rechnitzer [219].

(4) The results in Section 7.4 have been extended to SAWs. It is shown in Janse van Rensburg, Orlandini and Whittington [194] that the same phase diagram as in Fig. 7.13 holds, with ζ_c^+ the critical value for the halfspace SAW. Most features of the phase diagram are proved rigorously, some are

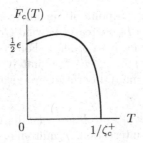

Fig. 7.16. Plot of $T \mapsto F_c(T)$ for SAW.

conjectured. In Janse van Rensburg, Orlandini, Owczarek, Rechnitzer and Whittington [192] and Martin, Orlandini, Owczarek, Rechnitzer and Whittington [237] these features are substantiated via exact enumeration and Monte Carlo simulation.

(5) Janse van Rensburg [189], [190], [191], and Janse van Rensburg and Ye [197], study directed polymers in two-dimensional wedges constrained to begin and end on the same side of the wedge, compute the free energy, and find the repulsive entropic force the polymer exerts on the wedge as the angle of the wedge is varied. In all cases the force diverges logarithmically with the inverse of the angle as the latter tends to zero, a property that appears to be robust. The problem is delicate, because without the constraint on the endpoints the free energy does not depend on the wedge angle (Hammersley and Whittington [144]) and the force is zero. In [189] an attractive interaction with one side of the wedge is added and the critical value for the wetting transition is determined.

(6) Whittington [315] considers wetting of a copolymer performing a partially directed self-avoiding walk and carrying two types of monomers, occurring in alternating order, of which only one type interacts with the interface. The critical threshold for adsorption is computed, and turns out to be strictly less than twice the critical threshold for the homopolymer. Moghaddam, Vrbová and Whittington [241] look at more general forms of periodicity and compare the associated critical thresholds. They also do a Monte Carlo study when the copolymer is a SAW, and find that the numerical differences with directed models tend to be small.

(7) Caravenna and Pétrélis [59], [60] consider a multi-interface medium, consisting of an infinite array of layers with width $L_n \in 2\mathbb{N}$, at which pinning occurs with energy $-\zeta$. The polymer follows a ballot path and for $\zeta > 0$ (attracting case) displays three regimes, depending on how fast L_n grows with n: (1) $L_n - C(\zeta) \log n \to -\infty$: infinitely many interfaces are visited; (2) $L_n - C(\zeta) \log n$ bounded: finitely many interfaces are visited; (3) $L_n - C(\zeta) \log n \to \infty$: only one interface is visited. Here, $C(\zeta)$ is a certain explicit function of ζ. As L_n increases from 1 to $C(\zeta) \log n$, the scaling of the

vertical displacement of the right endpoint of the polymer decreases smoothly from order $\sqrt{n}$ to order $\log n$. As L_n increases beyond $C(\zeta) \log n$, the vertical displacement jumps from order $\log n$ to order 1. For $\zeta < 0$ (repulsive case) there are three regimes also: (1) $L_n \ll n^{1/3}$: infinitely many interfaces are visited; (2) $L_n \asymp n^{1/3}$: finitely many interfaces are visited; (3) $L_n \gg n^{1/3}$: only one interface is visited.

(8) Vrbová and Whittington [306] and Janse van Rensburg [187] consider self-avoiding polygons with an interaction Hamiltonian that has both self-attraction (energy $-\gamma$ for monomers that are self-touching) and attraction to a linear wall (energy $-\zeta$ for monomers on the wall), which is a combination of the models studied in Chapter 6 and Section 7.2. Based on earlier work in the physics literature (Foster [107], Foster and Yeomans [109], Foster, Orlandini and Tesi [108]) and on simulations (Vrobová and Whittington [307], [308]), the phase diagram in the (γ, ζ)-plane for $d = 3$ is believed to have the shape drawn in Fig. 7.17. In [306] it is proved that there is an adsorption transition (between the (DE+DC)-phase and the (AE+AC)-phase) for all values of γ, with $\zeta_c^+(\gamma) \in [0, \log(\mu(3)/\mu(2)) + 2\gamma]$ for $\gamma \geq 0$. It is also proved that *if* there is a collapse transition in the desorbed phase (between the DE-phase and the DC-phase), *then* the associated critical curve is linear, i.e., the critical threshold γ_c for collapse does not depend on ζ (a similar result was found in [107] and [109] for a directed version of the model). In [187] it is proved that $\zeta_c^+(\gamma) > 0$ for all $\gamma \in \mathbb{R}$. There is strong evidence that the phase transitions DE/AE, DE/DC and AE/AC are all second order. The phase transitions DC/AC and DC/AE appear to be first order. Simulations further suggest that in $d = 2$ the AC-phase is absent. This has been proved for directed models (Foster and Yeomans [109]). More refined simulations appear in Krawczyk, Owczarek, Prellberg and Rechnitzer [220]. For a recent overview, see Owczarek and Whittington [261].

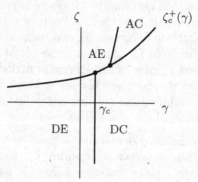

Fig. 7.17. Conjectured phase diagram for adsorption and collapse of self-avoiding polygons in $d = 3$. There are four phases: DE (= desorbed-extended), DC (= desorbed-collapsed), AE (= adsorbed-extended) and AC (= adsorbed-collapsed). Only the critical curve between DE+DC and AE+AC has been shown to exist. Note the occurrence of two tricritical points.

Similar behavior is expected for walks instead of polygons. However, for walks the free energy has only been shown to exists for $\zeta \in \mathbb{R}$ and $\gamma \leq 0$, i.e., for repulsive self-touchings, in which case it is equal to that for polygons (see [306]). It should be equal for all $\zeta, \gamma \in \mathbb{R}$.

(9) Velenik [305] provides an overview of pinning and wetting for interfaces, which are higher-dimensional analogues of polymers. The interaction Hamiltonian for these interfaces is rather general, allowing for a flexible description of a number of associated physical phenomena, such as the "roughening transition".

(10) Caputo, Martinelli and Toninelli [49] study the pinned SRW-polymer in the adsorbed phase subject to a Metropolis dynamics w.r.t. the Hamiltonian defined in (7.1–7.2) in which two successive increments $\nearrow\searrow$ can be replaced by $\searrow\nearrow$, and vice versa. The average transition time between a typical desorbed and a typical adsorbed configuration is computed. Dynamic models of polymers are important and interesting, but are mathematically largely unexplored. They fall outside the scope of the present monograph.

7.7 Challenges

(1) Prove that the phase transitions in the pinning and the wetting model of SAW, as described in Section 7.5, are second order. Prove that $\zeta_c = 0$. (It is already known that $\zeta_c^+ > 0$.) Show that $\zeta \mapsto f(\zeta)$ and $\zeta \mapsto f^+(\zeta)$ are infinitely differentiable on (ζ_c, ∞), respectively, (ζ_c^+, ∞).

(2) Prove the qualitative properties of the force-temperature diagram in Fig. 7.16 from Section 7.6, Extension (3).

(3) Prove the qualitative properties of the phase diagram in Fig. 7.17 from Section 7.6, Extension (8). In particular, show that there is a phase transition between AE and AC and between DE and DC. Determine the shape of the critical curve between AE and AC, and show that this curve does not meet the critical curve between DE+DC and AE+AC at the same point as the vertical critical curve between DE and DC does (i.e., show that there are two tricritical points as drawn). Try to determine the order of the phase transitions. All these are hard challenges.

(4) Extend the analysis in Sections 7.1–7.3 to the situation where the substrate has a finite width, i.e., in (7.2) the indicator $1_{\{w_i=0\}}$ is replaced by $1_{\{0\leq w_i\leq N\}}$ for some $N \in \mathbb{N}$. It is conjectured in Caravenna, Giacomin and Zambotti [56] that the scaling limits of the path measure are the same for all $N \in \mathbb{N}$, both for pinning and wetting.

(5) What can be said about the multi-interface medium in Section 7.6, Extension (7) when the widths of the layers are random, i.e., when $\mathbb{Z}^2$ is sliced

into rows of random sizes $H_k L_n$, $k \in \mathbb{Z}$, with $(H_k)_{k \in \mathbb{Z}}$ i.i.d. random variables taking values in $[C_1, C_2]$ for some $0 < C_1 < C_2 < \infty$?

(6) For all tractable models of wetting based on SRW it is known that the desorption transition is second order in the temperature without a force and first order in the temperature with a force, no matter how small (see Orlandini, Tesi and Whittington [254]). Investigate to what extent this property is robust.

(7) Prove the robustness of the property that "the force is proportional to the logarithm of the inverse of the wedge angle" as described in Section 7.6, Extension (5).

Polymers in Random Environment

OUTLINE:

In Part B we look at five models of polymers in random environment.

- In Chapter 8, we consider a polymer with positive and negative *charges*, arranged randomly along the chain, resulting in repulsion and attraction at the locations where the polymer intersects itself. This is a model for a polymer in an *electrolyte*: the charges on the polymer are surrounded by an ionic atmosphere that screens the charges so that they are felt "on site" only. Our main result will be that the *annealed* charged polymer is *subdiffusive* in any $d \geq 1$. We obtain the full scaling behavior. The annealed model is appropriate for polymers consisting of *amphoteric* monomers, i.e., monomers containing both acidic and basic functional groups, since such monomers may change their charge as the pH of the solution in which they are immersed is varied.

- In Chapter 9, we look at a copolymer, consisting of a random concatenation of *hydrophobic* and *hydrophilic* monomers, interacting with two solvents, *oil* and *water*, separated by a *linear* interface. If the interaction between monomers and solvents is strong and unbiased enough, then the polymer *localizes* near the interface, otherwise it *delocalizes* away from the interface. Our main result is a detailed analysis of the *quenched* critical curve separating the two phases, the order of the phase transition, and the path behavior in each of the two phases.

- In Chapter 10, we extend the results of Chapter 9 to a model where the linear interface is replaced by a *percolation-type* interface. This is a primitive model for a copolymer in an *emulsion* consisting of *mesoscopically* large droplets of oil floating in water. Various forms of (de)localization are possible, depending on whether the oil droplets percolate or not.

- In Chapter 11, the polymer interacts with a linear substrate composed of different types of atoms or molecules, occurring in a random order. Each time the polymer hits the substrate it picks up a reward or a penalty according to the type of atom or molecule it encounters. If the reward to hit the substrate is strong enough, then the polymer *localizes*, otherwise it *delocalizes*. Both the *pinning* and the *wetting* version of the model are of interest. Our main result is a detailed analysis of the *quenched* critical curve separating the two phases, the order of the phase transition, and the path behavior in each of the two phases. Moreover, we investigate when the quenched and the annealed critical curve are different, referred to as *relevant disorder*, and when they coincide, referred to as *irrelevant disorder*. The wetting version of the model can be used to describe the *denaturation transition* of DNA, where the two strands detach with increasing temperature, as well as the *unzipping transition* of DNA under the action of a force.

- In Chapter 12, finally, we consider a polymer interacting with a *random potential* field. An example is a hydrophobic homopolymer living in a *microemulsion* consisting of *microscopically* small droplets of oil and water. Depending on the dimension and on the strength of the interaction, the polymer can be in a phase of *weak disorder* or a phase of *strong disorder*, where the polymer is diffusive, respectively, superdiffusive. Our main result is a classification of these two phases. Another application is a polymer in a *gel* with different sizes of pockets.

Except in Chapter 8, we will consider only *directed paths*, i.e., paths that cannot backtrack (already used in Section 6.2 and Chapter 7). This restriction is necessary in order to make the models mathematically tractable.

The presence of the random environment adds on a considerable layer of complication, and necessitates the development of new ideas and techniques. Some of these are successful in elucidating the behavior of the polymer, others only partially resolve key questions.

Polymers in random environment is a challenging and captivating area, with problems that are driven by specific applications. Many of these problems are "easy to state but hard to solve", and frequently cannot be tackled at full power with non-rigorous methods (such as mean-field approximations and replica computations), in the best of the tradition of mathematical physics.

WARNING: In Part B we use the symbol ω to denote the random environment. Be careful to distinguish this from the symbol w that is used to denote the polymer path.

8

Charged Polymers

In this chapter we consider a model introduced in Kantor and Kardar [203], where each monomer carries a *random charge*, and each self-intersection of the polymer is rewarded when the two charges of the associated monomers have opposite sign and is penalized when they have the same sign. This model is a variation on the weakly self-avoiding walk described in Chapters 3 and 4, with a random self-interaction driven by the charges.

We will focus on the *annealed* path measure, of the type defined in (1.5). We will show that the annealed charged polymer is in a *collapsed phase*, irrespective of its overall charge distribution, and is *subdiffusive* with a scaling limit that can be computed explicitly, namely, Brownian motion conditioned to stay inside a finite ball. The free energy will be different for neutral and for non-neutral charged polymers, even though the scaling limit is the same. Once more *local times* will prove to be useful. In particular, the large deviation behavior of the local times of SRW will play a crucial role in the identification of the scaling limit.

DNA and proteins are polyelectrolytes, with each monomer in a specific charged state that depends on the pH of the solution in which they are immersed. Our annealed analysis applies to polymers consisting of *amphoteric* monomers, i.e., monomers containing both acidic and basic functional groups. An amphoteric monomer can be positively charged at low pH and negatively charged at high pH. In equilibrium the individual monomers do not retain their charge: the charges fluctuate in time, with the overall charge of the polymer being more or less constant. For polyelectrolytes with amphoteric monomers there is a particular value of the pH, called the *iso-electric point*, at which the overall charge is zero, separating the regimes of positive and negative overall charge.

In Section 8.1 we define the model, in Section 8.2 we compute the annealed free energy, in Section 8.3 we prove the subdiffusive scaling, while in Section 8.4 we make a link with the parabolic Anderson model.

F. den Hollander, *Random Polymers*,
Lecture Notes in Mathematics 1974, DOI: 10.1007/978-3-642-00333-2_8,
© Springer-Verlag Berlin Heidelberg 2009

8.1 A Polymer with Screened Random Charges

Let

$$\omega = (\omega_i)_{i\in\mathbb{N}_0} \text{ is i.i.d. with } \mathbb{P}_p(\omega_0 = +1) = p \text{ and } \mathbb{P}_p(\omega_0 = -1) = 1-p, \quad (8.1)$$

with $p \in (0,1)$. This random sequence labels the charges of the monomers along the polymer (see Fig. 8.1), with p the density of the positive charges and $1-p$ the density of the negative charges. Throughout the sequel, $\mathbb{P}_p$ denotes the law of ω.

Our set of paths and Hamiltonian are

$$\mathcal{W}_n = \{w = (w_i)_{i=0}^n \in (\mathbb{Z}^d)^{n+1} \colon w_0 = 0, \|w_{i+1} - w_i\| = 1 \ \forall 0 \le i < n\},$$
$$H_n^\omega(w) = \beta I_n^\omega(w), \tag{8.2}$$

where $\beta \in (0,\infty)$ and (compare with (3.2))

$$I_n^\omega(w) = \sum_{\substack{i,j=0 \\ i<j}}^n \omega_i\omega_j \, 1_{\{w_i=w_j\}}. \tag{8.3}$$

For fixed ω, the quenched path measure is

$$P_n^{\beta,\omega}(w) = \frac{1}{Z_n^{\beta,\omega}} \, e^{-\beta I_n^\omega(w)} \, P_n(w), \qquad w \in \mathcal{W}_n, \tag{8.4}$$

where we recall that P_n is the projection onto $\mathcal{W}_n$ of the law P of SRW. What this path measure does is reward self-intersections when the charges are opposite ($\omega_i\omega_j = -1$) and penalize self-intersections when the charges are the same ($\omega_i\omega_j = 1$). Note that the interaction between the charges is felt "on site" only, which is why we speak of *screened random charges*. Think of this as modeling a polymer in an electrolyte. Each charge gets surrounded by an ionic atmosphere, which reduces the spatial range of the charge-charge interaction (see Section 8.6, Challenge (4), for other options).

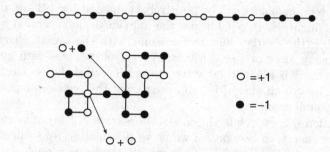

Fig. 8.1. A polymer carrying random charges. The path may or may not be self-avoiding. The charges only interact at self-intersections.

The cases $p = 0$ and $p = 1$ correspond to the weakly self-repellent random walk. Indeed, if $\omega \equiv +1$ or $\omega \equiv -1$, then all the charges are the same and there are only penalties.

We will be focusing on the *annealed model* with path measure (recall (1.5))

$$\mathbb{P}_n^{\beta,p}(w) = \frac{1}{\mathbb{Z}_n^{\beta,p}} \left[\int e^{-\beta I_n^\omega(w)} \mathbb{P}_p(\mathrm{d}\omega) \right] P_n(w), \qquad w \in \mathcal{W}_n. \tag{8.5}$$

This describes the situation in which the charges can vary along the polymer and can take part in the equilibration. Intuitively, we expect:

(a) $p = \frac{1}{2}$: The total charge is on average *neutral*, the repulsion and the attraction balance out, and the polymer behaves qualitatively like SRW, i.e., has a diffusive scaling.

(b) $p \neq \frac{1}{2}$: The total charge is on average *non-neutral*, but makes a large deviation under the annealed measure so that it becomes neutral. The price for this large deviation is an entropy factor that shows up in the free energy. Apart from this factor, the polymer behaves as if $p = \frac{1}{2}$.

In Sections 8.2–8.3 we will show that the above intuition is correct, except that actually the attraction wins from the repulsion as soon as the total charge is neutral, resulting in the polymer being *subdiffusive*.

8.2 Scaling of the Free Energy

8.2.1 Variational Characterization

The annealed partition sum equals

$$\mathbb{Z}_n^{\beta,p} = \mathbb{E}_p\left(Z_n^{\beta,\omega}\right) \quad \text{with} \quad Z_n^{\beta,\omega} = E\left(e^{-\beta I_n^\omega(w)}\right). \tag{8.6}$$

(Note that $\log \mathbb{Z}_n^{\beta,p}$ is different from $\mathbb{E}_p(\log Z_n^{\beta,\omega})$, which is the average of the quenched free energy.) The following result is due to Biskup and König [21] for $p = \frac{1}{2}$. Note that β plays no role in the scaling behavior.

Theorem 8.1. *For every $d \geq 1$, $\beta \in (0, \infty)$ and $p \in (0, 1)$,*

$$\lim_{n \to \infty} \frac{(\alpha_n)^2}{n} \left\{ \log \mathbb{Z}_n^{\beta,p} - n \tfrac{1}{2} \log[4p(1-p)] \right\} = -\chi \tag{8.7}$$

with $\alpha_n = (n/\log n)^{1/(d+2)}$ and $\chi \in (0, \infty)$ given by (8.26) below.

Proof. Let (similarly as in (3.8))

$$\widehat{I}_n^\omega = \sum_{i,j=0}^n \omega_i \omega_j \, 1\{w_i = w_j\}. \tag{8.8}$$

Then $\widehat{I}_n^\omega = 2I_n^\omega + (n+1)$. Hence, in (8.5) we may as well put $\widehat{I}_n^\omega$ in the exponential weight factor (which only changes β to 2β). We will do so and henceforth drop the overscript.

We begin by carrying out the first expectation in (8.6). To that end, define

$$G^{\beta,p}(l) = \log \mathbb{E}_p \left(\exp\left[-\beta \left(\sum_{k=1}^{l} \omega_k \right)^2 \right] \right), \qquad l \in \mathbb{N}_0. \tag{8.9}$$

Then, writing

$$I_n^\omega(w) = \sum_{i,j=0}^{n} \omega_i \omega_j \sum_{x \in \mathbb{Z}^d} 1\{w_i = x\} 1\{w_j = x\} = \sum_{x \in \mathbb{Z}^d} \left(\sum_{i=0}^{n} \omega_i 1\{w_i = x\} \right)^2, \tag{8.10}$$

using Fubini's theorem to interchange expectations, noting that ω is i.i.d., and using (8.9), we may rewrite (8.6) as

$$\begin{aligned}
Z_n^{\beta,p} &= E\left(\mathbb{E}_p \left(e^{-\beta I_n^\omega(w)} \right) \right) \\
&= E\left(\mathbb{E}_p \left(\prod_{x \in \mathbb{Z}^d} \exp\left[-\beta \left(\sum_{k=1}^{\ell_n(x)} \omega_{\tau_k(x)} \right)^2 \right] \right) \right) \\
&= E\left(\exp\left[\sum_{x \in \mathbb{Z}^d} G^{\beta,p}(\ell_n(x)) \right] \right),
\end{aligned} \tag{8.11}$$

where $\tau_k(x)$ is the time of the k-th visit of w to site x and $\ell_n(x)$ is the local time of w at site x defined in (3.9).

The following lemma allows us to reduce the proof to the case $p = \frac{1}{2}$, and shows that $G^{\beta,\frac{1}{2}}(l)$ tends to $-\infty$ at a very slow rate, which will serve us later on.

Lemma 8.2. *For every $\beta \in (0,1)$ and $p \in (0,1)$,*

$$G^{\beta,p}(l) = A(p)l - \tfrac{1}{2}\log l + O(1) \qquad \text{as } l \to \infty \tag{8.12}$$

with $A(p) = \frac{1}{2}\log[4p(1-p)] \le 0$.

Proof. Abbreviate $\Sigma_l = \sum_{k=0}^{l} \omega_k$, $l \in \mathbb{N}$. A change of measure from $\mathbb{P}_p$ to $\mathbb{P}_{\frac{1}{2}}$ gives

$$G^{\beta,p}(l) = Al + G_*^{\beta,p}(l) \tag{8.13}$$

with

$$G_*^{\beta,p}(l) = \log \mathbb{E}_{\frac{1}{2}} \left(\exp\left[-\beta(\Sigma_l)^2 + B\Sigma_l \right] \right), \tag{8.14}$$

where $A = A(p) = \frac{1}{2}\log[4p(1-p)]$ and $B = B(p) = \frac{1}{2}\log[p/(1-p)]$. To prove (8.12), we pick $N \in \mathbb{N}$ large and estimate

$$e^{G_*^{\beta,p}(l)} \geq e^{-\beta N^2 - |B|N} \, \mathbb{P}_{\frac{1}{2}}(|\Sigma_l| \leq N),$$

$$e^{G_*^{\beta,p}(l)} \leq e^{-\beta N^2 + |B|N} + e^{B^2/4\beta} \, \mathbb{P}_{\frac{1}{2}}(|\Sigma_l| \leq N). \tag{8.15}$$

Here, we use that $-\beta x^2 + Bx$ is maximal at $x = B/2\beta$, where it assumes the value $B^2/4\beta$, and is non-increasing in $|x|$ for $|x|$ large enough. By the local central limit theorem we have, for fixed N,

$$\mathbb{P}_{\frac{1}{2}}(|\Sigma_l| \leq N) \sim C(N)\, l^{-1/2} \quad \text{as } l \to \infty \tag{8.16}$$

for some $C(N) \in (0,\infty)$. Substituting this into (8.15) and letting $l \to \infty$ followed by $N \to \infty$, we obtain

$$G_*^{\beta,p}(l) = -\tfrac{1}{2}\log l + O(1) \quad \text{as } l \to \infty, \tag{8.17}$$

with the dependence on β and p sitting in the error term only. Combining (8.13) and (8.17), we get the claim in (8.12). Note that the above argument shows that for large l the term $B\Sigma_l$ in (8.14) is negligible, implying that $G_*^{\beta,p}(l) = G^{\beta,\frac{1}{2}}(l) + O(1)$ as $l \to \infty$. $\square$

Because $\sum_{x \in \mathbb{Z}^d} \ell_n(x) = n + 1$, it follows from (8.11) and (8.13) that

$$Z_n^{\beta,p} = e^{A(n+1)}\, E\left(\exp\left[\sum_{x \in \mathbb{Z}^d} G_*^{\beta,p}(\ell_n(x))\right]\right). \tag{8.18}$$

The frontfactor in (8.18) explains the shift of the free energy in (8.7). Thus, it remains to identify the scaling behavior of the expectation in (8.18) with the help of (8.17).

8.2.2 Heuristics

For all β and p,

$$[G^{\beta,p}(l+1) - A] \geq [G^{\beta,p}(l) - A] + [G^{\beta,p}(1) - A] \quad \forall l \in \mathbb{N}_0. \tag{8.19}$$

Indeed, from (8.9) we have

$$e^{G^{\beta,p}(l+1)} = \mathbb{E}_p\left(e^{-\beta(\Sigma_{l+1})^2}\right) = \mathbb{E}_p\left(e^{-\beta(\Sigma_l)^2}\, e^{-\beta}\left[pe^{-2\beta\Sigma_l} + (1-p)e^{2\beta\Sigma_l}\right]\right)$$
$$\geq e^{G^{\beta,p}(l)}\, e^{-\beta}\, [4p(1-p)]^{1/2} = e^{G^{\beta,p}(l)}\, e^{G^{\beta,p}(1)}\, e^A, \tag{8.20}$$

where we use that $pe^{-2\beta x} + (1-p)e^{2\beta x} \geq [4p(1-p)]^{1/2}$ for all $\beta, x \in \mathbb{R}$. Because $G^{\beta,p}$ is negative, we may refer to (8.19) as sublinearity. This property is inherited by $G_*^{\beta,p}$ defined in (8.13). A consequence of this sublinearity is that the local times in (8.18), which are subject to the restriction $\sum_{x \in \mathbb{Z}^d} \ell_n(x) = n+1$, have a tendency to "pile up" rather than to "spread out". This explains

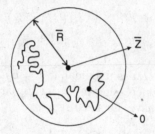

Fig. 8.2. A random walk starting at 0 conditioned to stay inside the ball with radius $R_n \sim \bar{R}\alpha_n$ and center $Z_n \sim \bar{Z}\alpha_n$.

why the polymer has a tendency to be subdiffusive. The following argument, which makes this intuition precise, is sketchy. We will indicate later how to fill in the details.

The maximal contribution to $\mathbb{Z}_n^{\beta,p}$ comes from a strategy where the polymer stays inside $B_{R_n}(0) \subset \mathbb{R}^d$, the closed ball of radius R_n (to be determined later) centered at 0, and distributes its local times evenly over this ball (see Fig. 8.2). The cost for staying inside $B_{R_n}(0)$ up to time n is

$$\exp\left[-\frac{\lambda_d n}{(R_n)^2}[1 + o(1)]\right] \qquad \text{as } n \to \infty, \tag{8.21}$$

where λ_d is the principal Dirichlet eigenvalue of $-\Delta_{\mathbb{R}^d}/2d$ on $B_1(0)$, with $\Delta_{\mathbb{R}^d}$ the continuous Laplacian. Conditionally on the polymer staying inside $B_{R_n}(0)$ up to time n, the local time per site is $(n/\omega_d(R_n)^d)[1 + o(1)]$, with $\omega_d = |B_1(0)|$. Since, by (8.17), $l \mapsto G_*^{\beta,p}(l)$ is almost constant for large l, the contribution of this strategy to (8.18) is

$$\exp\left[-\frac{1}{2}\omega_d(R_n)^d \log\left(\frac{n}{\omega_d(R_n)^d}\right)[1 + o(1)]\right] \qquad \text{as } n \to \infty. \tag{8.22}$$

The two exponentials in (8.21–8.22) are of the same order when we choose R_n such that

$$\frac{n}{(R_n)^2} \asymp (R_n)^d \log\left(\frac{n}{(R_n)^d}\right), \tag{8.23}$$

i.e., $R_n \asymp \alpha_n = (n/\log n)^{1/(d+2)}$ (where $\asymp$ means that the ratio of the two sides is bounded above and below by strictly positive and finite constants). Therefore, putting $R_n = \bar{R}\alpha_n$, we get (8.7) with

$$\chi = \inf_{R \in (0,\infty)}\left[\frac{\lambda_d}{R^2} + \frac{1}{d+2}\omega_d R^d\right], \tag{8.24}$$

where the scaled radius R is to be optimized over. The infimum is achieved at $R = \bar{R}$ with

$$\bar{R} = \left[\frac{2(d+2)\lambda_d}{d\omega_d}\right]^{1/(d+2)}, \tag{8.25}$$

and equals

$$\chi = \left[\frac{\omega_d}{2}\right]^{2/(d+2)} \left[\frac{(d+2)\lambda_d}{d}\right]^{d/(d+2)}. \tag{8.26}$$

8.2.3 Large Deviations

The above argument can be made rigorous with the help of large deviation techniques. For background on what follows next, we refer the reader to the Saint-Flour lectures by Bolthausen [28], Sections 2.5–2.6. First, the strategy of staying inside a large ball and distributing the local times evenly clearly provides a *lower bound* for the free energy. Second, the fact that no other strategy does better, is proved in three steps:

(1) *Compactification*: The random walk is "wrapped around" a large box containing $B_{R_n}(0)$, i.e., is periodized w.r.t. this box. By the sublinearity of $l \mapsto G_*^{\beta,p}(l)$, this wrapping increases the contribution to the expectation in (8.18), thus providing an *upper bound* for the free energy.

(2) *Scaling*: For the wrapped random walk, after scaling the large box down to a finite box, the scaled local times are close to those of a "mollified" Brownian motion. The latter satisfy an LDP, with a rate function given by the standard Donsker-Varadhan rate function for the empirical measure of Brownian motion.

(3) *Optimization*: By optimizing over the profile of the scaled local times, we get (8.7) with χ given by a functional variational problem involving empirical measures on a finite box. Here it is important that, because $l \mapsto G_*^{\beta,p}(l)$ is almost constant for large l, only the *order* of the local times is relevant, not the precise form of their spatial profile. Accordingly, the functional variational problem for χ reduces to the scalar variational problem stated in (8.24), in which only λ_d and ω_d enter. The fact that the empirical measures take their support in a ball is due to radial symmetry in combination with a standard isoperimetric inequality.

The full argument requires a number of smoothing and approximation techniques. It is easy to show that the optimal path spends a "negligible" amount of time outside the ball $B_{R_n}(0)$. It is harder to show that the optimal path never leaves the ball. For the latter we refer to Bolthausen [27]. □

8.3 Subdiffusive Behavior

The following theorem shows that the annealed charged polymer is in the *collapsed phase*, with subdiffusive scaling, irrespective of its overall charge.

Theorem 8.3. *For every $d \geq 1$, $\beta \in (0,\infty)$ and $p \in (0,1)$, under the path measure $\mathbb{P}_n^{\beta,p}$,*

$$\left(\frac{1}{a_n} S_{\lfloor nt \rfloor}\right)_{0 \le t \le 1} \Longrightarrow (\Theta_t)_{0 \le t \le 1} \qquad as \; n \to \infty \qquad (8.27)$$

with $(\Theta_t)_{t \ge 0}$ Brownian motion on $\mathbb{R}^d$ conditioned on not leaving the ball with radius $\tilde{R}$ given by (8.25) and with a randomly shifted center $\bar{Z}$.

Proof. The proof of Theorem 8.1 shows that the annealed charged polymer up to time n lives in a ball of radius R_n (see Fig. 8.2). The profile of its scaled local times is given by the eigenfunction associated with λ_d. Its scaled endpoint is distributed as the endpoint of a Brownian motion on a ball with radius 1, with a drift towards the center of this ball according to this eigenfunction. The center of the ball is not 0, but is randomly shifted according to that same eigenfunction. We again refer to Bolthausen [27] for details; the behavior is the same as for the range of SRW. □

8.4 Parabolic Anderson Equation

An interesting connection is the following. Let $S = (S_t)_{t \ge 0}$ be continuous-time SRW on $\mathbb{Z}^d$ jumping at rate 1, and let $(\ell_t)_{t \ge 0}$ be its associated local time process defined by $\ell_t(x) = \int_0^t 1\{S_t = x\}dt$, $x \in \mathbb{Z}^d$, $t \ge 0$. Define

$$\mathbb{Z}_t^{\beta, p} = E\left(\exp\left[\sum_{x \in \mathbb{Z}^d} G^{\beta, p}(\ell_t(x))\right]\right), \qquad t \ge 0, \qquad (8.28)$$

which is the continuous-time analogue of (8.11). Then

$$\mathbb{Z}_t^{\beta, p} = \mathbb{E}(u(0, t)), \qquad (8.29)$$

with $u \colon \mathbb{Z}^d \times [0, \infty) \to \mathbb{R}$ the solution of the parabolic Anderson equation

$$\begin{cases} \frac{\partial u}{\partial t}(x, t) = \Delta_{\mathbb{Z}^d} u(x, t) + \xi(x) u(x, t), & x \in \mathbb{Z}^d, \; t \ge 0, \\ u(x, 0) = 1, & x \in \mathbb{Z}^d, \end{cases} \qquad (8.30)$$

where $\Delta_{\mathbb{Z}^d}$ is the discrete Laplacian and $\xi = \{\xi(x) \colon x \in \mathbb{Z}^d\}$ is an i.i.d. field of non-positive random variables whose probability law is given by the moment generating function

$$\mathbb{E}\left(e^{l \xi(0)}\right) = e^{G^{\beta, p}(l)}, \qquad l \in \mathbb{N}_0, \qquad (8.31)$$

where we write $\mathbb{E}$ to denote expectation w.r.t. ξ. The relation between (8.28) and (8.29–8.31) is an immediate consequence of the Feynman-Kac representation of the solution of (8.30) given ξ, which reads

$$u(x, t) = E\left(\exp\left[\int_0^t \xi(S_s)\, ds\right]\right) = E\left(\exp\left[\sum_{x \in \mathbb{Z}^d} \xi(x)\, \ell_t(x)\right]\right). \qquad (8.32)$$

Indeed, if (8.32) is substituted into (8.29), Fubini's theorem is used to interchange the two expectations, and (8.31) is used afterwards, then the result is precisely (8.28). In Biskup and König [21] the annealed Lyapunov exponents of (8.30), i.e., the growth rates of the successive moments of $u(0,t)$, are computed for several choices of $\xi(0)$ whose moment generating function satisfies an appropriate scaling assumption. The quantity of interest in Theorem 8.1 corresponds to the first Lyapunov exponent for the special case where $\xi(0)$ is given by (8.31).

Remark: The moment generating function in (8.31) actually is not completely monotone as a function of l and therefore does not properly define the random variable $\xi(0)$. However, this can be repaired by perturbing $G^{\beta,p}(l)$ slightly, without affecting its asymptotics in (8.17).

The parabolic Anderson equation is used to model the evolution of reactant particles in the presence of a catalyst. For an overview, see Gärtner and König [113].

8.5 Extensions

(1) The argument in Sections 8.2–8.3 is easily extended to random charges that take values in $\mathbb{R}$, provided they have a finite moment generating function.

(2) Chen [66] studies the annealed scaling behavior of $I_n^\omega(w)$ defined in (8.3) for random charges that have a symmetric distribution with a finite moment generating function (so that the total charge is on average neutral). A central limit theorem is derived, a law of the iterated logarithm, as well as moderate and large deviation estimates. The behavior is different in $d = 1$, $d = 2$ and $d \geq 3$. The moderate and large deviation estimates are further refined in Asselah [11]. Chen [67] contains an extensive study of the large deviation properties of random walk intersections. This is useful also for the self-repellent polymer studied in Chapters 3–4.

(3) For the quenched model, with path measure $P_n^{\beta,\omega}$ given by (8.4), the two main questions are: (i) Is the free energy self-averaging in ω? (ii) Is there a phase transition from a collapsed state to an extended state at some $\beta_c \in (0,\infty)$? Both these questions remain unresolved. For $d = 1$, partial results have been obtained in a different direction, as we describe in Extensions (4), (5) and (6) below.

(4) Derrida, Griffiths and Higgs [86] and Derrida and Higgs [88] consider the case where the steps of the random walk are drawn from $\{0,1\}$ rather than $\{-1,+1\}$, which makes the model a bit more tractable, both analytically and numerically. In [86] the charge disorder is binary, and numerical evidence is found for the free energy to be self-averaging and to exhibit a "freezing transition" at a critical threshold $\beta_c \in (0,\infty)$, i.e., the quenched charged

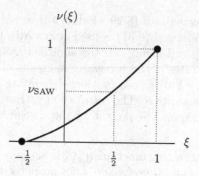

Fig. 8.3. Qualitative shape of $\xi \mapsto \nu(\xi)$ for the one-sided charged polymer conditioned on the overall charge.

polymer is ballistic when $\beta < \beta_c$ and subballistic when $\beta > \beta_c$. In the latter regime it seems that the end-to-end distance scales like n^ν, with $\nu = \nu(\beta)$ depending on β. In this regime, long and rare stretches of the polymer that are globally neutral find it energetically favorable to collapse onto single sites. Numerical simulation gives that $\beta_c \geq 0.48$. In [88] the charge disorder is standard normal and the total charge $\sum_{i=1}^n \omega_i$ is *conditioned* to grow like n^ξ, $\xi \in [-\frac{1}{2}, 1]$. It is found numerically that the end-to-end distance scales like n^ν, with $\nu = \nu(\xi)$ depending on ξ and growing roughly linearly from $\nu(-\frac{1}{2}) = 0$ to $\nu(1) = 1$, with $\nu(\frac{1}{2}) \approx 0.574$ (see Fig. 8.3). The latter is the exponent for the quenched charged polymer when the charges are typical.

(5) Martinez and Petritis [238] choose the charge distribution to be given by the increments of a SRW *conditioned* to return to 0 after n steps, which guarantees that the total charge is neutral. Instead of looking at the free energy conditioned on the charges, they study the free energy *conditioned on the path and averaged over the disorder*, i.e.,

$$\frac{1}{n} \log \mathbb{E}\left(e^{-\beta I_n^\omega(w)} \mid w\right), \qquad w \in \mathcal{W}_n, \tag{8.33}$$

and show that this is self-averaging in w in the limit as $n \to \infty$, and has no phase transition in β.

(6) A continuum version of the model in [238] is studied in Buffet and Pulé [46]. Here, the path of the polymer is given by a Brownian motion $B = (B_t)_{t \in [0,1]}$, the charge distribution is given by a Brownian bridge $b = (b_t)_{t \in [0,1]}$, while the interaction is no longer "on site" but is given by a two-body potential $h: \mathbb{R} \to \mathbb{R}$ of the form

$$h(x) = \int_{\mathbb{R}} g(x - y)g(y) \, dy, \qquad x \in \mathbb{R}, \tag{8.34}$$

for some $g: \mathbb{R} \to \mathbb{R}$ that is square integrable, so that h is even, positive definite, bounded and integrable. The interaction Hamiltonian is taken to be

$$H_t^{\beta,b}(B) = \frac{\beta t}{2} \int_0^1 \mathrm{d}b_u \int_0^1 \mathrm{d}b_v\, h\big(t(B_u - B_v)\big), \tag{8.35}$$

where t is the length of the polymer, and time is scaled down to $[0,1]$ for convenience: tB_u is the position of the polymer at time u, $\sqrt{t}b_u$ is the total charge carried by the polymer up to time u. The object of interest is the free energy conditioned on B and averaged over b, i.e.,

$$f(\beta; B) = \lim_{t \to \infty} \frac{1}{t} \log Z_t^\beta(B) \qquad B - a.s. \tag{8.36}$$

with

$$Z_t^\beta(B) = \mathbb{E}\left(e^{-H_t^{\beta,b}(B)} \mid B\right), \tag{8.37}$$

where $\mathbb{E}$ denotes expectation over b. The latter partition sum is computed in terms of a series involving the local times of B. It turns out that, in contrast with the discrete model in [238], the free energy is *not* self-averaging, i.e., $f(\beta; B)$ depends on B. However, $\beta \mapsto f(\beta; B)$ is shown to be analytic on $[0, \infty)$, which rules out a phase transition. The analysis indicates that there is a collapse transition at some $\beta_c = \beta_c(B) \in (-\infty, 0)$, which corresponds to the "unphysical" regime where charges of the same sign attract each other and charges of the opposite sign repel each other. For $\beta < \beta_c(B)$ the free energy diverges (because $\log Z_t^\beta(B)$ scales differently with t, similarly as in (6.7) for the polymer with self-intersections and self-touchings considered in Chapter 6). The analysis gives an explicit upper bound for $\beta_c(B)$, namely,

$$\beta_c(B) \leq -\left[L(B) \int_{\mathbb{R}} h(x)\mathrm{d}x\right]^{-1} \tag{8.38}$$

with $L(B)$ the maximal local time of B on $\mathbb{R}$.

(7) Almost nothing is known rigorously for SAW with random charges. The appropriate modifications of (8.2–8.3), relevant only for $d \geq 2$, read

$$\mathcal{W}_n = \big\{w = (w_i)_{i=0}^n \in (\mathbb{Z}^d)^{n+1}:\ w_0 = 0,\ \|w_{i+1} - w_i\| = 1\ \forall 0 \leq i < n,$$
$$w_i \neq w_j\ \forall 0 \leq i < j \leq n\big\},$$
$$H_n^\omega(w) = \beta I_n^\omega(w), \tag{8.39}$$

and

$$I_n^\omega(w) = \frac{1}{2d} \sum_{\substack{i,j=0 \\ i<j}}^n \omega_i \omega_j\, 1_{\{|w_i - w_j|=1\}}. \tag{8.40}$$

For $d = 2$ and $d = 3$, a number of results have been obtained with the help of exact enumeration, series analysis and Monte Carlo techniques. Monari and Stella [242] argue that for $p = \frac{1}{2}$ there is a collapse transition as β *increases* from 0 through a critical value $\beta_c^+(\frac{1}{2}) \in (0, \infty)$. This transition seems to be in the same universality class as the θ-transition in homopolymers mentioned

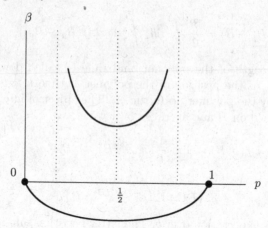

Fig. 8.4. Conjectured phase diagram for SAW with random charges. The top curve is $p \mapsto \beta_c^+(p)$, the lower curve is $p \mapsto \beta_c^-(p)$. The polymer is collapsed both above the top curve and below the bottom curve, and is extended elsewhere.

in Section 6.1.3. Kantor and Kardar [204], Grassberger and Hegger [127] and Golding and Kantor [126] find that, for $p > \frac{1}{2}$ but close to $\frac{1}{2}$, a similar collapse transition occurs, but at a value $\beta_c^+(p) \in (0, \infty)$ that increases with p, while for p close to 1 there is no collapse transition, i.e., $\beta_c^+(p) = \infty$, and the behavior is like SAW for all $\beta \in (0, \infty)$ (see Fig. 8.4). As β *decreases* from 0, there appears to be a collapse transition at a critical value $\beta_c^-(p) \in (-\infty, 0)$ for all $p \in [\frac{1}{2}, 1)$. This corresponds to the "unphysical" regime where charges of the same sign attract each other and charges of the opposite sign repel each other. In this regime, for $p = 1$ the model is the same as that of the SAW-version of the homopolymer in Section 6.1 with $\gamma = -\beta$ (recall (6.1–6.2)), and should be qualitatively similar for $p \in [\frac{1}{2}, 1)$ as well (see Fig. 8.4).

For further background we refer to the overview paper by Soteros and Whittington [283], Sections 2.3, 3.2.3 and 4.2.3. The "unphysical" regime mentioned in Extensions (6) and (7) can be reinterpreted as describing a copolymer consisting of hydrophobic and hydrophilic monomers, a topic that will be treated in Chapter 9.

8.6 Challenges

(1) Recall (1.3–1.5). Find the growth rate of the *quenched* free energy $\log Z_n^{\beta,\omega}$, the partition sum appearing in (8.4). Is this rate ω-a.s. constant or is it sample dependent? What about the *average quenched* free energy $\mathbb{E}_p(\log Z_n^{\beta,\omega})$?

(2) For $p = \frac{1}{2}$, find out whether or not the polymer is ω-a.s. subdiffusive under the quenched law $P_n^{\beta,\omega}$ as $n \to \infty$. The fluctuations of the charges in ω are expected to push the polymer farther apart.

(3) For $p \neq \frac{1}{2}$, show that, under the quenched law $P_n^{\beta,\omega}$ as $n \to \infty$, the charged polymer ω-a.s. has the same qualitative behavior as the soft polymer (which corresponds to the case $p = 0$ and $p = 1$). In particular, in $d = 1$ show that the charged polymer ω-a.s. is ballistic and that its speed is smaller than that of the soft polymer studied in Chapter 3.

(4) Investigate whether it is possible to deal with charges whose interaction extends beyond the "on site" interaction in (8.3), like a Coulomb potential (polynomial decay) or a Yukawa potential (exponential decay). A Yukawa potential arises from a Coulomb potential via screening of the charges.

Copolymers Near a Linear Selective Interface

A *copolymer* is a polymer consisting of different types of monomer. In this chapter we consider a two-dimensional *directed* copolymer, consisting of a random concatenation of *hydrophobic* and *hydrophilic* monomers, near a linear interface separating two immiscible solvents, *oil* and *water* (see Fig. 9.1). We will be interested in the *quenched* path measure (of the type defined in (1.3)). We will show that, as a function of the strength and the bias of the interaction between the monomers and the solvents, this model has a phase transition between a *localized* phase, where the copolymer stays near the interface, and a *delocalized* phase, where the copolymer wanders away from the interface. The critical curve separating the two phases has interesting properties, some of which remain to be clarified. The main techniques used are the *subadditive ergodic theorem*, the *method of excursions*, *large deviations* and *partial annealing estimates*.

In Section 9.1 we define the model, in Section 9.2 we study the free energy, in Section 9.3 we prove the existence of the phase transition, in Section 9.4 we identify the qualitative properties of the critical curve, while in Section 9.5 we describe the qualitative properties of the two phases.

In Chapter 10 we will turn to a model where the linear interface is replaced by a random interface, coming from large blocks of oil and water arranged in a percolation-type fashion.

The order of the monomers is determined by the *polymerization process* through which the copolymer is grown (see Section 1.1). Since the monomers cannot reconfigure themselves without some chemical reaction occurring, a copolymer is an example of a system with *quenched* disorder. Different copolymers typically have different orderings of monomers. Therefore, in order to determine the physical properties of a system of copolymers, an average must be taken over the possible monomer orderings. This is why it is natural to consider the monomers as being drawn *randomly* according to some appropriate probability distribution, i.i.d. in our case.

Copolymers near liquid-liquid interfaces are of interest due to their extensive application as surfactants, emulsifiers, and foaming or antifoaming agents.

F. den Hollander, *Random Polymers*,
Lecture Notes in Mathematics 1974, DOI: 10.1007/978-3-642-00333-2_9,
© Springer-Verlag Berlin Heidelberg 2009

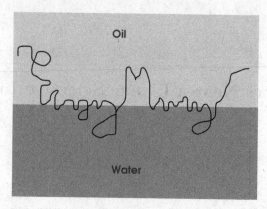

Fig. 9.1. An undirected copolymer near a linear interface.

Many fats contain stretches of hydrophobic and hydrophilic monomers, arranged "in some sort of erratic manner", and therefore are examples of random copolymers. The transition between a localized and a delocalized phase has been observed experimentally e.g. in neutron reflection studies of copolymers consisting of blocks of ethylene oxide and propylene oxide near a hexane-water interface (Phipps, Richardson, Cosgrove and Eaglesham [266]). Here, a thin layer of hexane, approximately 10^{-5} m thick, is spread on water. In the localized phase, the copolymer is found to stretch itself along the interface in a band of width approximately 20 Å.

Soteros and Whittington [283], Sections 2.2, 3.2.2 and 4.2.2, describes a number of rigorous, approximate, numerical and heuristic results for copolymers near linear interfaces. An earlier reference is Whittington [316]. Giacomin [116], Chapters 6–9, gives an overview of rigorous and numerical results based on the method of excursions.

9.1 A Copolymer Interacting with Two Solvents

For $n \in \mathbb{N}_0$, let

$$\mathcal{W}_n = \left\{ w = (i, w_i)_{i=0}^n \colon w_0 = 0,\ w_{i+1} - w_i = \pm 1\ \forall\, 0 \le i < n \right\} \qquad (9.1)$$

denote the set of all n-step directed paths in $\mathbb{N}_0 \times \mathbb{Z}$ that start from the origin and at each step move either north-east or south-east (called ballot paths; recall Fig. 1.4). This is the set of configurations of the copolymer. An example of a path in $\mathcal{W}_n$ is drawn in Fig. 9.2. Let

$$\omega = (\omega_i)_{i \in \mathbb{N}} \text{ be i.i.d. with } \mathbb{P}(\omega_1 = +1) = \mathbb{P}(\omega_1 = -1) = \tfrac{1}{2}. \qquad (9.2)$$

This random sequence labels the order of the monomers along the copolymer (see Fig. 9.2). Throughout the sequel, $\mathbb{P}$ denotes the law of ω. As Hamiltonian we pick, for fixed ω,

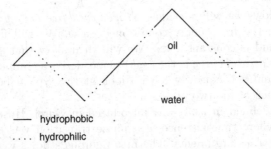

Fig. 9.2. A 14-step directed copolymer near a linear interface. Oil is above the interface, water is below. The drawn edges are hydrophobic monomers, the dashed edges are hydrophilic monomers.

$$H_n^\omega(w) = -\lambda \sum_{i=1}^{n} (\omega_i + h)\, \mathrm{sign}(w_{i-1}, w_i), \qquad w \in \mathcal{W}_n, \tag{9.3}$$

with $\mathrm{sign}(w_{i-1}, w_i) = \pm 1$ depending on whether the edge (w_{i-1}, w_i) lies above or below the horizontal axis, and $\lambda, h \in \mathbb{R}$. Without loss of generality we may restrict the interaction parameters to the quadrant

$$\mathrm{QUA} = \{(\lambda, h) \in [0, \infty)^2\}. \tag{9.4}$$

The choice in (9.3) has the following interpretation. Think of $w \in \mathcal{W}_n$ as a directed copolymer on $\mathbb{N}_0 \times \mathbb{Z}$, consisting of n monomers represented by the edges in the path (rather than the sites). The lower halfplane is water, the upper halfplane is oil. The monomers are labeled by ω, with $\omega_i = +1$ meaning that monomer i is hydrophobic and $\omega_i = -1$ that it is hydrophilic. The term $\mathrm{sign}\,(w_{i-1}, w_i)$ equals $+1$ or -1 depending on whether monomer i lies in the oil or in the water. Thus, the Hamiltonian rewards matches and penalizes mismatches of the chemical affinities between the monomers and the solvents. The parameter λ is the *disorder strength*. The parameter h plays the role of an *disorder bias*: $h = 0$ corresponds to the hydrophobic and hydrophilic monomers interacting equally strongly, while $h = 1$ corresponds to the hydrophilic monomers not interacting at all. Only the regime $h \in [0, 1]$ is of interest (because for $h > 1$ both types of monomers prefer to be in the oil and the copolymer is always delocalized).

The law of the copolymer given ω is denoted by P_n^ω and is defined as in (1.3), i.e.,

$$P_n^\omega(w) = \frac{1}{Z_n^\omega}\, e^{-H_n^\omega(w)}\, P_n(w), \qquad w \in \mathcal{W}_n, \tag{9.5}$$

where P_n is the law of the n-step directed random walk, which is the uniform distribution on $\mathcal{W}_n$. This law is the projection on $\mathcal{W}_n$ of the law P of the infinite directed walk whose vertical steps are SRW. Note that we have suppressed λ, h from the notation.

In what follows we will consider the *quenched free energy* $f(\lambda, h)$ of the copolymer, i.e., the free energy conditioned on ω. We will first show that $f(\lambda, h)$ exists and is constant ω-a.s. We will then show that $f(\lambda, h)$ is non-analytic along a *critical curve* in the (λ, h)-plane, derive a number of properties of this curve, and subsequently have a look at the typical behavior of the copolymer in each of the two phases.

The model defined in (9.3) was introduced in Garel, Huse, Leibler and Orland [111], where the existence of a phase transition was argued on the basis of non-rigorous arguments. The first rigorous studies were carried out in Sinai [274] and in Bolthausen and den Hollander [31]. The latter paper proved the existence of a phase transition, derived a number of properties of the critical curve, and raised a number of questions. Since then, several papers have appeared in which these questions have been settled and various aspects of the model have been further elucidated, leading not only to many interesting results, but also to challenging open problems. We will refer to these papers as we go along.

In Giacomin [116], Chapters 6–9, the more general situation is considered where the ω_i are $\mathbb{R}$-valued with a finite moment generating function and the $w_{i+1} - w_i$ are the increments of a random walk in the domain of attraction of a stable law. We will comment on this extension in Section 9.6. The special situation treated below (SRW and binary disorder) already collects the main ideas.

A key observation is the following. The energy of a path is a sum of contributions coming from its *successive excursions* away from the interface (see Fig. 9.2). What is relevant for the energy of these excursions is when they start and end, and whether they are above or below the interface. This simplifying feature, together with the directed nature of the path, is of great help when we want to do explicit calculations.

9.2 The Free Energy

Our starting point is the following self-averaging property, which is taken from Bolthausen and den Hollander [31].

Theorem 9.1. *For every* $\lambda, h \in \text{QUA}$,

$$f(\lambda, h) = \lim_{n \to \infty} \frac{1}{n} \log Z_n^\omega \qquad (9.6)$$

exists ω-*a.s. and in* $\mathbb{P}$-*mean, and is constant* ω-*a.s.*

Proof. The proof consists of two parts. In Lemma 9.2 we prove that the claim holds when the random walk is constrained to return to the origin at time $2n$. In Lemma 9.3 we show how to remove this constraint.

Fix $\lambda, h \in \text{QUA}$. Abbreviate

$$\Delta_i = \text{sign}(S_{i-1}, S_i) \qquad (9.7)$$

and define

$$Z_{2n}^{\omega,0} = E\left(\exp\left[\lambda\sum_{i=1}^{2n}(\omega_i + h)\Delta_i\right]1_{\{S_{2n}=0\}}\right),\tag{9.8}$$

where we recall that P is the law of SRW, $S = (S_i)_{i\in\mathbb{N}_0}$.

Lemma 9.2. $\lim_{n\to\infty}\frac{1}{2n}\log Z_{2n}^{\omega,0}$ exists ω-a.s. and in $\mathbb{P}$-mean, and is constant ω-a.s.

Proof. We need the following three properties:

(I) $Z_{2n}^{\omega,0} \geq Z_{2m}^{\omega,0}Z_{2n-2m}^{T^{2m}\omega,0}$ for all $0 \leq m \leq n$, with T the left-shift acting on ω as $(T\omega)_i = \omega_{i+1}$, $i \in \mathbb{N}$.

(II) $n \to \frac{1}{2n}\mathbb{E}(\log Z_{2n}^{\omega,0})$ is bounded from above.

(III) $\mathbb{P}(T\omega \in \cdot) = \mathbb{P}(\omega \in \cdot)$.

Property (I) follows from (9.8) by inserting an extra indicator $1_{\{S_{2m}=0\}}$ and using the Markov property of S at time $2m$. Property (II) holds because

$$\mathbb{E}\left(\log Z_{2n}^{\omega,0}\right) \leq \log\mathbb{E}\left(Z_{2n}^{\omega,0}\right)$$
$$= \log E\left((\cosh\lambda)^{2n}\exp\left[\lambda h\sum_{i=1}^{2n}\Delta_i\right]1_{\{S_{2n}=0\}}\right)\tag{9.9}$$
$$\leq 2n(\log\cosh\lambda + \lambda h).$$

Property (III) is trivial. Thus, $\omega \mapsto (\log Z_{2n}^{\omega,0})_{n\in\mathbb{N}_0}$ is a *superadditive random process* (which is the analogue of a superadditive deterministic sequence mentioned in Section 1.3). It therefore follows from the *subadditive ergodic theorem* (Kingman [214]) that $\lim_{n\to\infty}\frac{1}{2n}\log Z_{2n}^{\omega,0}$ converges ω-a.s. and in $\mathbb{P}$-mean, and is measurable w.r.t. the tail sigma-field of ω. Since the latter is trivial, i.e., all events not depending on finitely many coordinates have probability 0 or 1, the limit is constant ω-a.s. $\square$

Our original partition sum at time $2n$ was

$$Z_{2n}^{\omega} = E\left(\exp\left[\lambda\sum_{i=1}^{2n}(\omega_i + h)\Delta_i\right]\right),\tag{9.10}$$

which is (9.8) but without the indicator. Thus, in order to prove Theorem 9.1 we must show that this indicator is harmless as $n \to \infty$. Since $|\log(Z_{2n}^{\omega}/Z_{2n+1}^{\omega})| \leq \lambda(1+h)$, it will suffice to consider time $2n$.

Lemma 9.3. There exists a $C < \infty$ such that $Z_{2n}^{\omega,0} \leq Z_{2n}^{\omega} \leq CnZ_{2n}^{\omega,0}$ for all $n \in \mathbb{N}$ and ω.

Proof. The lower bound is obvious. The upper bound is proved as follows (compare with the proof of Theorem 7.1). For $1 \leq k \leq n$, consider the events

$$
\begin{aligned}
A_{2n,2k}^+ &= \{S_i > 0 \text{ for } 2n - 2k + 1 \leq i \leq 2n\}, \\
B_{2n,2k}^+ &= \{S_i > 0 \text{ for } 2n - 2k + 1 \leq i < 2n, S_{2n} = 0\},
\end{aligned}
\tag{9.11}
$$

and similarly for $A_{2n,2k}^-, B_{2n,2k}^-$ when the excursion is below the interface. By conditioning on the last hitting time of 0 prior to time $2n$, we may write

$$
\begin{aligned}
Z_{2n}^\omega = Z_{2n}^{\omega,0} &+ \sum_{k=1}^n Z_{2n-2k}^{\omega,0} \, E\left(\exp\left[\lambda \sum_{i=2n-2k+1}^{2n} (\omega_i + h)\Delta_i \right] \right. \\
&\left. \times 1\{A_{2n,2k}^+ \cup A_{2n,2k}^-\} \mid S_{2n-2k} = 0 \right) \\
= Z_{2n}^{\omega,0} &+ \sum_{k=1}^n Z_{2n-2k}^{\omega,0} \frac{a(2k)}{b(2k)} E\left(\exp\left[\lambda \sum_{i=2n-2k+1}^{2n} (\omega_i + h)\Delta_i \right] \right. \\
&\left. \times 1\{B_{2n,2k}^+ \cup B_{2n,2k}^-\} \mid S_{2n-2k} = 0 \right).
\end{aligned}
\tag{9.12}
$$

The reason for the second equality in (9.12) is that $\Delta_i = +1$ for all $2n-2k+1 \leq i \leq 2n$ on the events $A_{2n,2k}^+, B_{2n,2k}^+$ and $\Delta_i = -1$ for all $2n - 2k + 1 \leq i \leq 2n$ on the events $A_{2n,2k}^-, B_{2n,2k}^-$ (ω is fixed). We have

$$
\begin{aligned}
P(A_{2n,2k}^+ | S_{2n-2k} = 0) &= P(A_{n,k}^- \mid S_{2n-2k} = 0) = a(2k), \\
P(B_{2n,2k}^+ | S_{2n-2k} = 0) &= P(B_{n,k}^- \mid S_{2n-2k} = 0) = b(2k),
\end{aligned}
\tag{9.13}
$$

where (compare with (7.4))

$$
\begin{aligned}
a(2k) &= P(S_i > 0 \text{ for } 1 \leq i \leq 2k \mid S_0 = 0), \\
b(2k) &= P(S_i > 0 \text{ for } 1 \leq i < 2k, S_{2k} = 0 \mid S_0 = 0).
\end{aligned}
\tag{9.14}
$$

Moreover, there exist $0 < C_1, C_2 < \infty$ such that $a(2k) \sim C_1/k^{1/2}$ and $b(2k) \sim C_2/k^{3/2}$ as $k \to \infty$ (see Spitzer [284], Section 1). Hence $a(2k)/b(2k) \leq Ck$, $k \in \mathbb{N}$, for some $C < \infty$. Finally, without the factor $a(2k)/b(2k)$ the last sum in (9.12) is precisely $Z_{2n}^{\omega,0}$. Hence we get

$$
Z_{2n}^\omega \leq (1 + Cn) Z_{2n}^{\omega,0},
\tag{9.15}
$$

which proves the claim. □

Lemmas 9.2–9.3 complete the proof of Theorem 9.1. □

9.3 The Critical Curve

Now that we have proved the existence of the quenched free energy, we proceed to study its properties, in particular, we proceed to look for a phase transition in QUA. In Section 9.3.1 we define the localized and the delocalized phases, while in Section 9.3.2 we prove the existence of a non-trivial critical curve separating the two. Sections 9.4–9.5 will be devoted to establishing the qualitative properties of the critical curve and the two phases.

Grosberg, Izrailev and Nechaev [133] and Sinai and Spohn [276] study the annealed version of the model, in which Z_n^ω is averaged over ω. The free energy and the critical curve can in this case be computed exactly, but they provide little information on what the quenched model does. Nevertheless, in Section 9.4.1 we will use the annealed model to obtain an upper bound on the quenched critical curve.

9.3.1 The Localized and Delocalized Phases

The quenched free energy $f(\lambda, h)$ is continuous, nondecreasing and convex in each variable (convexity follows from Hölder's inequality, as shown in Section 1.3). Moreover, we have

$$f(\lambda, h) \geq \lambda h \qquad \forall (\lambda, h) \in \text{QUA}. \tag{9.16}$$

Indeed, since $P(\Delta_i = +1 \; \forall 1 \leq i \leq n) \sim C/n^{1/2}$ for some $C > 0$ as $n \to \infty$ (see Spitzer [284], Section 7), it follows from (9.3–9.5) that

$$
\begin{aligned}
Z_n^\omega &= E \left(\exp \left[\lambda \sum_{i=1}^n (\omega_i + h)\Delta_i \right] \right) \\
&\geq \left(\exp \left[\lambda \sum_{i=1}^n (\omega_i + h)\Delta_i \right] 1_{\{\Delta_i = +1 \; \forall 1 \leq i \leq n\}} \right) \\
&= \exp \left[\lambda \sum_{i=1}^n (\omega_i + h) \right] P(\Delta_i = +1 \; \forall 1 \leq i \leq n) \\
&= \exp[\lambda h n + o(n) = O(\log n)] \qquad \omega - a.s.,
\end{aligned}
\tag{9.17}
$$

where in the last line we use the strong law of large numbers for ω. Thus, we see that the lower bound in (9.16) corresponds to the strategy where the copolymer wanders away from the interface in the upward direction. This leads us to the following definition (see Fig. 9.3).

Definition 9.4. *We say that the copolymer is:*
(i) *localized if $f(\lambda, h) > \lambda h$,*
(ii) *delocalized if $f(\lambda, h) = \lambda h$.*
We write

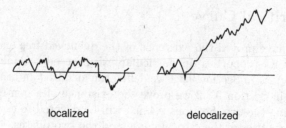

localized   delocalized

Fig. 9.3. Expected path behavior in the two phases.

$$\mathcal{L} = \{(\lambda, h) \in \text{QUA}: f(\lambda, h) > \lambda h\},$$
$$\mathcal{D} = \{(\lambda, h) \in \text{QUA}: f(\lambda, h) = \lambda h\},\tag{9.18}$$

to denote the localized, respectively, the delocalized region in QUA.

In case (i), the copolymer is able to beat on an exponential scale the trivial strategy of moving upward. It is intuitively clear that this is only possible by crossing the interface at a positive frequency, but a proof of the latter requires work. In case (ii), the copolymer is not able to beat the trivial strategy on an exponential scale. In principle it could do better on a smaller scale, but it actually does not, which also requires work. Path properties are dealt with in Section 9.5.1.

Albeverio and Zhou [1] prove that if $\lambda > 0$ and $h = 0$, then $\log Z_n^\omega$ satisfies a law of large numbers and a central limit theorem (as a random variable in ω). This result readily extends to the entire localization regime.

9.3.2 Existence of a Non-trivial Critical Curve

The following theorem, taken from Bolthausen and den Hollander [31], shows that $\mathcal{L}$ and $\mathcal{D}$ are separated by a non-trivial critical curve (see Fig. 9.4).

Theorem 9.5. *For every* $\lambda \in [0, \infty)$ *there exists an* $h_c(\lambda) \in [0, 1)$ *such that the copolymer is*

$$\begin{array}{ll} localized & if\ 0 \le h < h_c(\lambda), \\ delocalized & if\ h \ge h_c(\lambda). \end{array}\tag{9.19}$$

Moreover, $\lambda \mapsto h_c(\lambda)$ *is continuous and strictly increasing on* $[0, \infty)$*, with* $h_c(0) = 0$ *and* $\lim_{\lambda \to \infty} h_c(\lambda) = 1$.

Proof. Let

$$g_n^\omega(\lambda, h) = \frac{1}{n} \log E\left(\exp\left[\lambda \sum_{i=1}^n (\omega_i + h)(\Delta_i - 1)\right]\right).\tag{9.20}$$

Then, because $\lambda \sum_{i=1}^n (\omega_i + h) = \lambda h + o(n)$ ω-a.s., Theorem 9.1 says that

$$\lim_{n \to \infty} g_n^\omega(\lambda, h) = g(\lambda, h) \qquad \omega - a.s.\tag{9.21}$$

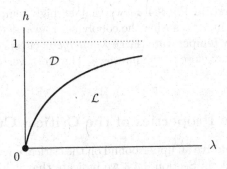

Fig. 9.4. Plot of $\lambda \mapsto h_c(\lambda)$.

with

$$g(\lambda, h) = f(\lambda, h) - \lambda h \tag{9.22}$$

the *excess free energy*. In terms of g, (9.18) becomes

$$\mathcal{L} = \{(\lambda, h) \in \text{QUA: } g(\lambda, h) > 0\},$$
$$\mathcal{D} = \{(\lambda, h) \in \text{QUA: } g(\lambda, h) = 0\}. \tag{9.23}$$

For $\lambda \in [0, \infty)$, define

$$h_c(\lambda) = \inf\{h \in [0, \infty) \colon (\lambda, h) \in \mathcal{D}\}. \tag{9.24}$$

To study this function, it is expedient to change variables by putting

$$\theta = \lambda h, \qquad \theta_c(\lambda) = \lambda h_c(\lambda), \qquad \bar{g}(\lambda, \theta) = g(\lambda, h). \tag{9.25}$$

Because λ and θ appear linearly in the Hamiltonian in (9.3), we know that $(\lambda, \theta) \mapsto \bar{g}(\lambda, \theta)$ is convex (see Section 1.3). Moreover, since $\bar{g} \geq 0$, we have $\mathcal{D} = \{(\lambda, \theta) \colon \bar{g}(\lambda, \theta) \leq 0\}$, i.e., $\mathcal{D}$ is a level set of the function $\bar{g}$. Since $\bar{g}$ is a convex function, it follows that $\mathcal{D}$ is a convex set, implying in turn that $\mathcal{L}$ and $\mathcal{D}$ are separated by a *single* critical curve

$$\theta_c(\lambda) = \inf\{\theta \in [0, \infty) \colon (\lambda, \theta) \in \mathcal{D}\} \tag{9.26}$$

that is itself a convex function.

We know that $h_c(0) = 0$. Now, Theorems 9.6–9.7 below will imply that

$$0 < \liminf_{\lambda \downarrow 0} \frac{1}{\lambda} h_c(\lambda) \leq \limsup_{\lambda \downarrow 0} \frac{1}{\lambda} h_c(\lambda) < \infty. \tag{9.27}$$

Moreover, Theorem 9.7 will imply that $\lim_{\lambda \to \infty} h_c(\lambda) = 1$. All that therefore remains to be done is to show that h_c is strictly increasing on $[0, \infty)$. To that end, note that $\theta_c(0) = 0$, while (9.27) shows that $\theta_c(\lambda)$ is of order λ^2 near 0. Consequently, $\lambda \mapsto \theta_c(\lambda)/\lambda$ is strictly increasing $[0, \infty)$ (with limiting value 0 at $\lambda = 0$). $\quad \Box$

The critical curve in Fig. 9.4 shows that at high temperature (small λ) *entropic* effects dominate, causing the copolymer to wander away from the interface, while at low temperature (large λ) *energetic* effects dominate, causing the copolymer to stay close to the interface. The crossover value of λ depends on the value of h.

9.4 Qualitative Properties of the Critical Curve

In Section 9.4.1 we derive an upper bound on the critical curve, in Section 9.4.2 a lower bound, while in Section 9.4.3 we indicate that it has a (positive and finite) slope at 0.

9.4.1 Upper Bound

The next result, also taken from Bolthausen and den Hollander [31], provides an upper bound on h_c. This bound is the critical curve for the *annealed* model.

Theorem 9.6. $h_c(\lambda) \leq (2\lambda)^{-1} \log \cosh(2\lambda)$ *for all* $\lambda \in (0, \infty)$.

Proof. Estimate (recall (9.20–9.22))

$$
\begin{aligned}
g(\lambda, h) &= \lim_{n \to \infty} \frac{1}{n} \mathbb{E} \left(\log E \left(\exp \left[\lambda \sum_{i=1}^{n} (\omega_i + h)(\Delta_i - 1) \right] \right) \right) \\
&\leq \lim_{n \to \infty} \frac{1}{n} \log E \left(\mathbb{E} \left(\exp \left[\lambda \sum_{i=1}^{n} (\omega_i + h)(\Delta_i - 1) \right] \right) \right) \qquad (9.28) \\
&= \lim_{n \to \infty} \frac{1}{n} \log E \left(\prod_{i=1}^{n} \left[\tfrac{1}{2} e^{-2\lambda(1+h)} + \tfrac{1}{2} e^{-2\lambda(-1+h)} \right]^{1\{\Delta_i = -1\}} \right).
\end{aligned}
$$

The first equality comes from the fact that Kingman's subadditive ergodic theorem holds in $\mathbb{P}$-mean. The inequality follows from Jensen's inequality and Fubini's theorem. The second equality uses (9.2). The right-hand side is ≤ 0 as soon as the term between square brackets is ≤ 1. Consequently,

$$
(2\lambda)^{-1} \log \cosh(2\lambda) > h \quad \Longrightarrow \quad g(\lambda, h) = 0, \qquad (9.29)
$$

which yields the desired upper bound on $h_c(\lambda)$. $\square$

Note that the proof of Theorem 9.6 is a *partial annealing estimate* in disguise, because to transform (9.18) into (9.23) we have used the law of large numbers for ω.

9.4.2 Lower Bound

The following counterpart of Theorem 9.6, due to Bodineau and Giacomin [22], provides a lower bound on $h_c(\lambda)$.

Theorem 9.7. $h_c(\lambda) \geq (\frac{4}{3}\lambda)^{-1} \log \cosh(\frac{4}{3}\lambda)$ *for all* $\lambda \in (0, \infty)$.

Proof. As we will see, the lower bound comes from strategies where the copolymer dips below the interface during rare long stretches in ω where the empirical mean is sufficiently biased.

We begin with some notation. Pick $l \in 2\mathbb{N}$. For $j \in \mathbb{N}$, let

$$I_j = ((j-1)l, jl] \cap \mathbb{N}, \qquad \Omega_j = \sum_{i \in I_j} \omega_i. \tag{9.30}$$

Pick $\delta \in (0, 1]$. Define recursively

$$i_0^\omega = 0, \qquad i_j^\omega = \inf\{k \geq i_{j-1}^\omega + 2 \colon \Omega_k \leq -\delta l\}, \; j \in \mathbb{N}, \tag{9.31}$$

and abbreviate $\tau_j^\omega = i_j^\omega - i_{j-1}^\omega - 1$, $j \in \mathbb{N}$. (In (9.31) we skip at least 2 to make sure that $\tau_j^\omega \geq 1$, which is needed below.) For $n \in \mathbb{N}$, put

$$t_{n,l,\delta}^\omega = \sup\{j \in \mathbb{N} \colon i_j^\omega \leq \lfloor n/l \rfloor\}, \tag{9.32}$$

and consider the set of paths (see Fig. 9.5)

$$\mathcal{W}_{n,l,\delta}^\omega = \Big\{ w \in \mathcal{W}_n \colon w_i \leq 0 \text{ for } i \in \cup_{j=1}^{t_{n,l,\delta}^\omega} I_{i_j^\omega},$$
$$w_i \geq 0 \text{ for } i \in \{(0, n] \cap \mathbb{N}\} \setminus \cup_{j=1}^{t_{n,l,\delta}^\omega} I_{i_j^\omega} \Big\}. \tag{9.33}$$

By restricting the path to $\mathcal{W}_{n,l,\delta}^\omega$, we find that the quantity in (9.20) can be bounded from below as

$$g_n^\omega(\lambda, h) = \frac{1}{n} \log E \left(\exp\left[\lambda \sum_{i=1}^n (\omega_i + h)(\Delta_i - 1) \right] \right)$$

$$\geq \frac{1}{n} \log E \left(\exp\left[\lambda \sum_{i=1}^n (\omega_i + h)(\Delta_i - 1) \right] 1_{\mathcal{W}_{n,l,\delta}^\omega} \right)$$

$$\geq \frac{1}{n} \log \left\{ \left(\prod_{j=1}^{t_{n,l,\delta}^\omega} b(\tau_j^\omega l) \, b(l) \, e^{-2\lambda(-\delta+h)l} \right) a\left(n - i_{t_{n,l,\delta}^\omega} l \right) \right\}$$

$$\geq \frac{1}{n} \sum_{j=1}^{t_{n,l,\delta}^\omega} \log b(\tau_j^\omega l) + \frac{t_{n,l,\delta}^\omega}{n} \left[\log b(l) + 2\lambda(\delta - h)l \right] + \frac{1}{n} \log a(n),$$
$$\tag{9.34}$$

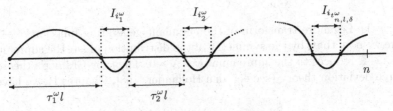

Fig. 9.5. A path in $\mathcal{W}_{n,l,\delta}^\omega$.

where $a(k)$ and $b(k)$ are defined in (9.13), the second inequality uses that $\Omega_j \leq -\delta l$ for $j = 1, \ldots, t^\omega_{n,l,\delta}$, and the third inequality uses that $k \mapsto a(k)$ is non-increasing. We need to compute the ω-a.s. limit of the right-hand side of (9.34) as $n \to \infty$.

The last term vanishes as $n \to \infty$, because $a(n) \asymp n^{-1/2}$. Furthermore, since there exists a $C > 0$ such that $b(k) \geq Ck^{-3/2}$ for $k \in 2\mathbb{N}$, we have

$$\frac{1}{n} \sum_{j=1}^{t^\omega_{n,l,\delta}} \log b(\tau^\omega_j l) \geq \frac{1}{n} \sum_{j=1}^{t^\omega_{n,l,\delta}} \log \left[C(\tau^\omega_j l)^{-3/2} \right]$$

$$\geq -\frac{3t^\omega_{n,l,\delta}}{2n} \log \left[\frac{\sum_{j=1}^{t^\omega_{n,l,\delta}} \tau^\omega_j l}{t^\omega_{n,l,\delta}} \right] + \frac{t^\omega_{n,l,\delta}}{n} \log C, \tag{9.35}$$

where the second inequality uses Jensen's inequality. Moreover, applying the ergodic theorem to $(\Omega_j)_{j \in \mathbb{N}}$ while recalling (9.31–9.32), we have

$$\lim_{n \to \infty} \frac{t^\omega_{n,l,\delta}}{n} = p_{l,\delta} \qquad \omega - a.s., \tag{9.36}$$

where

$$p_{l,\delta} = \frac{1}{l} q_{l,\delta}(1 - q_{l,\delta}) \text{ with } q_{l,\delta} = \mathbb{P}(\Omega_1 \leq -\delta l). \tag{9.37}$$

Since $\sum_{j=1}^{t^\omega_{n,l,\delta}} \tau^\omega_j l \leq n - t^\omega_{n,l,\delta} l$, it follows from (9.36) that

$$\limsup_{n \to \infty} \frac{\sum_{j=1}^{t^\omega_{n,l,\delta}} \tau^\omega_j l}{t^\omega_{n,l,\delta}} \leq \lim_{n \to \infty} \frac{n - t^\omega_{n,l,\delta} l}{t_{n,l,\delta}} = p_{l,\delta}^{-1} - l \qquad \omega - a.s. \tag{9.38}$$

Combining (9.34–9.38) and recalling (9.21), we arrive at

$$g(\lambda, h) \geq -\tfrac{3}{2} p_{l,\delta} \log \left(p_{l,\delta}^{-1} - l \right) + p_{l,\delta} \left[2 \log C - \tfrac{3}{2} \log l + 2\lambda(\delta - h)l \right]. \tag{9.39}$$

This inequality is valid for all $l \in 2\mathbb{N}$ and $\delta \in (0, 1]$.

Next, let

$$\Sigma(\delta) = \sup_{\lambda > 0} \left[\lambda\delta - \log \mathbb{E} \left(e^{-\lambda\omega_1} \right) \right], \qquad \delta \in (0, 1], \tag{9.40}$$

denote the Legendre transform of the cumulant generating function of $-\omega_1$, where we note that, by the symmetry of the distribution of ω_1, the supremum over $\lambda \in \mathbb{R}$ reduces to the supremum over $\lambda > 0$. Then, by Cramér's theorem of large deviation theory (see e.g. den Hollander [168], Chapter I), we have

$$\lim_{l \to \infty} \frac{1}{l} \log q_{l,\delta} = -\Sigma(\delta) \qquad \forall \delta \in (0, 1]. \tag{9.41}$$

Hence, letting $l \to \infty$ in (9.39) and using (9.37), we obtain

$$\exists \delta \in (0,1]: \quad -\tfrac{3}{2}\Sigma(\delta) + 2\lambda(\delta - h) > 0 \quad \Longrightarrow \quad g(\lambda, h) > 0. \tag{9.42}$$

Taking the inverse Legendre transform in (9.40), we have

$$\sup_{\delta \in (0,1]} \left[\tfrac{4}{3}\lambda(\delta - h) - \Sigma(\delta) \right] = -\tfrac{4}{3}\lambda h + \log \mathbb{E}\left(e^{-\tfrac{4}{3}\lambda \omega_1} \right) \tag{9.43}$$

$$= -\tfrac{4}{3}\lambda h + \log \cosh\left(\tfrac{4}{3}\lambda \right), \qquad \lambda > 0.$$

Combining (9.42–9.43), we get

$$\left(\tfrac{4}{3}\lambda \right)^{-1} \log \cosh\left(\tfrac{4}{3}\lambda \right) > h \quad \Longrightarrow \quad g(\lambda, h) > 0, \tag{9.44}$$

which yields the desired lower bound on $h_c(\lambda)$. $\quad \square$

9.4.3 Weak Interaction Limit

The upper and lower bounds in Theorems 9.6–9.7 are sketched in Fig. 9.6. Numerical work in Caravenna, Giacomin and Gubinelli [55] indicates that $\lambda \mapsto h_c(\lambda)$ lies somewhere halfway between these bounds (see also Garel and Monthus [244]).

The following weak interaction limit is proved in Bolthausen and den Hollander [31].

Theorem 9.8. *There exists a $K_c \in (0, \infty)$ such that*

$$\lim_{\lambda \downarrow 0} \frac{1}{\lambda} h_c(\lambda) = K_c. \tag{9.45}$$

Proof. The idea behind the proof is that, as $\lambda, h \downarrow 0$, the excursions away from the interface become longer and longer (i.e., entropy gradually takes over from energy). As a result, both w and ω can be approximated by Brownian motions. In essence, (9.45) follows from the scaling property

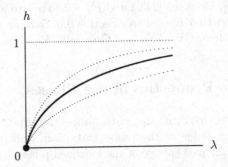

Fig. 9.6. Upper and lower bounds on $\lambda \mapsto h_c(\lambda)$.

$$\lim_{a \downarrow 0} a^{-2} f(a\lambda, ah) = \widetilde{f}(\lambda, h), \qquad \lambda, h \geq 0, \tag{9.46}$$

where $\widetilde{f}(\lambda, h)$ is the free energy of a space-time continuous version of the copolymer model, with Hamiltonian

$$H_t^b(B) = -\lambda \int_0^t (\mathrm{d}b_s + h \,\mathrm{d}s) \,\mathrm{sign}(B_s) \tag{9.47}$$

replacing (9.3), and with path measure given by the Radon-Nikodym derivative

$$\frac{\mathrm{d}P_t^b}{\mathrm{d}P}(B) = \frac{1}{Z_t^b} \,\mathrm{e}^{-H_t^b(B)} \tag{9.48}$$

replacing (9.5). Here, $B = (B_s)_{s \geq 0}$ is a path drawn from Wiener space, replacing (9.1), P is the Wiener measure (i.e., the law of standard Brownian motion on Wiener space), and $(b_s)_{s \geq 0}$ is a standard Brownian motion, replacing (9.2), playing the role of the quenched randomness. The proof of (9.46) is based on a *coarse-graining* argument. Due to the presence of exponential weight factors, (9.46) is a much more delicate property than the standard invariance principle relating simple random walk and Brownian motion. Moreover, (9.45) follows from (9.46) only after the latter has been shown to be "stable against perturbations" in λ, h. For details we refer to [31].

In the continuum model, the quenched critical curve turns out to be *linear* with slope K_c. The value of K_c is not known.   □

The proof of (9.45) shows that the same scaling property holds for the model in which the h-dependence sits in the probability law of ω rather than in the Hamiltonian, i.e., $\mathbb{P}(\omega_i = \pm 1) = \frac{1}{2}(1 \pm h)$ and $H_n^\omega(w) = \lambda \sum_{i=1}^n \omega_i \Delta_i(w)$ instead of (9.2–9.3). This describes a copolymer where the monomers occur with different densities but interact equally strongly. Alternatively, we could allow for more general ω, assuming values in $\mathbb{R}$ according to a symmetric distribution with a finite exponential moment (see Section 9.6). Thus, K_c has a certain degree of *universality*. See also Section 9.6, Extension (1).

The bounds in Theorems 9.6–9.7 give $K_c \in [\frac{2}{3}, 1]$. There have been various papers in the literature arguing in favor of $K_c = \frac{2}{3}$ (Stepanow, Sommer and Erukhimovich [285], Monthus [243]) and $K_c = 1$ (Trovato and Maritan [298]). Toninelli [295] proves that $K_c < 1$ (see Section 9.6, Extension (2)). The numerical work in Caravenna, Giacomin and Gubinelli [55] gives $K_c \in [0.82, 0.84]$.

9.5 Qualitative Properties of the Phases

In Section 9.5.1 we look at the path properties in the two phases, in Section 9.5.2 at the order of the phase transition, and in Section 9.5.3 at the smoothness of the free energy in the localized phase.

9.5.1 Path Properties

We next state two theorems identifying the path behavior in the two phases. The first is due to Biskup and den Hollander [20], the second to Giacomin and Toninelli [120]. These theorems confirm the naive view put forward in Section 9.3 when we defined $\mathcal{L}$ and $\mathcal{D}$ (recall Fig. 9.3).

Theorem 9.9. ω-a.s. under P_n^ω as $n \to \infty$:
(a) If $(\lambda, h) \in \mathcal{L}$, then the path intersects the interface with a strictly positive density, while the lengths and the heights of its excursions away from the interface are exponentially tight.
(b) If $(\lambda, h) \in \mathrm{int}(\mathcal{D})$, then the path intersects the interface with a zero density.

Theorem 9.10. ω-a.s. under P_n^ω as $n \to \infty$, if $(\lambda, h) \in \mathrm{int}(\mathcal{D})$, then the path intersects the interface $O(\log n)$ times.

The idea behind Theorem 9.9 is that in the localized regime, where the excess free energy $g = g(\lambda, h)$ defined in (9.22) is strictly positive, the probability for the copolymer to be away from the interface during a time l is roughly e^{-gl} as $l \to \infty$. This is because the copolymer contributes an amount gl to the free energy when it stays near the interface, but contributes 0 when it moves away. The idea behind Theorem 9.10 is that strictly inside the delocalized regime, where the excess free energy is zero, the probability for the copolymer to return to the interface a large number of times is small. This idea is exploited with the help of concentration inequalities.

It is believed that strictly inside the delocalized regime the number of intersections with the interface is in fact $O(1)$. This has only been proved deep inside the delocalized phase, namely, above the annealed upper bound in Theorem 9.6, where ideas similar to those that went into the proof of Theorem 7.3 can be exploited (Giacomin and Toninelli [120]). See Section 9.7, Challenge (3).

9.5.2 Order of the Phase Transition

The next theorem, due to Giacomin and Toninelli [123], shows that the phase transition is *at least* of second order.

Theorem 9.11. *For every* $\lambda \in (0, \infty)$,

$$0 \le g(\lambda, h) = O\left([h_c(\lambda) - h]^2\right) \qquad as \ h \uparrow h_c(\lambda). \tag{9.49}$$

Proof. We give the proof for the case where $\omega = (\omega_i)_{i \in \mathbb{N}}$ is an i.i.d. sequence of standard normal random variables, rather than binary random variables as in (9.2). At the end of the proof we indicate how to adapt the argument.

Recall the definitions in the proof of Theorem 9.7. Consider the set of paths (see Fig. 9.7)

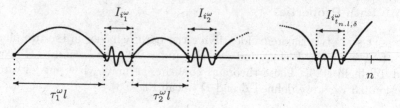

Fig. 9.7. A path in $\widehat{\mathcal{W}}_{n,l,\delta}^{\omega}$.

$$\widehat{\mathcal{W}}_{n,l,\delta}^{\omega} = \left\{ w \in \mathcal{W}_n \colon\; w_i = 0 \text{ for } i \in \cup_{j=1}^{t_{n,l,\delta}^{\omega}} \partial I_{i_j^{\omega}}, \right.$$
$$\left. w_i \geq 0 \text{ for } i \in \{(0,n] \cap \mathbb{N}\} \setminus \cup_{j=1}^{t_{n,l,\delta}^{\omega}} I_{i_j^{\omega}} \right\} \tag{9.50}$$

with $\partial I_j = \{(j-1)l, jl\}$, $j \in \mathbb{N}$. By restricting the path to $\widehat{\mathcal{W}}_{n,l,\delta}^{\omega}$, we get

$$\widetilde{Z}_n^{\omega} \geq \left(\prod_{j=1}^{t_{n,l,\delta}^{\omega}} b(\tau_j^{\omega} l)\; \widetilde{Z}_l^{T^{i_j^{\omega} l} \omega} \right) a\left(n - i_{t_{n,l,\delta}^{\omega}} l \right), \tag{9.51}$$

where T is the left-shift acting on ω, and the tilde is used to indicate that the partition sums are taken with $\Delta_i - 1$ instead of Δ_i, as in (9.20) and (9.34). Letting $n \to \infty$, using that $k \mapsto a(k)$ is non-increasing with $\lim_{n\to\infty} \frac{1}{n} \log a(n) = 0$, and noting that the convergence in (9.21) holds in $\mathbb{P}$-mean as well, we obtain

$$g(\lambda, h) \geq \liminf_{n\to\infty} \frac{1}{n} \mathbb{E}\left(\sum_{j=1}^{t_{n,l,\delta}^{\omega}} \log b(\tau_j^{\omega} l) \right) + \liminf_{n\to\infty} \frac{1}{n} \mathbb{E}\left(\sum_{j=1}^{t_{n,l,\delta}^{\omega}} \log \widetilde{Z}_l^{T^{i_j^{\omega} l} \omega} \right). \tag{9.52}$$

The first term in the right-hand side of (9.52) was computed in (9.35–9.38) and equals $-\frac{3}{2} p_{l,\delta} \log(p_{l,\delta}^{-1} - l)$. The expectation under the limit in the second term equals (recall (9.31–9.32))

$$\frac{1}{n} \mathbb{E}\left(\sum_{j=1}^{t_{n,l,\delta}^{\omega}} \log \widetilde{Z}_l^{T^{i_j^{\omega} l} \omega} \right) = \mathbb{E}\left(\frac{t_{n,l,\delta}^{\omega}}{n} \right) \mathbb{E}\left(\log \widetilde{Z}_l^{\omega} \mid \Omega_1 \leq -\delta l \right). \tag{9.53}$$

Conditioning l i.i.d. normal random variables with mean 0 and variance 1 to have sum equal to $-\delta l$ is the same as taking l i.i.d. normal random variables with mean $-\delta$ and variance 1. Hence, the effect of the conditioning in (9.53) is that, in (9.20), ω_i becomes $\omega_i - \delta$, so that

$$\mathbb{E}\left(\log \widetilde{Z}_l^{\omega}(\lambda, h) \mid \Omega_1 \leq -\delta l \right) = \mathbb{E}\left(\log \widetilde{Z}_l^{\omega}(\lambda, h - \delta) \right), \tag{9.54}$$

where we add the parameters λ, h as arguments to exhibit the shift in h. Inserting (9.54) into (9.53), picking $h = h_c(\lambda)$ in (9.52), noting that $g(\lambda, h_c(\lambda)) = 0$, and recalling (9.36), we arrive at

$$0 \geq -\tfrac{3}{2}\, p_{l,\delta} \log\left(p_{l,\delta}^{-1} - l\right) + p_{l,\delta}\, \mathbb{E}\left(\log \widetilde{Z}_l^\omega\left(\lambda, h_c(\lambda) - \delta\right)\right). \tag{9.55}$$

This inequality is valid for all $l \in 2\mathbb{N}$ and $\delta \in (0, 1]$.

Dividing (9.55) by l, letting $l \to \infty$ and using (9.41), we obtain

$$0 \geq -\tfrac{3}{2}\Sigma(\delta) + g\left(\lambda, h_c(\lambda) - \delta\right), \tag{9.56}$$

where Σ is the Cramér rate function for standard normal random variables, i.e., $\Sigma(\delta) = \tfrac{1}{2}\delta^2$. Hence we conclude that

$$g\left(\lambda, h_c(\lambda) - \delta\right) \leq \tfrac{3}{4}\delta^2, \tag{9.57}$$

which proves the claim.

It is easy to extend the proof to binary ω. All that is needed is to show that (9.54) holds asymptotically as $l \to \infty$ and to use that $\Sigma(\delta) \sim \tfrac{1}{2}\delta^2$ as $\delta \downarrow 0$. $\quad\square$

9.5.3 Smoothness of the Free Energy in the Localized Phase

We conclude with a theorem by Giacomin and Toninelli [124] showing that the free energy is smooth throughout the localized phase. Consequently, our critical curve is the *only* location where a phase transition of *finite order* occurs.

Theorem 9.12. $(\lambda, h) \mapsto f(\lambda, h)$ *is infinitely differentiable on* $\mathcal{L}$.

Proof. The main idea behind the proof is that, for $(\lambda, h) \in \mathcal{L}$, the correlation between any pair of events that depend on the values of the path w of the copolymer in finite and disjoint subsets of $\{1, \ldots, n\}$ decays exponentially with the distance between these sets. This is formulated in the following lemma.

Lemma 9.13. *Let $\mathcal{K}$ be an arbitrary compact subset of $\mathcal{L}$. Then there exist $0 < c_1, c_2 < \infty$ (depending on $\mathcal{K}$) such that, for all $(\lambda, h) \in \mathcal{K}$, all $n \in \mathbb{N}$, all integers $1 \leq a_1 < b_1 < a_2 < b_2 \leq n$, and all pairs of events A and B that are measurable w.r.t. the sigma-fields generated by $(w_{a_1}, \ldots, w_{b_1})$, respectively, $(w_{a_2}, \ldots, w_{b_2})$,*

$$\mathbb{E}\left(\left| P_n^\omega(A \cap B) - P_n^\omega(A)\, P_n^\omega(B)\right|\right) \leq c_1 e^{-c_2(a_2 - b_1)}. \tag{9.58}$$

Proof. The proof of Lemma 9.13 is based on the following lemma. For $n \in \mathbb{N}$, let $P_n^{\omega, \otimes 2}$ be the joint law of two independent copies w^1 and w^2 of w of length n.

Lemma 9.14. *There exist $0 < c_1, c_2 < \infty$ (depending on $\mathcal{K}$) such that for all $(\lambda, h) \in \mathcal{K}$, all $n \in \mathbb{N}$, and all integers $1 \leq a < b \leq n$,*

$$\mathbb{E}\left(P_n^{\omega, \otimes 2}(E_{a,b})\right) \leq c_1 e^{-c_2(b - a)}, \tag{9.59}$$

where $E_{a,b} = \{\#i \in \{a+1, \ldots, b-1\}: w_i^1 = w_i^2 = 0\}$.

Proof. The main idea is that, for $(\lambda, h) \in \mathcal{L}$, the lengths of the excursions of the copolymer away from the interface are exponentially tight. This forces w^1 and w^2 to hit the interface in a set of sites with a strictly positive density. Therefore the probability that w^1 and w^2 hit the interface at the same site at least once in $\{a+1, \ldots, b-1\}$ tends to 1 exponentially fast as $b - a \to \infty$. For details we refer the reader to Giacomin and Toninelli [124]. □

Now note that the l.h.s. of (9.58) is equal to

$$\mathbb{E}\left(\left| E_n^{\omega, \otimes 2} \left[\{ 1_A(w^1) 1_B(w^1) - 1_A(w^1) 1_B(w^2) \} 1_E(w^1, w^2) \right] \right| \right). \qquad (9.60)$$

Together with (9.59), this completes the proof of Lemma 9.13. □

We use Lemma 9.13 to complete the proof of Theorem 9.12. This can be done by appealing to the theorem of Arzela-Ascoli, according to which it is enough to prove that for every $k_1, k_2 \in \mathbb{N}$ the partial derivative

$$\frac{\partial^{k_1 + k_2}}{\partial \lambda^{k_1} \partial h^{k_2}} \left(\frac{1}{n} \mathbb{E} \log Z_n^{\omega} \right) \qquad (9.61)$$

is bounded from above, uniformly in $n \in \mathbb{N}$ and $(\lambda, h) \in \mathcal{K}$.

In order to achieve the latter, we define, for $k \in \mathbb{N}$ and for $\{f_1, \ldots, f_k\}$ any family of bounded functions on the path space $\mathcal{W}_n$ whose supports are finite,

$$E_n^{\omega}[f_1(w); \ldots; f_k(w)] = \sum_{P \in \mathcal{P}} (-1)^{|P|-1} (|P| - 1)! \prod_{p=1}^{|P|} E_n^{\omega} \left[\prod_{l \in P_p} f_l(w) \right], \qquad (9.62)$$

where $\mathcal{P}$ denotes the set of all partitions $P = (P_p)_{p \in |P|}$ of $\{1, \ldots, k\}$, with $|P|$ the number of sets in the partition P. We need the following lemma for this quantity.

Lemma 9.15. *For all $k \in \mathbb{N}$ there exist $0 < c_1, c_2 < \infty$ (depending on k and $\mathcal{K}$) such that, for all $(\lambda, h) \in \mathcal{K}$,*

$$\mathbb{E}\left[\left| E_n^{\omega}[f_1; \ldots; f_k] \right| \right] \leq c_1 \|f_1\|_\infty \cdots \|f_k\|_\infty \, e^{-c_2 (|\mathcal{I}| - [|\mathcal{S}(f_1)| + \cdots + |\mathcal{S}(f_k)|])}, \qquad (9.63)$$

where $\mathcal{S}(f_1), \ldots, \mathcal{S}(f_k)$ are the supports of $f_1, \ldots, f_k$, and $\mathcal{I}$ is the smallest interval containing $\mathcal{S}(f_1), \ldots, \mathcal{S}(f_k)$.

Proof. The proof uses induction on k. We skip the details, noting only that the case $k = 1$ is trivial, while the case $k = 2$ is a direct consequence of Lemma 9.13 after we write bounded functions with finite support as linear combinations of indicators of events. □

The point of the notation introduced in (9.62) is that (9.61) can be written in the form

$$\frac{1}{n} \sum_{1 \le i_1,\dots,i_{k_1} \le n} \sum_{1 \le j_1,\dots,j_{k_2} \le n} \mathbb{E}\Big[(\omega_{i_1} + h)\dots(\omega_{i_{k_1}} + h)$$

$$\times E_n^{\omega}\big[\Delta_{i_1}(w);\dots;\Delta_{i_{k_1}}(w);\Delta_{j_1}(w);\dots;\Delta_{j_{k_2}}(w)\big]\Big]. \tag{9.64}$$

Since the ω_i's are bounded, the proof will therefore be complete once we show that

$$\frac{1}{n} \sum_{1 \le i_1,\dots,i_k \le n} \mathbb{E}\Big[\big|E_n^{\omega}\big[\Delta_{i_1}(w);\dots;\Delta_{i_k}(w)\big]\big|\Big], \qquad k = k_1 + k_2, \tag{9.65}$$

is bounded uniformly in $n \in \mathbb{N}$ and $(\lambda, h) \in \mathcal{K}$. For this we use Lemma 9.15, to bound (9.65) from above by

$$\frac{1}{n} \sum_{m=1}^{n-1} \sum_{(i_1,\dots,i_k) \in T_n^m} c_1 e^{-c_2(m-k)}, \tag{9.66}$$

where

$$T_n^m = \big\{(i_1,\dots,i_k):\ 1 \le i_1,\dots,i_k \le n,$$
$$\max\{i_1,\dots,i_k\} - \min\{i_1,\dots,i_k\} = m\big\}. \tag{9.67}$$

Since $|T_n^m| \le nm^k$, (9.66) is at most $c_1 e^{c_2 k} \sum_{m \in \mathbb{N}} m^k e^{-c_2 m} < \infty$. □

9.6 Extensions

(1) With the exception of Theorem 9.11, all the results described in Sections 9.2–9.5 extend to the situation where the ω_i are $\mathbb{R}$-valued with a finite moment generating function. For instance, define

$$h^*(\lambda) = \frac{1}{2\lambda} \log \mathbb{E}\big(e^{2\lambda\omega_1}\big), \qquad \lambda \in (0,\infty), \tag{9.68}$$

and assume that h^* is finite on $(0,\infty)$. Then the proofs of Theorems 9.6–9.7 show that

$$h^*(\tfrac{2}{3}\lambda) \le h_c(\lambda) \le h^*(\lambda), \qquad \lambda \in (0,\infty). \tag{9.69}$$

Moreover, if the random walk is replaced by a renewal process whose return times to the interface have distribution $P(S_i > 0 \ \forall 0 < i < n, S_n = 0) \sim n^{-1-a}L(n)$ as $n \to \infty$ for some $a > 0$ and some function L that is slowly varying at infinity (as in (7.5)), then the same proofs yield

$$h^*(\tfrac{1}{1+a}\lambda) \le h_c(\lambda) \le h^*(\lambda), \qquad \lambda \in (0,\infty). \tag{9.70}$$

We refer the reader to Giacomin [116], Chapters 6–8, for a full account of these extensions. The reason why the generalization from random walks to renewal processes works is precisely that the Hamiltonian in (9.3) decomposes into contributions coming from single excursions away from the interface. Therefore only the asymptotics of the law of these excursions matters. It also allows for the inclusion of $(1 + d)$-dimensional excursions away from a d-dimensional flat interface with $d \geq 2$.

Theorem 9.11 has been extended to bounded disorder, and also to continuous disorder subject to a mild entropy condition that is satisfied e.g. for Gaussian disorder (see Giacomin and Toninelli [122], [123]).

(2) Toninelli [294] shows that the upper bound in Theorem 9.6 is strict for unbounded disorder and large λ. This result is extended by Bodineau, Giacomin, Lacoin and Toninelli [23] to arbitrary disorder (subject to h^* in (9.68) being finite) and arbitrary λ. The proofs are based on *fractional moment estimates* of the partition sum. Toninelli [295] further refines the latter technique to show that $K_c < 1$ for arbitrary disorder, thereby ruling out $K_c = 1$ (recall the discussion at the end of Section 9.4.3). Bodineau, Giacomin, Lacoin and Toninelli [23] also show that the lower bound in Theorem 9.7 is strict for arbitrary disorder and small λ, at least for a large subclass of excursion return time distributions. The proof is based on finding appropriate *localization strategies*, in the spirit of the computation in Section 9.4.2.

(3) The difficulty behind improving the upper bound is that the typical length of the excursions diverges as the critical curve is approached from below. Consequently, any attempt to do a higher order partial annealing in the hope to improve the first order partial annealing argument in (9.28) is doomed to fail. This fact was observed in Orlandini, Rechnitzer and Whittington [253] and in Iliev, Rechnitzer and Whittington [182], and was subsequently proved for a large class of models in Caravenna and Giacomin [54]. In the latter paper, the setting is an arbitrary Hamiltonian $w \mapsto H_n^\omega(w)$ with the property that there exists a sequence $(D_n)_{n \in \mathbb{N}}$ of subsets of $\mathbb{Z}^d$ such that

$$\lim_{n \to \infty} \frac{1}{n} \log P(w_i \in D_i \,\forall\, 1 \leq i \leq n) = 0,$$
$$w_i \in D_i \,\forall\, 1 \leq i \leq n \implies H_n^\omega(w) = 0 \,\forall\, \omega, \tag{9.71}$$

where P is the reference path measure. Given an arbitrary local, bounded and measurable function $\omega \mapsto F(\omega)$ with $\mathbb{E}(F(\omega)) = 0$, it is shown

$$\liminf_{n \to \infty} \frac{1}{n} \log \mathbb{E}\big(E\big(\exp[-H_n^\omega(S)]\big)\big) > 0$$
$$\implies \liminf_{n \to \infty} \frac{1}{n} \log \mathbb{E}\Big(E\Big(\exp\Big[-H_n^\omega(S) - \sum_{i=1}^n F(T^i \omega)\Big]\Big)\Big) > 0, \tag{9.72}$$

where T is the left-shift acting on ω. What this says is that if the *annealed* free energy is strictly positive, then it remains strictly positive after adding the empirical average of a centered local function of the disorder. In other words, the two annealed free energies have the same critical curve. Note that the centering implies that the *quenched* free energies associated with H and $H+F$ are the same: the term $\sum_{i=1}^{n} F(T^i\omega)$ does not depend on S. Thus, in order to improve an annealed estimate of a critical curve one has to resort to adding non-local functions of the disorder, which are computationally unattractive.

(4) Caravenna, Giacomin and Gubinelli [55] carry out a detailed numerical analysis of the critical curve, with full statistical control on the errors. They exploit the superadditivity property noted in the proof of Theorem 9.1, which implies that

$$g(\lambda, h) = \sup_{n \in \mathbb{N}} \frac{1}{2n} \log \mathbb{E}(\log Z_{2n}^{*,\omega}). \tag{9.73}$$

Hence, for any given (λ, h), if $\mathbb{E}(\log Z_{2n}^{*,\omega})$ for some finite n, then $(\lambda, h) \in \mathcal{L}$. This leads to a sharp lower bound for the critical curve. It is much harder to get a decent upper bound. Computations are carried out up to $n = 2 \times 10^8$, and concentration inequalities are used to estimate the expectation $\mathbb{E}$ with only relatively few samples of ω. Giacomin and Sohier (private communication) have pushed the computation up to $n = 10^{12}$, thereby improving the lower bound for the critical curve even further.

Interestingly, the results show that to a remarkable degree of accuracy

$$h_c(\lambda) \approx h^*(K_c \lambda), \qquad \lambda \in (0, \infty), \tag{9.74}$$

with K_c the constant in (9.45), both for binary and standard normal disorder. Nevertheless, Bodineau, Giacomin, Lacoin and Toninelli [23] prove that equality in (9.74) cannot hold for all λ, by showing that the critical curve depends on the fine details of the excursion return time distribution and not only on the universal constant K_c.

(5) The restriction to the quadrant in (9.4) can be trivially removed. Indeed, because $f(\lambda, h)$ is antisymmetric under the transformations $\lambda \to -\lambda$ and $h \to -h$ (recall (9.2–9.3)), the full phase diagram consists of four critical curves, one in each quadrant of $\mathbb{R}^2$, which are images of the critical curve in the first quadrant (drawn in Fig. 9.4) under reflection in the horizontal and the vertical axes. For instance, below the critical curve in the fourth quadrant (i.e., for $h \leq -h_c(\lambda)$ and $\lambda \geq 0$) the copolymer is delocalized into the water rather than into the oil.

A further observation is the following. By subtracting $\lambda n + \lambda h \sum_{i=1}^{n} \omega_i$ from the Hamiltonian H_n^ω in (9.3), we obtain a new Hamiltonian

$$\widehat{H}_n^\omega(w) = -2\lambda(1+h) \sum_{i=1}^{n} \frac{1+\omega_i}{2} \frac{1+\Delta_i}{2} - 2\lambda(1-h) \sum_{i=1}^{n} \frac{1-\omega_i}{2} \frac{1-\Delta_i}{2}, \tag{9.75}$$

where we recall (9.7). In terms of the reparameterization $\alpha = 2\lambda(1+h)$ and $\beta = 2\lambda(1-h)$, this reads

$$\widehat{H}_n^\omega(w) = -\alpha \sum_{i=1}^n 1_{\{\omega_i = \Delta_i = 1\}} - \beta \sum_{i=1}^n 1_{\{\omega_i = \Delta_i = -1\}}, \qquad (9.76)$$

in which energy $-\alpha$ is assigned to the hydrophobic monomers in the oil, $-\beta$ to the hydrophilic monomers in the water, and 0 to the other two combinations. By the strong law of large numbers for ω, we have $H_n^\omega - \widehat{H}_n^\omega = \lambda n + o(n)$ ω-a.s. Therefore the quenched free energies associated with H_n^ω and $\widehat{H}_n^\omega$ differ by a term $\log \lambda$, which allows for a direct link between the respective phase diagrams. The form in (9.76) is used in a number of papers (see Extensions (7–10) below), and also in Chapter 10, where we look at a copolymer near a random selective interface.

In the (α, β)-model the quenched critical curve $\alpha \mapsto \beta_c(\alpha)$ takes the form in Fig. 9.8. This curve is continuous, strictly increasing and concave as a function of α, with a finite asymptote as $\alpha \to \infty$, and with a curvature as $\alpha \downarrow 0$ that tends to K_c, the universal constant in Theorem 9.8. The former comes from the bound $1 - h_c(\lambda) \leq C/\lambda$, $\lambda \to \infty$, $C < \infty$, implied by Theorem 9.7. The latter comes from the observation that, by (9.45), as $\lambda, \alpha \downarrow 0$,

$$\alpha - \beta_c(\alpha) = 4\lambda h_c(\lambda) \sim 4K_c\lambda^2 = \tfrac{1}{4}K_c\big(\alpha + \beta_c(\alpha)\big)^2 \sim \tfrac{1}{4}K_c(2\alpha)^2 = K_c\alpha^2. \tag{9.77}$$

(6) Bolthausen and Giacomin [29] look at the version of the model where the disorder is *periodic* (earlier work can be found in Grosberg, Izrailev and Nechaev [133]). For this case the free energy can be expressed in terms of a variational formula. However, it turns out to be delicate to deal with large periods, since computations quickly become prohibitive. In particular, the fact that random disorder arises from periodic disorder as the period tends to infinity seems hard to implement when probing the fine details of the critical curve. Caravenna, Giacomin and Zambotti [57], [58] look at the path properties for periodic disorder, showing that under the law P_∞^ω, i.e., the weak limit

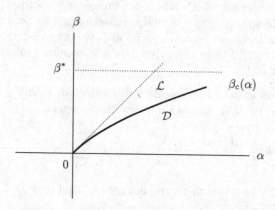

Fig. 9.8. Plot of $\alpha \mapsto \beta_c(\alpha)$.

of the law P_n^ω, the path is transient in the interior of $\mathcal{D}$, null-recurrent at the boundary $\partial\mathcal{D}$, and positive recurrent in $\mathcal{L}$. In each of the three cases they identify the scaling limit of the path.

(7) Den Hollander and Wüthrich [173] consider a version of the copolymer model with an *infinite array* of linear interfaces stacked on top of each other, equally spaced at distance L_n with n the length of the copolymer. If $\lim_{n\to\infty} L_n = \infty$, then this model has the same free energy as the single interface model, because the copolymer "sees only one interface at a time". Under the assumption that $\lim_{n\to\infty} L_n/\log n = 0$ and $\lim_{n\to\infty} L_n/\log\log n = \infty$, they show that the copolymer hops between neighboring interfaces on a time scale of order $\exp[\chi L_n]$, for some $\chi = \chi(\lambda, h) > 0$, when $(\lambda, h) \in \mathcal{L}$. The reason is that in the localized phase the excursions away from the interface are exponentially unlikely in their length and height (recall Theorem 9.9(a)). See also Extension (7) in Section 7.6.

(8) Martin, Causo and Whittington [236] consider a version of the single interface model in which the path is self-avoiding, the sites (rather than the edges) in the path represent the monomers, while the Hamiltonian is given by (9.76). They identify the qualitative properties of the phase diagram in the (α, β)-plane with the help of exact enumeration methods, series analysis techniques and rigorous bounds. It turns out that there are two critical curves (which are images of each other w.r.t. the line $\{\alpha = \beta\}$ when both monomer types occur with density $\frac{1}{2}$). Both curves lie in the first and in the third quadrant, with horizontal and vertical asymptotes, and meet each other at the origin. Below the lower curve the copolymer is delocalized into the water, above the upper curve it is delocalized into the oil, while in between the two curves it is localized near the interface. Madras and Whittington [233], building on earlier work by Maritan, Riva and Trovato [235], prove that inside the first quadrant the lower curve lies strictly inside the first octant except at the origin, and is a continuous, non-decreasing and concave function of α (and thus is similar to the one in Fig. 9.8, showing that the SAW-restriction does not alter the qualitative properties of the phase diagram). Causo and Whittington [64] analyze the phase diagram with the help of Monte Carlo techniques, and their results suggest that the phase transition is second order in the third quadrant but higher order in the first quadrant. If the latter were true, then this would imply that the origin is a tricritical point, i.e., a point where three phases meet. The results in [235], [236] and [233] apply to SAW's in dimensions $d \geq 2$.

(9) James, Soteros and Whittington [198], [199] generalize the SAW-version of the (α, β)-model by adding to the Hamiltonian an energy $-\gamma$ for each monomer that lies at the interface, irrespective of its type. It turns out that the value of γ affects the phase diagram (see Fig. 9.9). Indeed, as later shown in Madras and Whittington [233], for $\gamma < 0$ the phase diagram is qualitatively like that for $\gamma = 0$, but not for $\gamma > 0$. Indeed, there are *two critical values* $0 \leq \gamma_1 \leq \gamma_2 < \infty$ such that for $\gamma \in (\gamma_1, \gamma_2)$ the two critical curves are separated,

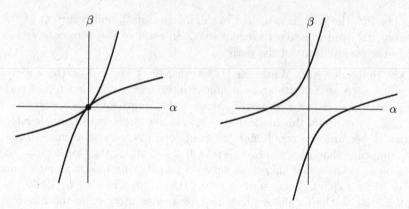

Fig. 9.9. Phase diagrams for the (α, β, γ)-model. The first curve is for $\gamma \in (-\infty, \gamma_1]$, the second curve for $\gamma \in (\gamma_1, \gamma_2)$. For $\gamma \in [\gamma_2, \infty)$ both curves have moved out to infinity.

i.e., they pass through the second and the fourth quadrant (resulting in a single localized phase), while for $\gamma \in [\gamma_2, \infty)$ the two critical curves have "moved out to infinity", i.e., the system is localized for all values of (α, λ). It is shown that $\gamma_2 < \infty$. It is believed that $\gamma_1 = 0$ and $\gamma_2 > 0$, but this remains open. For $\alpha = \beta = 0$ the model reduces to the homogeneous pinning model described in Section 7.1. Also the results in [198] and [199] apply to SAW's in dimensions $d \geq 2$. Exact enumeration is done for $d = 3$.

Habibzadah, Iliev, Saguia and Whittington [140] analyze the (α, β, γ)-model for generalized ballot paths rather than ballot paths (recall Fig. 1.4). The qualitative properties of the phase diagram are the same as for self-avoiding paths.

(10) Iliev, Orlandini and Whittington [181] study the effect of a force in the (α, β)-model with $\alpha < 0$ and $\beta > 0$, i.e., when the copolymer is delocalized below the interface. The force is applied to the endpoint of the copolymer and is perpendicular to the interface in the upward direction (in the spirit of the models considered in Section 7.3). Both for ballot paths and for generalized ballot paths the quenched free energy is computed in the first order Morita approximation, i.e., Z_n^ω is averaged over ω conditioned on $n^{-1} \sum_{i=1}^n \omega_i = 0$. Even though the Morita free energy is only an upper bound for the quenched free energy, it is argued that both free energies lead to the same critical force, which can therefore be computed explicitly. The critical curve in the force-temperature diagram is strictly increasing and so the phase transition is *not re-entrant* (compare with Section 7.3).

The situation is different when $\beta \geq \alpha > 0$. In that case the phase transition is *re-entrant* for $\beta > \alpha$, with the critical curve in the force-temperature diagram hitting 0 at some finite temperature, and *not re-entrant* for $\beta = \alpha$, with the critical curve being strictly increasing. This fits with the observations made in Extension (4), where the critical curve in the phase diagram of the (α, β)-plane without force was found to touch the line $\{\alpha = \beta\}$ (see Fig. 9.8). It is argued that the critical curve in the force-temperature diagram in the Morita approximation is an upper bound for the one of the quenched model.

(11) For the directed version of the (α, β, γ)-model, Orlandini, Tesi and Whittington [254] and Orlandini, Rechnitzer and Whittington [253] analyze the phase diagram in the Morita approximation, i.e., they compute the annealed free energy subject to restrictions on the first and second moments of ω. This gives only limited rigorous information on the properties of the critical curves, but it provides considerable insight into the mechanisms driving the phase transition. The analysis was subsequently extended by Iliev, Rechnitzer and Whittington [182], who found evidence that the lower critical curve is non-analytic at the origin when $\gamma = 0$, at a point in the third quadrant when $\gamma < 0$, and at a point in the first quadrant when $\gamma > 0$. The nature of these tricritical points remains unclear.

(12) Brazhnyi and Stepanow [40] consider a model of a random copolymer in $\mathbb{R}^3$ where the Hamiltonian includes elasticity, self-repulsion and self-attraction, thereby combining aspects from Chapters 3–6 and 9. Obviously, such models display highly complex behavior, but they are worthwhile to investigate when the aim is to model particular experimental situations. With a combination of analytical and numerical techniques it is shown that localization is disfavored by self-repulsion, favored by self-attraction, and can be re-entrant as a function of the densities of the different monomer types.

9.7 Challenges

(1) Improve Theorem 9.11 by finding out whether the phase transition is second order or higher order. Numerical analysis seems to indicate that the order is higher (Trovato and Maritan [298], Causo and Whittington [64], Caravenna, Giacomin and Gubinelli [55]). Monthus [243] suggests that the order is infinite.

(2) We know from Theorem 9.12 that the quenched free energy is infinitely differentiable on $\mathcal{L}$. Find out whether or not it is analytic. At points where this fails the system is said to have a Griffiths-McCoy singularity (Griffiths [132], McCoy [239]). Such singularities are known to occur is some disordered systems, e.g. the low-temperature dilute Ising model in zero magnetic field. The singular behavior is due to the occurrence of rare but arbitrarily large regions where the disorder is such that locally the system is almost at a phase transition. Is the critical curve, i.e., the boundary of $\mathcal{L}$, analytic?

(3) Improve Theorem 9.10 by showing that the path intersects the interface only finitely often. This has been proved by Giacomin and Toninelli [120] deep inside $\mathcal{D}$, namely, for (λ, h) on or above the annealed critical curve that is the upper bound in Theorem 9.6.

(4) Identify the analogue of the weak interaction limit in (9.46–9.48) when the reference random walk is not SRW (see Section 9.6, Extension (1)).

(5) The global properties of the copolymer do not depend on the fact that the monomer types are drawn in an i.i.d. rather than a stationary ergodic fashion, but the local properties do. Since in real polymers the order of the monomer types is at best Markov (as a result of the underlying polymerization process), it is interesting to try and extend the results in Sections 9.2–9.5 to the Markov setting.

(6) Investigate whether or not the critical curve for the SAW-version of the (α, β)-model defined in Section 9.6, Extension (8), has a positive and finite curvature as $\alpha \downarrow 0$. Is this curvature equal to K_c, as in the directed version of the (α, β)-model, or not? Prove that $\gamma_1 = 0$ in the SAW-version of the (α, β, γ)-model defined in Section 9.6, Extension (9).

10

Copolymers Near a Random Selective Interface

In this chapter we consider a different version of the directed copolymer model analyzed in Chapter 9, namely, one where the linear interface is replaced by a *random interface*. In particular, rather than putting the oil and the water in two halfplanes, we place them in large square blocks in a random percolation-type fashion. This is *a crude model of a copolymer in an emulsion*, consisting of oil droplets floating in water (see Fig. 10.1). We will see that this model exhibits a remarkably rich critical behavior, with *two phases* in a supercritical percolation regime and *four phases* in a subcritical percolation regime. *Large deviations* and *partial annealing estimates* will again play a central role, together with *coarse-graining*, entropy estimates and variational calculus.

The random interface model was introduced in den Hollander and Whittington [172] and the crude properties of the phase diagram were derived there. Finer details of the phase diagram were subsequently derived in den Hollander and Pétrélis [170], [171]. The text below follows these three papers.

One application of the model is the stabilization of milk by a protein called casein. This is a copolymer that adsorbs at the fat-water interface, wrapping itself around the tiny fat droplets that float in the water, thereby preventing them to coagulate. Another application is ink, which consists of carbon particles that are coated, for instance, with albumin (from egg white) or gum arabic (a polysaccharide obtained from the Acacia tree), with the same stabilizing effect.

In Section 10.1 we define the model. In Section 10.2 we derive a *variational formula* for the quenched free energy in terms of constituent quenched free energies at the block level and an underlying set of frequencies at which the copolymer can visit the blocks. In Section 10.3 we use the results from Section 10.2 to study the phase diagram in the supercritical percolation regime. In Section 10.4 we turn our attention to the subcritical percolation regime.

F. den Hollander, *Random Polymers*,
Lecture Notes in Mathematics 1974, DOI: 10.1007/978-3-642-00333-2_10,
© Springer-Verlag Berlin Heidelberg 2009

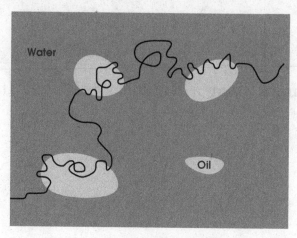

Fig. 10.1. An undirected copolymer in an emulsion.

10.1 A Copolymer Diagonally Crossing Blocks

Each positive integer is randomly labeled A or B, with probability $\frac{1}{2}$ each, independently for different integers. The resulting labeling is denoted by

$$\omega = \{\omega_i \colon i \in \mathbb{N}\} \in \{A, B\}^{\mathbb{N}} \tag{10.1}$$

and represents the *randomness of the copolymer*, with A denoting a hydrophobic monomer and B a hydrophilic monomer. Fix $p \in (0,1)$ and $L_n \in \mathbb{N}$. Partition $\mathbb{R}^2$ into square blocks of size L_n:

$$\mathbb{R}^2 = \bigcup_{x \in \mathbb{Z}^2} \Lambda_{L_n}(x), \qquad \Lambda_{L_n}(x) = xL_n + (0, L_n]^2. \tag{10.2}$$

Each block is randomly labeled A or B, with probability p, respectively, $1-p$, independently for different blocks, with A denoting oil and B denoting water. The resulting labeling is denoted by

$$\Omega = \{\Omega(x) \colon x \in \mathbb{Z}^2\} \in \{A, B\}^{\mathbb{Z}^2} \tag{10.3}$$

and represents the *randomness of the emulsion* (see Fig. 10.2).
  Let

- $\mathcal{W}_n =$ the set of n-step *directed self-avoiding paths* starting at the origin and being allowed to move *upwards, downwards and to the right* (recall Fig. 1.4):

$$\mathcal{W}_n = \{w = (w_i)_{i=0}^n \colon w_0 = 0,\ w_{i+1} - w_i \in \{\uparrow, \downarrow, \rightarrow\}\ \forall\, 0 \le i < n,$$
$$w_i \ne w_j\ \forall\, 0 \le i < j \le n\}. \tag{10.4}$$

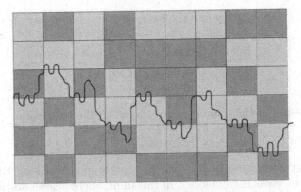

Fig. 10.2. A directed self-avoiding path crossing blocks of oil and water diagonally. The light-shaded blocks are oil, the dark-shaded blocks are water. Each block is L_n lattice spacings wide in both directions. The path carries hydrophobic and hydrophilic monomers on the lattice scale, which are not indicated.

- $\mathcal{W}_{n,L_n}$ = the subset of $\mathcal{W}_n$ consisting of those paths that enter blocks at a corner, exit blocks at one of the two corners *diagonally opposite* the one where it entered, and in between *stay confined* to the two blocks that are seen upon entering (see Fig. 10.2).

In other words, after the path reaches a site xL_n for some $x \in \mathbb{Z}^2$, it must make a step to the right, it must subsequently stay confined to the pair of blocks labeled x and $x + (0,-1)$, and it must exit this pair of blocks either at site $xL_n + (L_n, L_n)$ or at site $xL_n + (L_n, -L_n)$. This restriction – which is put in to make the model mathematically tractable – is unphysical. Nonetheless, as we will see in Sections 10.3–10.4, the model has physically very relevant behavior.

Given ω, Ω and n, with each path $w \in \mathcal{W}_{n,L_n}$ we associate an energy given by the Hamiltonian

$$H_{n,L_n}^{\omega,\Omega}(w) = -\sum_{i=1}^{n}\left(\alpha\, 1_{\{\omega_i = \Omega_{(w_{i-1},w_i)}^{L_n} = A\}} + \beta\, 1_{\{\omega_i = \Omega_{(w_{i-1},w_i)}^{L_n} = B\}}\right), \quad (10.5)$$

where (w_{i-1}, w_i) denotes the i-th step of the path and $\Omega_{(w_{i-1},w_i)}^{L_n}$ denotes the label of the block that the i-th step lies in. What this Hamiltonian does is count the number of AA-matches and BB-matches and assign them energy $-\alpha$ and $-\beta$, respectively, where $\alpha, \beta \in \mathbb{R}$. Note that, as in (9.3), the interaction is assigned to edges rather than to vertices, i.e., we identify the monomers with the steps of the path. We will see later that without loss of generality we may restrict the interaction parameters to the cone

$$\text{CONE} = \{(\alpha, \beta) \in \mathbb{R}^2 \colon \alpha \geq |\beta|\}. \quad (10.6)$$

Given ω, Ω and n, we define the *quenched free energy per step* as

$$f_{n,L_n}^{\omega,\Omega} = \frac{1}{n} \log Z_{n,L_n}^{\omega,\Omega},$$

$$Z_{n,L_n}^{\omega,\Omega} = \sum_{w \in \mathcal{W}_{n,L_n}} \exp\left[-H_{n,L_n}^{\omega,\Omega}(w)\right]. \tag{10.7}$$

We are interested in the limit $n \to \infty$ subject to the restriction

$$L_n \to \infty \qquad \text{and} \qquad \frac{1}{n}L_n \to 0. \tag{10.8}$$

This is a *coarse-graining* limit where the path spends a long time in each single block yet visits many blocks. In this limit, there is a separation between a *polymer scale* and an *emulsion scale*.

In Theorem 10.1 below we will see that

$$\lim_{n \to \infty} f_{n,L_n}^{\omega,\Omega} = f = f(\alpha, \beta; p) \quad \text{exists } \omega, \Omega - a.s., \tag{10.9}$$

is finite and non-random, and can be expressed as a variational problem involving the free energies of the polymer in each of the four possible pairs of adjacent blocks it may encounter and the frequencies at which the polymer visits each of these pairs of blocks on the coarse-grained block scale. This variational problem will be the starting point of our analysis.

10.2 Preparations

This section contains some preparatory definitions, lemmas and theorems that are needed to formulate our main results in Sections 10.3–10.4 for the phase diagram. We state the variational formula for the free energy, compute some path entropies, and identify the block free energies in terms of a *linear interface free energy*. Proofs of the lemmas and theorems stated below are elementary and can be found in den Hollander and Whittington [172].

10.2.1 Variational Formula for the Free Energy

We begin with two sets of definitions.

I. For $L \in \mathbb{N}$ and $a \geq 2$ (with aL integer), let $\mathcal{W}_{aL,L}$ denote the set of aL-step directed self-avoiding paths starting at $(0,0)$, ending at (L,L), and in between not leaving the two adjacent blocks of size L labeled $(0,0)$ and $(-1,0)$. For $k, l \in \{A, B\}$, let

$$\psi_{kl}^{\omega}(aL, L) = \frac{1}{aL} \log Z_{aL,L}^{\omega},$$

$$Z_{aL,L}^{\omega} = \sum_{w \in \mathcal{W}_{aL,L}} \exp\left[-H_{aL,L}^{\omega,\Omega}(w)\right] \text{ when } \Omega(0,0) = k \text{ and } \Omega(0,-1) = l,$$

$$\tag{10.10}$$

denote the free energy per step in a kl-block when the number of steps inside the block is a times the size of the block. Let

$$\lim_{L \to \infty} \psi_{kl}^{\omega}(aL, L) = \psi_{kl}(a) = \psi_{kl}(\alpha, \beta; a). \tag{10.11}$$

Note here that k labels the type of the block that is diagonally crossed, while l labels the type of the block that appears as its neighbor at the starting corner (see Fig. 10.3). We will see in Section 10.2.3 that the limit exists ω-a.s. and is non-random, and that ψ_{AA} and ψ_{BB} take on a simple form whereas ψ_{AB} and ψ_{BA} do not.

II. Let $\mathcal{W}$ denote the class of all *coarse-grained paths* $W = \{W_j : j \in \mathbb{N}\}$ that step diagonally from corner to corner (see Fig. 10.4, where each dashed line with arrow denotes a single step of W). For $n \in \mathbb{N}$, $W \in \mathcal{W}$ and $k, l \in \{A, B\}$, let

$$\rho_{kl}^{\Omega}(W, n) = \frac{1}{n} \sum_{j=1}^{n} 1_{E_{kl}^{\Omega}(j)} \tag{10.12}$$

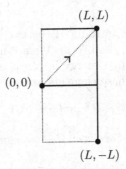

Fig. 10.3. Two neighboring blocks. The dashed line with the arrow indicates that the coarse-grained path makes a step diagonally upwards.

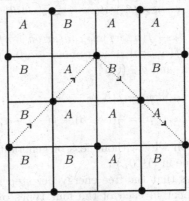

Fig. 10.4. W sampling Ω. The dashed lines with arrows indicate the steps of W.

with $E_{kl}^{\Omega}(j)$ the event that (W_{j-1}, W_j) diagonally crosses a k-block in Ω having an l-block in Ω as its neighbor at the starting corner. Abbreviate

$$\rho^{\Omega}(W, n) = \left(\rho_{kl}^{\Omega}(W, n)\right)_{k,l \in \{A,B\}}, \tag{10.13}$$

which is a 2×2 matrix with nonnegative elements that sum up to 1. Let $\mathcal{R}^{\Omega}(W)$ denote the set of all limits points of the sequence $\{\rho^{\Omega}(W, n) \colon n \in \mathbb{N}\}$, and put

$$\mathcal{R}^{\Omega} = \text{the closure of the set} \bigcup_{W \in \mathcal{W}} \mathcal{R}^{\Omega}(W). \tag{10.14}$$

Clearly, $\mathcal{R}^{\Omega}$ exists for all Ω. Moreover, since Ω has a trivial sigma-field at infinity (i.e., all events not depending on finitely many coordinates of Ω have probability 0 or 1) and $\mathcal{R}^{\Omega}$ is measurable with respect to this sigma-field, we have

$$\mathcal{R}^{\Omega} = \mathcal{R}(p) \qquad \Omega - a.s. \tag{10.15}$$

for some *non-random closed* set $\mathcal{R}(p)$. This set, which depends on the parameter p controlling Ω, is the set of all possible limit points of the frequencies at which the four pairs of adjacent blocks can be seen along an infinite coarse-grained path.

With I and II above we have introduced the two main quantities of the model, namely, ψ_{kl}, $k, l \in \{A, B\}$, and $\mathcal{R}(p)$. We are now ready to formulate the key theorem expressing the free energy in terms of these quantities. Let $\mathcal{A}$ be the set of 2×2 matrices whose elements are ≥ 2.

Theorem 10.1. (a) *For all $(\alpha, \beta) \in \mathbb{R}^2$ and $p \in (0, 1)$,*

$$\lim_{n \to \infty} f_{n, L_n}^{\omega, \Omega} = f = f(\alpha, \beta; p) \qquad \text{exists } \omega, \Omega - a.s. \text{ and in mean,} \tag{10.16}$$

is finite and non-random, and is given by

$$f = \sup_{(a_{kl}) \in \mathcal{A}} \sup_{(\rho_{kl}) \in \mathcal{R}(p)} \frac{\sum_{k,l} \rho_{kl} a_{kl} \psi_{kl}(a_{kl})}{\sum_{k,l} \rho_{kl} a_{kl}}. \tag{10.17}$$

(b) *The function $(\alpha, \beta) \mapsto f(\alpha, \beta; p)$ is convex on $\mathbb{R}^2$ for all $p \in (0, 1)$.*
(c) *The function $p \mapsto f(\alpha, \beta; p)$ is continuous on $(0, 1)$ for all $(\alpha, \beta) \in \mathbb{R}^2$.*
(d) *For all $(\alpha, \beta) \in \mathbb{R}^2$ and $p \in (0, 1)$,*

$$\begin{aligned}
f(\alpha, \beta; p) &= f(\beta, \alpha; 1 - p), \\
f(\alpha, \beta; p) &= \tfrac{1}{2}(\alpha + \beta) + f(-\beta, -\alpha; p).
\end{aligned} \tag{10.18}$$

(Part (d) is the reason why without loss of generality we may restrict the parameters to the cone in (10.6).)

Theorem 10.1 says that the free energy per step is obtained by keeping track of the times spent in each of the four types of blocks, summing the free energies of the four types of blocks given these times, and afterwards optimizing over these times and over the coarse-grained random walk. The latter carries no entropy because of (10.8). Thus, we see that *self-averaging* occurs both at the *polymer scale* and at the *emulsion scale*.

Theorem 10.1 is the starting point for our analysis of the phase diagram in Sections 10.3–10.4. What we need to do is collect the necessary information on the key ingredients of (10.17). This is done in Sections 10.2.2–10.2.3. We will see that, to get the qualitative properties of the phase diagram, only very little information is needed about $\mathcal{R}(p)$.

The behavior of f as a function of (α, β) is different for $p \geq p_c$ and $p < p_c$, where $p_c \approx 0.64$ is *the critical percolation density for directed bond percolation on the square lattice*. The reason is that the coarse-grained paths W, which determine the set $\mathcal{R}(p)$, sample Ω just like paths in directed bond percolation on the square lattice rotated by 45 degrees sample the percolation configuration (see Fig. 10.4). We will see in Sections 10.3–10.4 that, consequently, the phase diagrams in the supercritical and subcritical regimes are different.

10.2.2 Path Entropies

We next state two lemmas that identify the entropy of a path diagonally crossing a block or running along the interface in a block (the numbers appearing in these lemmas are model specific). We need these lemmas in Section 10.2.3. Their proofs are straightforward.

Let

$$\text{DOM} = \{(a, b)\colon a \geq 1 + b, b \geq 0\}. \tag{10.19}$$

For $(a, b) \in \text{DOM}$, let $N_L(a, b)$ denote the number of aL-step self-avoiding directed paths from $(0, 0)$ to (bL, L) whose vertical displacement stays within $(-L, L]$ (aL and bL are integer). Let

$$\kappa(a, b) = \lim_{L \to \infty} \frac{1}{aL} \log N_L(a, b). \tag{10.20}$$

Lemma 10.2. (i) $\kappa(a, b)$ *exists and is finite for all* $(a, b) \in \text{DOM}$.
(ii) $(a, b) \mapsto a\kappa(a, b)$ *is continuous and strictly concave on* DOM *and analytic on the interior of* DOM.
(iii) *For all* $a \geq 2$,

$$a\kappa(a, 1) = \log 2 + \tfrac{1}{2} \left[a \log a - (a - 2) \log(a - 2) \right]. \tag{10.21}$$

(iv) $\sup_{a \geq 2} \kappa(a, 1) = \kappa(a^*, 1) = \tfrac{1}{2} \log 5$ *with unique maximizer* $a^* = \tfrac{5}{2}$.
(v) $(\tfrac{\partial}{\partial a}\kappa)(a^*, 1) = 0$ *and* $a^*(\tfrac{\partial}{\partial b}\kappa)(a^*, 1) = \tfrac{1}{2} \log \tfrac{9}{5}$.

For $\mu \geq 1$, let $\hat{N}_L(\mu)$ denote the number of μL-step self-avoiding paths from $(0, 0)$ to $(L, 0)$ with no restriction on the vertical displacement (μL is integer). Let

$$\hat{\kappa}(\mu) = \lim_{L \to \infty} \frac{1}{\mu L} \log \hat{N}_L(\mu). \tag{10.22}$$

Lemma 10.3. (i) $\hat{\kappa}(\mu)$ *exists and is finite for all* $\mu \geq 1$.
(ii) $\mu \mapsto \mu\hat{\kappa}(\mu)$ *is continuous and strictly concave on* $[1, \infty)$ *and analytic on* $(1, \infty)$.
(iii) $\hat{\kappa}(1) = 0$ *and* $\mu\hat{\kappa}(\mu) \sim \log\mu$ *as* $\mu \to \infty$.
(iv) $\sup_{\mu \geq 1} \mu[\hat{\kappa}(\mu) - \frac{1}{2}\log 5] < \frac{1}{2}\log\frac{9}{5}$.

10.2.3 Free Energies per Pair of Blocks

In this section we identify the block free energies defined in (10.11). Because AA-blocks and BB-blocks have no interface, both ψ_{AA} and ψ_{BB} can be computed explicitly.

Lemma 10.4. *For all* $(\alpha, \beta) \in \mathbb{R}^2$ *and* $a \geq 2$, ω-*a.s. and in mean,*

$$\psi_{AA}(a) = \tfrac{1}{2}\alpha + \kappa(a, 1) \qquad and \qquad \psi_{BB}(a) = \tfrac{1}{2}\beta + \kappa(a, 1). \qquad (10.23)$$

To compute $\psi_{AB}(a)$ and $\psi_{BA}(a)$, we first consider the free energy per step when the path moves in the vicinity of a *single linear interface* $\mathcal{I}$ separating a solvent A in the upper halfplane from a solvent B in the lower halfplane including the interface itself (see Fig. 10.5). To that end, for $c \geq b > 0$, let $\mathcal{W}_{cL,bL}$ denote the set of cL-step directed self-avoiding paths starting at $(0,0)$ and ending at $(bL, 0)$. Define

$$\psi_L^{\omega,\mathcal{I}}(c, b) = \frac{1}{cL}\log Z_{cL,bL}^{\omega,\mathcal{I}} \qquad (10.24)$$

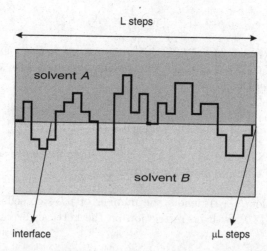

Fig. 10.5. Illustration of (10.24–10.25) for $c = \mu$ and $b = 1$.

with

$$Z_{cL,bL}^{\omega,\mathcal{I}} = \sum_{w \in \mathcal{W}_{cL,bL}} \exp\left[-H_{cL}^{\omega,\mathcal{I}}(w)\right],$$

$$H_{cL}^{\omega,\mathcal{I}}(w) \tag{10.25}$$

$$= -\sum_{i=1}^{cL} \left(\alpha\, 1_{\{\omega_i=A,(w_{i-1},w_i)>0\}} + \beta\, 1_{\{\omega_i=B,(w_{i-1},w_i)\leq 0\}}\right),$$

where $(w_{i-1}, w_i) > 0$ means that the i-th step lies in the upper halfplane and $(w_{i-1}, w_i) \leq 0$ means that the i-th step lies in the lower halfplane or in the interface.

Lemma 10.5. *For all $(\alpha, \beta) \in \mathbb{R}^2$ and $c \geq b > 0$,*

$$\lim_{L\to\infty} \psi_L^{\omega,\mathcal{I}}(c,b) = \phi^{\mathcal{I}}(c/b) = \phi^{\mathcal{I}}(\alpha,\beta;c/b) \tag{10.26}$$

exists ω-a.s. and in mean, and is non-random.

Lemma 10.6. *For all $(\alpha, \beta) \in \mathbb{R}^2$ and $a \geq 2$,*

$$a\psi_{AB}(a) = a\psi_{AB}(\alpha,\beta;a)$$
$$= \sup_{\substack{0\leq b\leq 1, c\geq b \\ a-c\geq 2-b}} \left\{c\phi^{\mathcal{I}}(c/b) + (a-c)[\tfrac{1}{2}\alpha + \kappa(a-c,1-b)]\right\}. \tag{10.27}$$

Lemma 10.7. *Let $k, l \in \{A, B\}$.*
(i) For all $(\alpha, \beta) \in \mathbb{R}^2$, $a \mapsto a\psi_{kl}(\alpha,\beta;a)$ is continuous and concave on $[2,\infty)$.
(ii) For all $a \in [2,\infty)$, $\alpha \mapsto \psi_{kl}(\alpha,\beta;a)$ and $\beta \mapsto \psi_{kl}(\alpha,\beta;a)$ are continuous and non-decreasing on $\mathbb{R}$.

Lemma 10.5 is an immediate consequence of the subadditive ergodic theorem. Indeed, $\phi^{\mathcal{I}}$ is a *single interface free energy* and can be treated as in Chapter 9. The idea behind Lemma 10.6 is that the polymer follows the AB-interface over a distance bL during cL steps and then wanders away from the AB-interface to the diagonally opposite corner over a distance $(1-b)L$ during $(a-c)L$ steps. The optimal strategy is obtained by maximizing over b and c (see Fig. 10.6). In den Hollander and Pétrélis [170] it is shown that the supremum in (10.27) is uniquely attained and that $a \mapsto \psi_{kl}$, $k, l \in \{A, B\}$, is strictly concave.

With (10.23) and (10.27) we have identified the key ingredients in the variational formula for the free energy in (10.17) in terms of the single interface free energy $\phi^{\mathcal{I}}$. This constitutes a *major simplification*, in view of the methods and techniques available from Chapter 9.

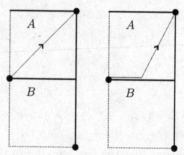

Fig. 10.6. Two possible strategies inside an *AB*-block: The path can either move straight across or move along the interface for awhile and then move across. Both strategies correspond to a coarse-grained step diagonally upwards as in Fig. 10.3. The dotted lines represent the motion of the path on the block level. On a microscopic scale the path is erratic, which is however not indicated.

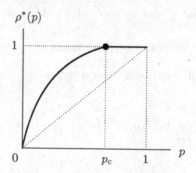

Fig. 10.7. Plot of $p \mapsto \rho^*(p)$.

10.2.4 Percolation

Let

$$\rho^*(p) = \sup_{(\rho_{kl}) \in \mathcal{R}(p)} [\rho_{AA} + \rho_{AB}], \qquad p \in (0, 1). \tag{10.28}$$

This is the maximal frequency of *A*-blocks crossed by an infinite coarse-grained path (recall (10.12–10.15)). The graph of $p \mapsto \rho^*(p)$ is sketched in Fig. 10.7.

The elements of $\mathcal{R}(p)$ are matrices

$$\begin{pmatrix} \rho_{AA} & \rho_{AB} \\ \rho_{BA} & \rho_{BB} \end{pmatrix} \tag{10.29}$$

whose elements are non-negative and sum up to 1. It is easy to see that $p \mapsto \mathcal{R}(p)$ is continuous in the Hausdorff metric. Moreover, for $p \geq p_c$, $\mathcal{R}(p)$ contains matrices of the form

$$\begin{pmatrix} 1 - \gamma & \gamma \\ 0 & 0 \end{pmatrix} \qquad \text{for } \gamma \text{ in a closed subset of } (0, 1). \tag{10.30}$$

The reason is that, when $p > p_c$, the A-blocks percolate and W may eventually either stay inside the infinite A-cluster or move along its boundary.

We have now gathered enough information on the ingredients of the variational formula in Theorem 10.1 for the free energy to be able to start addressing the phase diagram. In Section 10.3 we consider the supercritical regime $p \geq p_c$, where the oil blocks percolate, in Section 10.4 the subcritical regime, where the oil blocks do not percolate. These regimes display completely different behavior, since CONE defined in (10.6) favors the oil blocks over the water blocks.

10.3 Phase Diagram in the Supercritical Regime

The phase diagram is relatively simple in the supercritical regime. This is because the *oil blocks percolate*, and so the coarse-grained path can choose between moving into the oil or running along the interface between the oil and the water (see Fig. 10.8). We may therefore expect the behavior to be *qualitatively similar to that for a single linear interface*. In Section 10.4 we will see that the phase diagram is considerably more complex in the subcritical regime. This is because when the oil does not percolate there is no (!) infinite interface available.

In Section 10.3.1 we derive a criterion for localization in terms of the single interface free energy $\phi^\mathcal{I}$ defined in Lemma 10.5. In Section 10.3.3 we use this criterion to state a theorem identifying the qualitative properties of the critical curve. In Section 10.3.3 we give the proof of this theorem. Section 10.3.4 states finer details of the critical curve, the proofs of which are technical and are omitted.

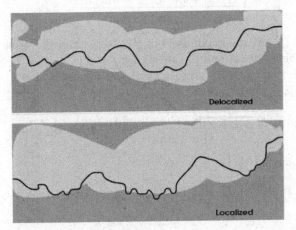

Fig. 10.8. Two possible strategies when the oil percolates.

10.3.1 Free Energy in the Two Phases

The following result gives a preliminary identification of the two phases.

Theorem 10.8. *Let $p \geq p_c$. Then*

$$
\begin{aligned}
\mathcal{D} &= \{(\alpha, \beta) \in \text{CONE} \colon S_{AB} = S_{AA}\}, \\
\mathcal{L} &= \{(\alpha, \beta) \in \text{CONE} \colon S_{AB} > S_{AA}\},
\end{aligned}
\tag{10.31}
$$

where

$$
S_{kl} = \sup_{a \geq 2} \psi_{kl}(a), \qquad k, l \in \{A, B\}. \tag{10.32}
$$

Proof. The proof uses (10.30) and Theorem 10.1(a), in combination with the inequalities

$$
S_{BB} \leq S_{AA}, \qquad S_{BA} \leq S_{AB}, \tag{10.33}
$$

which hold because $\beta \leq \alpha$. We prove that for $p \geq p_c$:

$$
\begin{aligned}
&\text{(i)} \quad S_{AB} = S_{AA} \Longrightarrow f = S_{AA}. \\
&\text{(ii)} \quad S_{AB} > S_{AA} \Longrightarrow f > S_{AA}.
\end{aligned}
\tag{10.34}
$$

(i) Suppose that $S_{AB} = S_{AA}$. Then, because $\psi_{kl}(a) \leq S_{kl}$ for all $a \geq 2$ and $k, l \in \{A, B\}$ by (10.32), Theorem 10.1(a) and (10.33) yield

$$
f \leq \sup_{(a_{kl}) \in \mathcal{A}} \sup_{(\rho_{kl}) \in \mathcal{R}(p)} \frac{\sum_{k,l} \rho_{kl} a_{kl} S_{kl}}{\sum_{k,l} \rho_{kl} a_{kl}} \leq \sup_{k,l} S_{kl} = S_{AB} = S_{AA}. \tag{10.35}
$$

On the other hand, (10.30) and Theorem 10.1(a) yield

$$
f \geq \frac{(1-\gamma)\bar{a}_{AA} S_{AA} + \gamma \bar{a}_{AB} S_{AB}}{(1-\gamma)\bar{a}_{AA} + \gamma \bar{a}_{AB}} = S_{AA}, \tag{10.36}
$$

where $\bar{a}_{AA}, \bar{a}_{AB}$ are any maximizers of S_{AA}, S_{AB} (and the value of γ is irrelevant). Combine (10.35) and (10.36) to get $f = S_{AA}$.

(ii) Suppose that $S_{AB} > S_{AA}$. Then (10.30) with $0 < \gamma < 1$ and Theorem 10.1(a) yield

$$
f \geq \frac{(1-\gamma)\bar{a}_{AA} S_{AA} + \gamma \bar{a}_{AB} S_{AB}}{(1-\gamma)\bar{a}_{AA} + \gamma \bar{a}_{AB}} > S_{AA}. \tag{10.37}
$$

Here it is important that $\gamma > 0$. $\square$

It follows from Lemma 10.2(iv), (10.23) and (10.32) that

$$
S_{AA} = \tfrac{1}{2}\alpha + \tfrac{1}{2}\log 5. \tag{10.38}
$$

Theorem 10.8, together with (10.34) and (10.38), yields:

Theorem 10.9. *Let $p \geq p_c$. Then $(\alpha, \beta) \mapsto f(\alpha, \beta; p)$ is non-analytic along the curve in* CONE *separating the two regions*

$$
\begin{aligned}
\mathcal{D} = \text{ delocalized phase } &= \left\{ (\alpha, \beta) \in \text{CONE}: f(\alpha, \beta; p) = \tfrac{1}{2}\alpha + \tfrac{1}{2}\log 5 \right\}, \\
\mathcal{L} = \text{ localized phase } &= \left\{ (\alpha, \beta) \in \text{CONE}: f(\alpha, \beta; p) > \tfrac{1}{2}\alpha + \tfrac{1}{2}\log 5 \right\},
\end{aligned}
$$
$$(10.39)$$

where $\lim_{n \to \infty} \frac{1}{n} \log |\mathcal{W}_{n,L_n}| = \frac{1}{2}\log 5$ is the entropy per step of the walk in a single block subject to (10.8).

The intuition behind Theorem 10.9 is as follows. Suppose that $p > p_c$. Then the A-blocks percolate. Therefore the polymer has the option of moving to the infinite cluster of A-blocks and staying in that infinite cluster forever, thus seeing only AA-blocks. In doing so, it loses an entropy of at most $o(n/L_n) = o(n)$ on the coarse-grained scale, it gains an energy $\frac{1}{2}\alpha n + o(n)$ on the lattice scale (because only half of its monomers are matched), and it gains an entropy $(\frac{1}{2}\log 5)n + o(n)$ on the lattice scale (see the top half of Fig. 10.8). Alternatively, the path has the option of following the boundary of the infinite cluster (at least part of the time), during which it sees AB-blocks and (when $\beta \geq 0$) gains more energy by matching more than half of its monomers (see the bottom half of Fig. 10.8). Consequently,

$$f(\alpha, \beta; p) \geq \tfrac{1}{2}\alpha + \tfrac{1}{2}\log 5. \tag{10.40}$$

The boundary between the two regimes in (10.39) corresponds to the crossover where one option takes over from the other. Note that the critical curve does *not* depend on p. Because $p \mapsto f(\alpha, \beta; p)$ is continuous (by Theorem 10.1(c) in Section 10.2.1), the same critical curve occurs at $p = p_c$. Note further that f does not depend on p in the delocalized phase, but does in the localized phase.

10.3.2 Criterion for Localization

The key result identifying the critical curve in the supercritical regime is the following criterion in terms of the free energy of the *single linear interface*.

Proposition 10.10. *Let $p \geq p_c$. Then $(\alpha, \beta) \in \mathcal{L}$ if and only if*

$$\sup_{\mu \geq 1} \mu[\phi^{\mathcal{I}}(\alpha, \beta; \mu) - \tfrac{1}{2}\alpha - \tfrac{1}{2}\log 5] > \tfrac{1}{2}\log \tfrac{9}{5}. \tag{10.41}$$

Proof. Combining (10.38) with Lemma 10.6, together with the reparameterization $\mu = c/b$ and $\nu = (a - c)/b$, we get

$$S_{AB} - S_{AA} = \sup_{\mu \geq 1, \nu \geq 1} \frac{\mu[\phi^{\mathcal{I}}(\mu) - S_{AA}] - \nu[\tfrac{1}{2}\log 5 - u(\nu)]}{\mu + \nu} \tag{10.42}$$

with

$$u(\nu) = \sup_{\frac{2}{\nu+1} \le b \le 1} \kappa(b\nu, 1-b), \qquad \nu \ge 1. \tag{10.43}$$

Abbreviate $v(\nu) = \nu[\frac{1}{2} \log 5 - u(\nu)]$. Below we will show that

$$\begin{aligned} &\text{(i)} \ v(\nu) > \tfrac{1}{2} \log \tfrac{9}{5} \text{ for all } \nu \ge 1, \\ &\text{(ii)} \ \lim_{\nu \to \infty} v(\nu) = \tfrac{1}{2} \log \tfrac{9}{5}. \end{aligned} \tag{10.44}$$

This will imply the claim as follows. If $\mu[\phi^{\mathcal{I}}(\mu) - S_{AA}] \le \frac{1}{2} \log \frac{9}{5}$ for all μ, then by (i) the numerator in (10.42) is strictly negative for all μ and ν, and so by (ii) the supremum is taken at $\nu = \infty$, resulting in $S_{AB} - S_{AA} = 0$. On the other hand, if $\mu[\phi^{\mathcal{I}}(\mu) - S_{AA}] > \frac{1}{2} \log \frac{9}{5}$ for some μ, then, for that μ, by (i) and (ii) the numerator is strictly positive for ν large enough, resulting in $S_{AB} - S_{AA} > 0$.

To prove (10.44), we will need the following inequality. Abbreviate $\chi(a,b) = a\kappa(a,b)$. Then by Lemma 10.2(ii) we have, for all $(s,t) \ne (u,v)$ in DOM,

$$\begin{aligned} \chi(s,t) - \chi(u,v) &= \int_0^1 dw \, \frac{\partial}{\partial w} \chi(u + w(s-u), v + w(t-v)) \\ &> \left[\frac{\partial}{\partial w} \chi(u + w(s-u), v + w(t-v)) \right]_{w=1} \\ &= (s-u) \left(\frac{\partial}{\partial a} \chi \right)(s,t) + (t-v) \left(\frac{\partial}{\partial b} \chi \right)(s,t). \end{aligned} \tag{10.45}$$

To prove (10.44)(i), put $b = a/\nu$ in (10.43) and use Lemma 10.2(iv–v) to rewrite the statement in (10.44)(i) as

$$\kappa\left(a, 1 - \frac{a}{\nu}\right) < \kappa(a^*, 1) - \frac{a^*}{\nu} \left(\frac{\partial}{\partial b} \kappa \right)(a^*, 1) \text{ for all } \nu \ge 1 \text{ and } \frac{2\nu}{\nu+1} \le a \le \nu. \tag{10.46}$$

But this inequality follows from (10.45) by picking $s = a^*$, $t = 1$, $u = a$, $v = 1 - \frac{a}{\nu}$, canceling a term $a^*\kappa(a^*, 1)$ on both sides, using that $(\frac{\partial}{\partial a}\kappa)(a^*, 1) = 0$, and afterwards canceling a common factor a on both sides.

To prove (10.44)(ii), we argue as follows. Picking $b = \frac{a^*}{\nu}$ in (10.43), we get from Lemma 10.2(iv) that

$$v(\nu) \le \nu \left[\kappa(a^*, 1) - \kappa\left(a^*, 1 - \frac{a^*}{\nu}\right) \right]. \tag{10.47}$$

Letting $\nu \to \infty$, we get from Lemma 10.2(v) that

$$\limsup_{\nu \to \infty} v(\nu) \le a^* \left(\frac{\partial}{\partial b} \kappa \right)(a^*, 1) = \tfrac{1}{2} \log \tfrac{9}{5}. \tag{10.48}$$

Combine this with (10.44)(i) to get (10.44)(ii). $\square$

Proposition 10.10 says that localization occurs if and only if the free energy per step for the single linear interface exceeds the free energy per step for an AA-block by a certain positive amount. This excess is needed to *compensate* for the loss of entropy that occurs when the path runs along the interface for awhile before moving upwards from the interface to end at the diagonally opposite corner (recall Fig. 10.3). The constants $\frac{1}{2}\log 5$ and $\frac{1}{2}\log\frac{9}{5}$ are special to our model.

10.3.3 Qualitative Properties of the Critical Curve

● **Statement of Qualitative Properties**

With the help of Proposition 10.10, we get the following theorem identifying the qualitative properties of the supercritical phase diagram (see Fig. 10.9). The proof is deferred to Section 10.3.3.

Theorem 10.11. *Let $p \geq p_c$.*
(a) *For every $\alpha \geq 0$ there exists a $\beta_c(\alpha) \in [0,\alpha]$ such that the copolymer is*

$$\text{delocalized if } -\alpha \leq \beta \leq \beta_c(\alpha),$$
$$\text{localized} \quad \text{if } \beta_c(\alpha) < \beta \leq \alpha. \tag{10.49}$$

(b) *The function $\alpha \mapsto \beta_c(\alpha)$ is independent of p, continuous, non-decreasing and concave on $[0,\infty)$. There exists an $\alpha^* \in (0,\infty)$ such that*

$$\beta_c(\alpha) = \alpha \text{ if } \alpha \leq \alpha^*,$$
$$\beta_c(\alpha) < \alpha \text{ if } \alpha > \alpha^*, \tag{10.50}$$

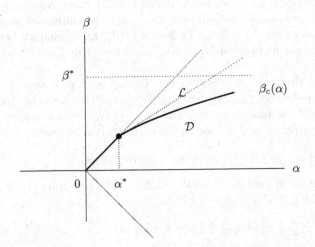

Fig. 10.9. Plot of $\alpha \mapsto \beta_c(\alpha)$ for $p \geq p_c$. Compare with Fig. 9.8.

and

$$\lim_{\alpha \downarrow \alpha^*} \frac{\alpha - \beta_c(\alpha)}{\alpha - \alpha^*} \in [0, 1). \tag{10.51}$$

Moreover, there exists a $\beta^* \in [\alpha^*, \infty)$ *such that*

$$\lim_{\alpha \to \infty} \beta_c(\alpha) = \beta^*. \tag{10.52}$$

The intuition behind Theorem 10.11 is as follows. Pick a point (α, β) inside $\mathcal{D}$. Then the polymer spends almost all of its time deep inside the A-blocks. Increase β while keeping α fixed. Then there will be a larger energetic advantage for the polymer to move some of its monomers from the A-blocks to the B-blocks by *crossing the interface inside the AB-block pairs*. There is some entropy loss associated with doing so, but if β is large enough, then the energy advantage will dominate, so that AB-localization sets in. The value at which this happens depends on α and is strictly positive. Since the entropy loss is finite, for α large enough the energy-entropy competition plays out not only below the diagonal, but also below a horizontal asymptote. On the other hand, for α small enough the loss of entropy dominates the energetic advantage, which is why the critical curve has a piece that lies on the diagonal. The larger the value of α the larger the value of β where AB-localization sets in. This explains why the critical curve moves to the right and up. The trivial inequality $\phi^{\mathcal{I}}(\mu) \leq \alpha + \hat{\kappa}(\mu)$ and the gap in Lemma 10.3(iv) are responsible for the linear piece of the critical curve. At the end of this linear piece the critical curve has a slope discontinuity, because the linear interface is already strictly inside its localized region.

• Proof of Qualitative Properties

In this section we prove Theorem 10.11. Recall (10.32). Since $S_{AA}(\alpha, \beta)$ does not depend on β by Lemma 10.4, and $\beta \mapsto S_{AB}(\alpha, \beta)$ is continuous and non-decreasing on $\mathbb{R}$ for every $\alpha \in \mathbb{R}$ by Lemma 10.7(ii), the boundary between $\mathcal{D}$ and $\mathcal{L}$ is a continuous function in CONE. We denote this function by $\alpha \mapsto \beta_c(\alpha)$.

We first show that the curve is concave. To that end, pick any $\alpha_1 < \alpha_2$, and consider the points $(\alpha_1, \beta_c(\alpha_1))$ and $(\alpha_2, \beta_c(\alpha_2))$ on the curve. Let (α_3, β_3) be the midpoint of the line connecting the two. We want to show that $\beta_3 \leq \beta_c(\alpha_3)$. By the convexity of f, stated in Theorem 10.1(a), we have

$$f(\alpha_3, \beta_3) \leq \tfrac{1}{2}[f(\alpha_1, \beta_c(\alpha_1)) + f(\alpha_2, \beta_c(\alpha_2))]. \tag{10.53}$$

Since the curve itself is part of $\mathcal{D}$ (recall (10.39)), it follows from (10.38) and (10.34)(i) that the right-hand side of (10.53) equals

$$\tfrac{1}{2} \left[\left(\tfrac{1}{2}\alpha_1 + \tfrac{1}{2}\log 5 \right) + \left(\tfrac{1}{2}\alpha_2 + \tfrac{1}{2}\log 5 \right) \right]$$
$$= \tfrac{1}{2} \left(\frac{\alpha_1 + \alpha_2}{2} \right) + \tfrac{1}{2}\log 5 = \tfrac{1}{2}\alpha_3 + \tfrac{1}{2}\log 5. \tag{10.54}$$

Thus, $f(\alpha_3, \beta_3) \leq \frac{1}{2}\alpha_3 + \frac{1}{2}\log 5$. But, by (10.34), the reverse inequality is true always, and so equality holds. Consequently, $(\alpha_3, \beta_3) \in \mathcal{D}$, which proves the claim that $\beta_3 \leq \beta_c(\alpha_3)$.

The concavity in combination with the lower bound in part (i) of the following lemma show that the curve is non-decreasing. This lemma settles most of Theorem 10.11.

Lemma 10.12. *Let $p \geq p_c$.*
(i) $\beta_c(\alpha) \geq \log(2 - e^{-\alpha})$ *for all $\alpha \geq 0$.*
(ii) $\beta_c(\alpha) < 8\log 3$ *for all $\alpha \geq 0$.*
(iii) $\beta_c(\alpha) = \alpha$ *for all $0 \leq \alpha \leq \alpha_0$, where $\alpha_0 \approx 0.125$ is implicitly defined by*

$$\sup_{\mu \geq 1} \mu \left[\hat{\kappa}(\mu) + \tfrac{1}{2}\alpha_0 - \tfrac{1}{2}\log 5 \right] = \tfrac{1}{2}\log \tfrac{9}{5}. \tag{10.55}$$

Proof. (i) We have, recalling (10.24–10.25),

$$\phi^{\mathcal{I}}(\mu) = \lim_{L \to \infty} \frac{1}{L} \log Z_L^{\omega, \mathcal{I}}(\mu) \quad \omega - a.s. \tag{10.56}$$

with

$$Z_L^{\omega, \mathcal{I}}(\mu) = \sum_{w \in \mathcal{W}_{\mu L, L}} \exp\left[-H_{\mu L}^{\omega, \mathcal{I}}(w) \right]$$

$$H_{\mu L}^{\omega, \mathcal{I}}(w) = -\sum_{i=1}^{\mu L} \left(\alpha \, 1_{\{\omega_i = A, (w_{i-1}, w_i) > 0\}} + \beta \, 1_{\{\omega_i = B, (w_{i-1}, w_i) \leq 0\}} \right). \tag{10.57}$$

We will derive an upper bound on $\phi^{\mathcal{I}}(\mu)$ by doing a so-called *first-order partial annealing estimate*, similar to the one we did in Section 9.4. This estimate is also referred to as a first-order Morita approximation (see Orlandini, Rechnitzer and Whittington [253] for a general account of Morita approximations). The estimate consists in writing

$$H_{\mu L}^{\omega, \mathcal{I}}(w) = -\sum_{i=1}^{\mu L} \alpha 1_{\{\omega_i = A\}}$$

$$-\sum_{i=1}^{\mu L} 1_{\{(w_{i-1}, w_i) \leq 0\}} \left(-\alpha 1_{\{\omega_i = A\}} + \beta 1_{\{\omega_i = B\}} \right), \tag{10.58}$$

using that the first term is $-\mu L \frac{1}{2}\alpha[1 + o(1)]$ ω-a.s. as $L \to \infty$ and is independent of w, substituting the latter into (10.57), and performing the expectation over ω. This gives

$$\mathbb{E}\left(\log Z_L^{\omega, \mathcal{I}}(\mu) \right)$$

$$\leq \mu L \tfrac{1}{2}\alpha[1 + o(1)] + \log \sum_{w \in \mathcal{W}_{\mu L, L}} \prod_{i=1}^{\mu L} 1_{\{(w_{i-1}, w_i) \leq 0\}} \left\langle e^{-\alpha 1_{\{\omega_i = A\}} + \beta 1_{\{\omega_i = B\}}} \right\rangle$$

$$\leq \mu L \tfrac{1}{2}\alpha[1 + o(1)] + \mu L[\hat{\kappa}(\mu) + o(1)] + \mu L \log\left(\tfrac{1}{2}e^{-\alpha} + \tfrac{1}{2}e^{\beta} \right), \tag{10.59}$$

where we use Jensen's inequality as well as the i.i.d. property of ω for the first inequality, and we use Lemma 10.3(i) for the second inequality. Consequently,

$$\phi^{\mathcal{I}}(\mu) = \lim_{L \to \infty} \frac{1}{\mu L} \mathbb{E}\left(\log Z_L^{\omega,\mathcal{I}}(\mu)\right) \leq \tfrac{1}{2}\alpha + \hat{\kappa}(\mu) + \log\left(\tfrac{1}{2}e^{-\alpha} + \tfrac{1}{2}e^{\beta}\right). \quad (10.60)$$

Suppose that

$$\log\left(\tfrac{1}{2}e^{-\alpha} + \tfrac{1}{2}e^{\beta}\right) \leq 0. \quad (10.61)$$

Then substitution of (10.60) into (10.42) gives

$$S_{AB} - S_{AA} \leq \sup_{\mu \geq 1, \nu \geq 1} \frac{\mu[\hat{\kappa}(\mu) - \tfrac{1}{2}\log 5] - \nu[\tfrac{1}{2}\log 5 - f(\nu)]}{\mu + \nu}. \quad (10.62)$$

But the right-hand side is the same as $S_{AB} - S_{AA}$ when $\alpha = \beta = 0$ (as can be seen from (10.42) because $\phi^{\mathcal{I}}(\mu) = \hat{\kappa}(\mu)$ when $\alpha = \beta = 0$), and therefore is equal to 0. Hence, recalling (10.14), we find that (10.61) implies that $(\alpha, \beta) \in \mathcal{D}$. Consequently,

$$\log\left(\tfrac{1}{2}e^{-\alpha} + \tfrac{1}{2}e^{\beta_c(\alpha)}\right) \geq 0 \qquad \text{for all } \alpha \geq 0, \quad (10.63)$$

which gives the lower bound that is claimed.

(ii) We will show that there exists a $\mu_0 > 1$ such that

$$\mu_0[\phi^{\mathcal{I}}(\mu_0) - S_{AA}] > \tfrac{1}{2}\log\tfrac{9}{5} \quad \text{for all } \alpha \geq 0 \text{ when } \beta \geq 8\log 3. \quad (10.64)$$

This will prove the claim via Proposition 10.10.

Consider the polymer along the single infinite interface $\mathcal{I}$. Fix ω. In ω, look for the strings of B's that are followed by a string of at least three A's. Call these B-strings "good", and call all other B-strings "bad". Let $w(\omega)$ be the path that starts at $(0,0)$, steps to $(0,1)$ and proceeds as follows. Each time a good B-string comes up, the path moves down from height 1 to height 0 during the step that carries the A just preceding the good B-string, moves at height 0 during the steps that carry the B's inside the string, moves up from height 0 to height 1 during the step that carries the first A after the string, and moves at height 1 during the step that carries the second A after the string. The third A can be used to either move from height 1 to height 0 in case the next good B-string comes up immediately, or to move at height 1 in case it is not. When a bad B-string comes up, the path stays at height 1.

Along $w(\omega)$, we have that all the A's lie in the upper halfplane, all the bad B-strings lie in the upper halfplane, while all the good B-strings lie in the interface. Asymptotically, the good B-strings contain $\tfrac{1}{4}$-th of the B's. Hence $\tfrac{1}{8}$-th of the steps carry a B that is in a good B-string. Moreover, the number of steps between heights 0 and 1 is $\tfrac{1}{8}$ times the number of steps at heights 0 and 1 (because the average length of a good B-string is 2), and so $w(\omega)$ travels a distance L in time $\tfrac{9}{8}L$ for L large, which corresponds to $\mu = \mu_0 = \tfrac{9}{8}$. Thus, the value of the Hamiltonian in (10.25) at the path $w(\omega)$ equals

$$H_{\mu_0 L}^{\omega,\mathcal{I}}(w(\omega)) = -\mu_0 L \left(\tfrac{1}{2}\alpha + \tfrac{1}{8}\beta\right) [1 + o(1)] \qquad \omega - a.s. \text{ as } L \to \infty. \quad (10.65)$$

Therefore, recalling (10.26), we have

$$\phi^{\mathcal{I}}(\mu_0) \geq \tfrac{1}{2}\alpha + \tfrac{1}{8}\beta. \qquad (10.66)$$

Via (10.38) this gives

$$\mu_0 [\phi^{\mathcal{I}}(\mu_0) - S_{AA}] \geq \mu_0 \left(\tfrac{1}{8}\beta - \tfrac{1}{2}\log 5\right). \qquad (10.67)$$

Consequently, the inequality in (10.64) holds as soon as

$$\tfrac{1}{8}\beta > \tfrac{8}{9}\log 3 + \tfrac{1}{18}\log 5. \qquad (10.68)$$

Since the right-hand side is strictly smaller than $\log 3$, this proves the claim.

(iii) Pick $\alpha = \beta$. Then $\phi^{\mathcal{I}}(\mu) \leq \alpha + \hat{\kappa}(\mu)$. If $\alpha \in [0, \alpha_0]$, with α_0 given by (10.55), then this bound in combination with (10.38) and Proposition 10.10 gives $S_{AB} = S_{AA}$. Thus, $\{(\alpha, \alpha) \colon \alpha \in [0, \alpha_0]\} \subset \mathcal{D}$. $\square$

Lemma 10.12, together with the concavity of $\alpha \mapsto \beta_c(\alpha)$ (shown prior to Lemma 10.12), proves Theorem 10.11, except for the slope discontinuity stated in (10.51). But the latter follows from the fact that if the piece of β_c on $[\alpha^*, \infty)$ is "analytically continued" outside CONE, then it hits the vertical axis at a strictly positive value, namely, α_0. Indeed, for $\alpha = 0$ we have $\phi^{\mathcal{I}}(\mu) = \tfrac{1}{2}\beta + \hat{\kappa}(\mu)$, because there is zero exponential cost for the path to stay in the lower halfplane (recall (10.24–10.26)). Consequently, the criterion for delocalization in Proposition 10.10, $S_{AB} = S_{AA}$, reduces to

$$\sup_{\mu \geq 1} \mu \left[\hat{\kappa}(\mu) + \tfrac{1}{2}\beta - \tfrac{1}{2}\log 5\right] \leq \tfrac{1}{2}\log \tfrac{9}{5}. \qquad (10.69)$$

This inequality holds precisely when $\beta \leq \alpha_0$ with α_0 defined in (10.55).

10.3.4 Finer Details of the Critical Curve

In den Hollander and Pétrélis [170] the following three theorems are proved, which complete the analysis of the phase diagram in Fig. 10.9.

Theorem 10.13. *Let $p \geq p_c$. Then $\alpha \mapsto \beta_c(\alpha)$ is strictly increasing on $[0, \infty)$.*

Theorem 10.14. *Let $p \geq p_c$. Then for every $\alpha \in (\alpha^*, \infty)$ there exist $0 < C_1 < C_2 < \infty$ and $\delta_0 > 0$ (depending on p and α) such that*

$$C_1 \delta^2 \leq f(\alpha, \beta_c(\alpha) + \delta; p) - f(\alpha, \beta_c(\alpha); p) \leq C_2 \delta^2 \qquad \forall \delta \in (0, \delta_0]. \quad (10.70)$$

Theorem 10.15. *Let $p \geq p_c$. Then $(\alpha, \beta) \mapsto f(\alpha, \beta; p)$ is infinitely differentiable throughout $\mathcal{L}$.*

Theorem 10.13 implies that the critical curve never reaches the horizontal asymptote, which in turn implies that $\alpha^* < \beta^*$ and that the slope in (10.51) is > 0. Theorem 10.14 shows that the phase transition along the critical curve in Fig. 10.9 is *second order off the diagonal*. In contrast, we know that the phase transition is *first order on the diagonal*. Indeed, the free energy equals $\frac{1}{2}\alpha + \frac{1}{2}\log 5$ on and below the diagonal segment between $(0,0)$ and (α^*, α^*), and equals $\frac{1}{2}\beta + \frac{1}{2}\log 5$ on and above this segment as is evident from interchanging α and β. Theorem 10.15 tells us that the critical curve in Fig. 10.9 is the *only* location in CONE where a phase transition of *finite* order occurs.

The proofs of Theorems 10.13–10.15 are technical. They rely on *perturbation arguments*, in combination with *exponential tightness of the excursions away from the interface* inside the localized phase, and are hard because we have two variational formulas to fight with: (10.17) and (10.27). For details we refer to [170].

Theorems 10.14–10.15 are the analogues of Theorems 9.11–9.12 for the single flat infinite interface. For the latter model, Theorem 9.11 says that the phase transition is *at least of second order*, i.e., only the quadratic upper bound is proved. As mentioned in Section 9.7, Challenge (1), numerical simulation seems to indicate that the order is higher, contrary to what we have here.

The mechanisms behind the phase transition in the two models are different. While for the single interface model the copolymer makes long excursions away from the interface and dips below the interface during a fraction of time that is at most of order δ^2, in our emulsion model the copolymer runs along the interface during a fraction of time that is of order δ, and in doing so stays close to the interface. Moreover, because near the critical curve for the emulsion model the single interface model is already strictly inside its localized phase, there is a variation of order δ in the single interface free energy as a constituent part of the emulsion free energy. Thus, the δ^2 in the emulsion model is the product of two factors δ, one coming from the time spent running along the interface and one coming from the variation of the constituent single interface free energy away from its critical curve.

The proof of Theorem 10.15 uses some of the ingredients of the proof of Theorem 9.12 in Giacomin and Toninelli [122]. However, in the emulsion model there is an extra complication, namely, the speed per step to move one unit of space forward may vary (because steps are up, down and to the right), while in the single interface model this is fixed at one (because steps are north-east and south-east). Consequently, the infinite differentiability with respect to this speed variable needs to be controlled too. This is handled by taking the Legendre transform w.r.t. this variable. Along the way we need to prove *uniqueness of maximizers* and *non-degeneracy of the Jacobian matrix at these maximizers*, in order to be able to apply implicit function theorems. For details we refer to den Hollander and Pétrélis [170].

10.4 Phase Diagram in the Subcritical Regime

In the subcritical regime the phase diagram is *much more complex* than in the supcritical regime. The reason is that *the oil does not percolate*, and so the copolymer no longer has the option of moving into the oil (in case it prefers to delocalize) nor of running along the interface between the oil and the water (in case it prefers to localize). Instead, it has to every now and then cross blocks of water, even though it prefers the oil. Below we list the main results, skipping the mathematics.

It turns out that there are *four phases*, separated by three critical curves, meeting at *two tricritical points*, all of which depend on p. The phase diagram is sketched in Fig. 10.10. For details on the derivation, we refer to den Hollander and Pétrélis [171]. The copolymer has the following behavior in the four phases of Fig. 10.10, as illustrated in Figs. 10.11–10.12:

- $\mathcal{D}_1$: *fully delocalized* into A-blocks and B-blocks, but never inside both blocks of a neighboring pair.
- $\mathcal{D}_2$: *fully delocalized* into A-blocks and B-blocks, and inside both blocks of a neighboring BA-pair.
- $\mathcal{L}_1$: *partially localized* near the interface in BA-pairs.
- $\mathcal{L}_2$: *partially localized* near the interface in both BA-pairs and AB-pairs.

This is to be compared with the much simpler behavior in the two phases of Fig. 10.9, as given by Fig. 10.8.

The intuition behind the phase diagram is as follows. In $\mathcal{D}_1$ and $\mathcal{D}_2$, β is not large enough to induce localization. In both types of block pairs, the reward for running along the interface is too small compared to the loss of entropy that comes with having to cross the block at a steeper angle. In $\mathcal{D}_1$, where α

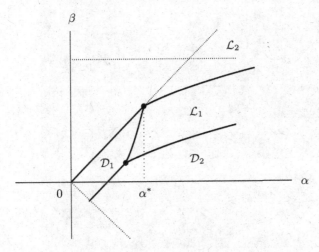

Fig. 10.10. Plot of the phase diagram for $p < p_c$. There are four phases, separated by three critical curves, meeting at two tricritical points.

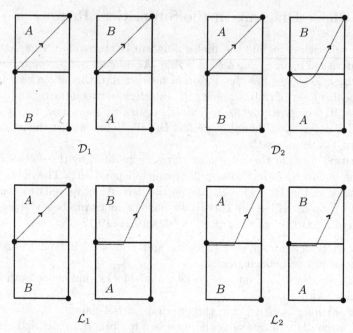

Fig. 10.11. Behavior of the copolymer, inside the four block pairs containing oil and water, for each of the four phases in Fig. 10.10.

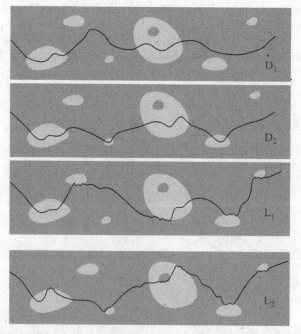

Fig. 10.12. Illustration of the four phases in Fig. 10.10 according to Fig. 10.11. Compare with Fig. 10.8.

and β are both small, the copolymer stays on one side of the interface in both types of block pairs. In $\mathcal{D}_2$, where α is larger, when the copolymer diagonally crosses a water block (which it has to do every now and then because the oil does not percolate), it dips into the oil block before doing the crossing. Since β is small, it still has no interest to localize. In $\mathcal{L}_1$ and $\mathcal{L}_2$, β is large enough to induce localization. In $\mathcal{L}_1$, where β is moderate, localization occurs in those block pairs where the copolymer crosses the water rather than the oil. This is because $\alpha > \beta$, making it more advantageous to localize away from water than away from oil. In $\mathcal{L}_2$, where β is large, localization occurs in both types of block pairs.

Note that the boundary between $\mathcal{D}_1$ and $\mathcal{D}_2$ is linear, as drawn in Fig. 10.10. This is because in $\mathcal{D}_1$ and $\mathcal{D}_2$ the free energy is a function of $\alpha - \beta$ only. This boundary extends above the horizontal because no localization can occur when $\beta \leq 0$. In $\mathcal{L}_1$ and $\mathcal{L}_2$ the free energy is no longer a function of $\alpha - \beta$. Note that there are *two tricritical points*, one that depends on p and one that does not.

Very little is known so far about the fine details of the four critical curves in the subcritical regime. The reason is that in none of the four phases does the free energy take on a simple form (contrary to what we saw in the supercritical regime, where the free energy is simple in the delocalized phase). In particular, in the subcritical regime there is no simple criterion like Proposition 10.10 to characterize the phases. In den Hollander and Pétrélis [171] it is shown that the phase transition between $\mathcal{D}_1$ and $\mathcal{D}_2$ has order ≤ 2, between $\mathcal{D}_1$ and $\mathcal{L}_1$ has order in $(1, 2]$, and between $\mathcal{D}_2$ and $\mathcal{L}_1$ has order at most that of the single linear interface (which has order ≥ 2 by the analogue of Theorem 9.11). It is further shown that the free energy is infinitely differentiable in the interior of $\mathcal{D}_1$ and $\mathcal{D}_2$. The same is believed to be true for $\mathcal{L}_1$ and $\mathcal{L}_2$, but a proof is missing.

It was argued in den Hollander and Whittington [172] that *the phase diagram is discontinuous at $p = p_c$*. Indeed, none of the three critical curves in the subcritical phase diagram in Fig. 10.10 converges to the critical curve in the supercritical phase diagram in Fig. 10.9. This is because *percolation or no percolation of the oil completely changes the character of the phase transition(s)*.

10.5 Extensions

(1) It is straightforward to extend the model to more than two types of monomers and solvents, say, $A, B, C, \ldots$, and to put parameters $\alpha, \beta, \gamma, \ldots$ in the Hamiltonian in (10.5) to reward the different matches. All that changes in the variational formula for the quenched free energy in (10.17) is that the sum over kl runs over all possible pairs drawn from $\{A, B, C, \ldots\}$. There are now more than two regimes of percolation, but the behavior is qualitatively similar.

(2) It is possible to relax the corner restriction in the definition of the path space $\mathcal{W}_{n,L_n}$ defined below (10.4), by allowing the copolymer to exit at the corner straight across the corner where it enters a pair of adjacent blocks. The result is that in (10.17) an additional block pair free energy ψ_{kl}^* is needed that records the contribution to f from such straight crossings. Also this free energy can be expressed in terms of crossing entropies and the linear interface free energy $\phi^{\mathcal{I}}$, just like the free energies ψ_{kl} were in Lemmas 10.4 and 10.6. The coarse-grained path W defined in Section 10.2.1 will now be able to make three rather than two steps, and the link with oriented percolation is therefore somewhat different. The resulting phase diagram becomes even richer. In particular, in the supercritical regime a second critical curve appears that is associated with the phase transition of the single linear interface.

10.6 Challenges

(1) Prove the analogues of Theorems 9.9–9.10 for a single block in the supercritical regime. This ought to be straightforward. Use these analogues to prove that the typical path indeed behaves as sketched in Figures 10.8 and 10.11.

(2) In the subcritical regime, the first and the second critical curve are understood only qualitatively. For instance, it has not been shown that these curves are monotone in α and p. For the third critical curve (between $\mathcal{L}_1$ and $\mathcal{L}_2$) the situation is worse: we do not even have a rigorous proof that the free energy is non-analytic at this curve, for the simple reason that it is not given by an explicit expression on either side of the curve. The challenge is to provide such a proof and thereby establish the existence of the phase transition.

(3) We know from Theorem 10.15 that in the supercritical regime the quenched free energy is infinitely differentiable on $\mathcal{L}$. Find out whether or not it is analytic. This is the analogue of Challenge (2) in Section 9.7. In the subcritical regime infinite differentiability has only been proved on the interior of $\mathcal{D}_1$ and $\mathcal{D}_2$ (see the end of Section 10.4), and the same challenge applies here.

(4) What happens when the block sizes become random? For instance, we may slice $\mathbb{Z}^2$ into rows and columns of random sizes $H_k L_n$, $k \in \mathbb{Z}$, and $V_k L_n$, $k \in \mathbb{Z}$, with $(H_k)_{k\in\mathbb{Z}}$ and $(V_k)_{k\in\mathbb{Z}}$ i.i.d. random variables taking values in $[C_1, C_2]$ for some $0 < C_1 < C_2 < \infty$. In the resulting blocks we may, as before, choose to place oil with probability p and water with probability $1 - p$, and ask what happens in the limit (10.8). We expect a qualitatively similar phase diagram, but the order of the phase transition is likely to depend on the law of H_1 and V_1 near C_1 and C_2. The challenge is to prove or disprove this. Compare with Section 7.7, Challenge (5).

(5) Investigate to what extent the behavior in the subcritical phase is qualitatively similar to that of the polymer in a random potential treated in Chapter 12. In both models the path has to hunt for "energetically favorable" locations in a random environment that are "not too far way".

(6) Attempt to remove the corner restriction in the definition of the path space $\mathcal{W}_{n,L_n}$ below (10.4) altogether. In the supercritical regime, does the phase diagram still have only one critical curve?

11

Random Pinning and Wetting of Polymers

In this chapter we look at *pinning* and *wetting* of a polymer by a *random substrate*. To that end, we think of the substrate as being composed of different types of atoms or molecules, occurring in a *random order*, and each time the polymer visits the substrate it picks up a reward or a penalty according to the type it encounters (see Fig. 11.1). Alternatively, we may think of a random copolymer consisting of different types of monomers, which interact differently with a homogeneous substrate. We will again be focusing on the phase transition between a *localized phase* and a *delocalized phase*. The results to be described below are an extension of those obtained for the homogeneous pinning model treated in Chapter 7.

The random pinning model is close in nature to the model of a copolymer near a linear selective interface described in Chapter 9. Therefore we will be able to re-use many of the ideas and techniques that were exposed in Chapter 9, allowing for a shortening of the proofs. In fact, in some sense the random pinning model is mathematically easier than the copolymer model. On the other hand, it has distinctive properties. One such property, which we will describe below, is that in some cases the disorder is *relevant*, meaning that the quenched critical curve is distinct from the annealed critical curve, while in other cases the disorder is *irrelevant*, meaning that the two critical curves coincide. This dichotomy does not occur for the copolymer model, where the disorder is always relevant. Another distinctive property is that the weak interaction limit of the random pinning model is trivial, in the sense that the disorder is not felt, whereas we saw that this is not the case for the copolymer model.

In Sections 11.1–11.5 we define the random pinning model, the quenched free energy and the quenched critical curve, and identify the qualitative properties of the critical curve and the two phases. This part runs parallel to Sections 9.1–9.5 for the copolymer near a linear selective interface. In Section 11.6 we focus on the issue of relevant vs. irrelevant disorder. In Section 11.7 we take a brief look at the extension of the results in Sections 11.1–11.6 to wetting. The wetting analogue may be used to describe the *denaturation transition*

F. den Hollander, *Random Polymers*,
Lecture Notes in Mathematics 1974, DOI: 10.1007/978-3-642-00333-2_11,
© Springer-Verlag Berlin Heidelberg 2009

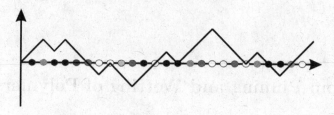

Fig. 11.1. A directed homopolymer near a linear interface with random disorder. Different shades of white, grey and black represent different values of the disorder.

of DNA with increasing temperature. This is done via the so-called Poland-Sheraga model, which views the two strands of DNA as polymers adsorbed onto each other. In Section 11.8 we consider what happens in the presence of a force, both for pinning and wetting, thus providing an extension of the results obtained in Section 7.3 for the homopolymer. The wetting version may be used to describe the *unzipping* of DNA under the action of a force: one strand of the DNA is pulled off the other strand as if it were a polymer adsorbed onto a substrate. Since the binding energies in an AT-pair and a CG-pair are different, and the order in which these pairs occur is more or less random, DNA can be modeled as random wetting with *binary disorder*.

The main techniques in the present chapter are the *subadditive ergodic theorem*, the *method of excursions* and *partial annealing estimates*.

11.1 A Polymer Near a Linear Penetrable Random Substrate: Pinning

For $n \in \mathbb{N}_0$, let

$$\mathcal{W}_n = \big\{w = (i, w_i)_{i=0}^n \colon w_0 = 0, \ w_i \in \mathbb{Z} \ \forall \, 0 \le i \le n\big\}, \qquad (11.1)$$

denote the set of all n-step directed paths in $\mathbb{N}_0 \times \mathbb{Z}$ starting from the origin. This set labels the configurations of the polymer where, as in (7.1), we allow for arbitrary increments. Let

$$\omega = (\omega_i)_{i\in\mathbb{N}} \text{ be i.i.d. with } \mathbb{P}(\omega_1 = +1) = \mathbb{P}(\omega_1 = -1) = \tfrac{1}{2}. \qquad (11.2)$$

This random sequence labels the order of the species along the interface $\mathbb{N} \times \{0\}$. Throughout the sequel, $\mathbb{P}$ denotes the law of ω. (In Section 11.9, Extension (3), we will comment on the extension to more general disorder.) As Hamiltonian we pick, for fixed ω,

$$H_n^\omega(w) = -\sum_{i=1}^n (\lambda \omega_i - h) \, 1_{\{w_i = 0\}}, \qquad w \in \mathcal{W}_n, \qquad (11.3)$$

with $(\lambda, h) \in \text{QUA}$ (recall (9.4)). Here, λ is the *disorder strength* and h is the *disorder bias*. Thus, a visit of the polymer to the substrate at time i contributes an energy $-(\lambda\omega_i - h)$. Note that, compared to (9.3), the parameter λ appears in front of the disorder rather than in front of the sum, and h appears with the opposite sign. This is the notation used in most of the literature.

The law of the copolymer given ω is denoted by P_n^ω and is defined as in (9.5), i.e.,

$$P_n^\omega(w) = \frac{1}{Z_n^\omega} e^{-H_n^\omega(w)} P_n(w), \qquad w \in W_n, \tag{11.4}$$

where, as in Section 7.1, P_n is the projection onto W_n of the path measure P of a directed irreducible random walk $S = (S_i)_{i \in \mathbb{N}_0}$. We will make the same assumptions on S as in (7.4–7.6), namely, that $b(k) = P(S_k = 0, S_i \neq 0 \; \forall 0 < i < k)$ satisfies $\sum_{k \in \mathbb{N}} b(k) = 1$ and

$$b(k) \sim k^{-1-a} L(k) \quad \text{as } k \to \infty \text{ through } \Lambda \tag{11.5}$$

for some $a \in (0, \infty)$ and some slowly varying function L, where $\Lambda = \{k \in \mathbb{N} \colon b(k) > 0\}$. As in Chapter 7, the fact that under the law P the successive visits of S to 0 form a *renewal process* will drive the computations, and the proofs will be valid for general renewals not necessarily arising from random walk as long as (11.5) is in force.

Comparing (11.1–11.3) with (7.1–7.2), we see that in the random pinning model the random sequence $(\lambda\omega_i - h)_{i \in \mathbb{N}}$ takes over the role of the parameter ζ in the homogeneous pinning model.

11.2 The Free Energy

Our starting point is the following analogue of Theorem 9.1.

Theorem 11.1. *For every $\lambda, h \in$ QUA,*

$$f(\lambda, h) = \lim_{n \to \infty} \frac{1}{n} \log Z_n^\omega \tag{11.6}$$

exists ω-a.s. and in $\mathbb{P}$-mean, and is constant ω-a.s.

Proof. The proof of Theorem 9.1 can be carried over almost verbatim. Indeed, consider the constrained partition sum

$$Z_{2n}^{\omega,0} = E\left(\exp\left[\sum_{i=1}^{2n}(\lambda\omega_i - h) 1_{\{S_i=0\}}\right] 1_{\{S_{2n}=0\}}\right). \tag{11.7}$$

Then $\omega \mapsto (\frac{1}{2n} \log Z_{2n}^\omega)_{n \in \mathbb{N}_0}$ is a superadditive random process, and so the analogue of Lemma 9.2 holds. Moreover, the same estimate as in Lemma 9.3 applies, showing that $Z_{2n}^{\omega,0}$ and Z_{2n}^ω differ by at most a polynomial factor. Finally, $|\log(Z_{2n}^\omega/Z_{2n+1}^\omega)| \leq \lambda + h$, which removes the restriction to even n. $\square$

11.3 The Critical Curve

11.3.1 The Localized and Delocalized Phases

The same argument as in (9.16–9.17) shows that

$$f(\lambda, h) \geq 0 \qquad \forall (\lambda, h) \in \text{QUA}, \tag{11.8}$$

with the lower bound corresponding to the polymer wandering away from the interface in either the upward or the downward direction. This leads us to the following analogue of Definition 9.4.

Definition 11.2. *We say that the polymer is:*
(i) *localized if $f(\lambda, h) > 0$,*
(ii) *delocalized if $f(\lambda, h) = 0$.*
We write

$$\begin{aligned} \mathcal{L} &= \{(\lambda, h) \in \text{QUA}\colon f(\lambda, h) > 0\}, \\ \mathcal{D} &= \{(\lambda, h) \in \text{QUA}\colon f(\lambda, h) = 0\}, \end{aligned} \tag{11.9}$$

to denote the localized and the delocalized region in QUA.

11.3.2 Existence of a Non-trivial Critical Curve

The quenched free energy $f(\lambda, h)$ is continuous and convex in (λ, h) (recall Section 1.3), and non-increasing in h for every λ. Moreover, $f(0, h)$ is the free energy of the homogeneous pinning model (with $\zeta = -h$).

The analogue of Theorem 9.5 reads as follows (see Fig. 11.2).

Theorem 11.3. *For every $\lambda \in [0, \infty)$ there exists an $h_c(\lambda) \in [0, \infty)$ such that the polymer is*

$$\begin{aligned} &\text{localized} &&\text{if } 0 \leq h < h_c(\lambda), \\ &\text{delocalized} &&\text{if } h \geq h_c(\lambda). \end{aligned} \tag{11.10}$$

Moreover, $\lambda \mapsto h_c(\lambda)$ is continuous, strictly increasing and convex on $[0, \infty)$, with $h_c(0) = 0$ and $\lim_{\lambda \to \infty} h_c(\lambda)/\lambda = 1$.

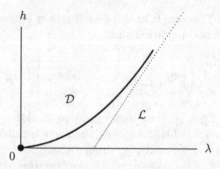

Fig. 11.2. Plot of $\lambda \mapsto h_c(\lambda)$.

Proof. The fact that $\mathcal{L}$ and $\mathcal{D}$ are separated by a single critical curve $\lambda \mapsto h_c(\lambda)$, defined by $h_c(\lambda) = \inf\{h \in \mathbb{R} \colon f(\lambda, h) = 0\}$, is immediate from the fact that $h \mapsto f(\lambda, h)$ is non-increasing for every $\lambda \in [0, \infty)$. Since the level sets of a convex function are convex, $\mathcal{D} = \{(\lambda, h) \in \mathrm{QUA} \colon f(\lambda, h) \leq 0\}$ is a convex set and h_c is a convex function.

Since the critical threshold for pinning in the homogeneous model is 0, we have that $h_c(0) = 0$. Theorem 11.5 below states that $h_c(\lambda) > 0$ for $\lambda > 0$. Together with the convexity this yields that $\lambda \mapsto h_c(\lambda)$ is strictly increasing on $[0, \infty)$. Theorem 11.4 below implies that $h_c(\lambda)/\lambda \leq 1$. It therefore remains to show that $\liminf_{\lambda \to \infty} h_c(\lambda)/\lambda \geq 1$.

To that end, note that

$$Z_n^\omega \geq E\left(\mathrm{e}^{(\lambda - h)|I_n^\omega|} 1_{\{S \in \mathcal{W}_n^\omega\}}\right), \tag{11.11}$$

where

$$\begin{aligned} I_n^\omega &= \{i \in \{1, \ldots, n\} \colon \omega_i = 1\}, \\ \mathcal{W}_n^\omega &= \{w \in \mathcal{W}_n \colon w_i = 0 \ \forall i \in I_n^\omega, \ w_i \neq 0 \ \forall i \in \{1, \ldots, n\} \backslash I_n^\omega\}. \end{aligned} \tag{11.12}$$

We have $\lim_{n \to \infty} \frac{1}{n} I_n^\omega = \frac{1}{2}$ ω-a.s. Moreover, $\lim_{n \to \infty} \frac{1}{n} \log P(\mathcal{W}_n^\omega) = \Sigma$ with $\Sigma = \sum_{k \in \Lambda} 2^{-(k+1)} \log b(k)$. It therefore follows from (11.6) and (11.11) that

$$f(\lambda, h) \geq \tfrac{1}{2}(\lambda - h) + \Sigma. \tag{11.13}$$

But $\Sigma \in (-\infty, 0)$ by (11.5). Therefore, for any $\delta > 0$, we have $f(\lambda, h) > 0$ when $\lambda > -2\Sigma/\delta$ and $h \leq (1-\delta)\lambda$. Consequently, $\liminf_{\lambda \to \infty} h_c(\lambda)/\lambda \geq 1-\delta$. Let $\delta \downarrow 0$ to get the claim. $\square$

The mean value of the disorder is $\mathbb{E}(\lambda \omega_1 - h) = -h$. Thus, we see from Fig. 11.2 that for the random pinning model localization may even occur for *moderately negative* mean values of the disorder, contrary to what we found in Section 7.2 for the homogeneous pinning model, where localization occurs only for $\zeta > 0$. In other words, even a globally repulsive random interface can pin the polymer: all that the polymer needs to do is to hit the positive values of the disorder and avoid the negative values as much as possible.

11.4 Qualitative Properties of the Critical Curve

11.4.1 Upper Bound

The analogue of Theorem 9.6 yields the following upper bound for h_c. This upper bound is the critical curve for the *annealed* model.

Theorem 11.4. $h_c(\lambda) \leq h_c^{\mathrm{ann}}(\lambda)$ *with* $h_c^{\mathrm{ann}}(\lambda) = \log \cosh(\lambda)$ *for all* $\lambda \geq 0$.

Proof. Compute, as in (9.28),

$$f(\lambda, h) = \lim_{n\to\infty} \frac{1}{n} \mathbb{E}\left(\log E\left(\exp\left[\sum_{i=1}^n (\lambda\omega_i - h) 1_{\{S_i = 0\}} \right] \right) \right)$$

$$\leq \lim_{n\to\infty} \frac{1}{n} \log E\left(\mathbb{E}\left(\exp\left[\sum_{i=1}^n (\lambda\omega_i - h) 1_{\{S_i = 0\}} \right] \right) \right) \qquad (11.14)$$

$$= \lim_{n\to\infty} \frac{1}{n} \log E\left(\prod_{i=1}^n \left[\tfrac{1}{2} e^{\lambda - h} + \tfrac{1}{2} e^{-\lambda - h} \right]^{1_{\{S_i = 0\}}} \right).$$

The right-hand side is ≤ 0 as soon as the term between square brackets is ≤ 1. Consequently,

$$h > \log\cosh(\lambda) \quad \Longrightarrow \quad f(\lambda, h) = 0, \qquad (11.15)$$

which yields the desired upper bound on $h_c(\lambda)$. □

It is natural to wonder whether it is possible to improve the upper bound in Theorem 11.4 by doing partial annealing estimates, i.e., by doing a first-order or higher-order Morita approximation (see e.g. Orlandini, Rechnitzer and Whittington [253]). The answer is no. As already noted in Extension (3) in Section 9.6, Caravenna and Giacomin [54] have shown that such approximations do not lead to an improvement. In fact, we will see in Section 11.2 that the upper bound in Theorem 11.4 is sharp in some cases.

11.4.2 Lower Bound

There is no analogue of Theorem 9.6. Indeed, this lower bound comes from strategies where the polymer dips below the interface during rare long stretches in ω where the empirical mean is sufficiently biased (see Fig. 9.5). Such strategies no longer work for the random pinning model, because the path only picks up a reward or a penalty *at* the interface. The following lower bound for h_c is proved in Alexander and Sidoravicius [6].

Theorem 11.5. $h_c(\lambda) > 0$ *for* $\lambda > 0$.

Proof. We follow the proof given in Giacomin [116], Section 5.2. This proof has the advantage that it quantifies the lower bound. In what follows, we first restrict ourselves to the subclass of excursion laws $b(\cdot)$ for which $b(1), b(2) > 0$ (recall (7.7)). At the end of the proof we indicate how to remove this restriction.

The key idea is to rewrite Z_n^ω as an expectation w.r.t. P of an exponential weight factor in which the disorder has a strictly positive expectation under $\mathbb{P}$. Once that is done, a homogeneous localization strategy will yield the desired result.

To formulate the rewrite, let

$$\chi = \frac{b(1)^2}{b(1)^2 + b(2)}, \qquad \widehat{\omega}_i = \widehat{\omega}_i(\lambda, h) = \log\left[\chi\, e^{\lambda\omega_i - h} + (1 - \chi)\right], \quad i \in \mathbb{N}.$$
(11.16)

Let $e(\lambda, h) = \mathbb{E}(\widehat{\omega}_1)$ and note that, because $\chi \in (0, 1)$,

$$\exists \delta > 0 \colon \forall \lambda > 0 \; \exists h_0 = h_0(\lambda, \delta) > 0 \colon \qquad e(\lambda, h) > \delta \; \forall h \le h_0. \qquad (11.17)$$

For $i \in \mathbb{N}$, abbreviate $\delta_i = 1_{\{S_i = 0\}}$ and $\widehat{\delta}_i = 1_{\{S_{i-1} = S_{i+1} = 0\}} 1_{\{i \text{ odd}\}}$.

Lemma 11.6. *For $n \in 2\mathbb{N}$,*

$$Z_n^\omega = E\left(\exp\left[\sum_{i=1}^n (\lambda\omega_i - h)\delta_i(1 - \widehat{\delta}_i) + \sum_{i=1}^n \widehat{\omega}_i\widehat{\delta}_i\right]\right). \qquad (11.18)$$

Proof. Fix $n \in 2\mathbb{N}$. Let

$$\mathcal{I}_n = \{i \in \{1, 3, \dots, n-3, n-1\} \colon S_{i-1} = S_{i+1} = 0\}. \qquad (11.19)$$

Number the points in $\{0\} \cup (\cup_{i \in \mathcal{I}_n}\{i-1, i+1\}) \cup \{n\}$ as $0 = n_1 \le n_2 \le \dots n_{k-1} \le n_k = n$ with $k = 2|\mathcal{I}_n| + 2$ (which includes repetitions). Note that $n_{2j-1} \le n_{2j}$ and $n_{2j+1} = n_{2j} + 2$ for $j = 1, \dots, |\mathcal{I}_n|$, and that $\mathcal{I}_n = \{n_{2j} + 1 \colon j = 1, \dots, |\mathcal{I}_n|\}$. Further note that all the n_j's are even, while all the elements of $\mathcal{I}_n$ are odd.

Partition the set of possible excursions according to the values taken by $\mathcal{I}_n$. Then, for $A \subset \{1, 3, \dots, n-3, n-1\}$, we have

$$E\left(\exp\left[\sum_{i=1}^n (\lambda\omega_i - h)\delta_i\right] 1_{\{\mathcal{I}_n = A\}}\right)$$

$$= \left(\prod_{j=1}^{|A|} \widehat{Z}_{n_{2j-1}, n_{2j}}^{\omega, 0}\left[b(1)^2\, e^{\lambda\omega_{n_{2j}+1} - h} + b(2)\right] e^{\lambda\omega_{n_{2j}+1} - h}\right) \widehat{Z}_{n_{2|A|+1}, n}^\omega,$$
(11.20)

where

$$\widehat{Z}_{u,v}^\omega = E\left(\exp\left[\sum_{i=u+1}^v (\lambda\omega_i - h)\delta_i\right] 1_{\{\widehat{\delta}_{u+1} = \dots = \widehat{\delta}_{v-1} = 0\}}\right), \quad u, v \in \mathbb{N}_0, \; u \le v,$$
(11.21)

and $\widehat{Z}_{u,v}^{\omega,0}$ is the same with an extra indicator $1_{\{\delta_n = 1\}}$ under the expectation. The term between square brackets in the r.h.s. of (11.20) comes from the fact that the sites in A may or may not contain a renewal time.

By (11.16), we have

$$b(1)^2\, e^{\lambda\omega_{n_{2j}+1} - h} + b(2) = [b(1)^2 + b(2)]\, e^{\widehat{\omega}_{n_{2j}+1}}. \qquad (11.22)$$

Substituting this into (11.20) and summing over A, we obtain the relation in (11.18). Indeed, the effect of the substitution in (11.22) is that, when $\widehat{\delta}_i = 1$, the contribution of the disorder at site $n_{2j} + 1$ changes from $e^{\lambda \omega_{n_{2j}+1} - h}$ to $e^{\widehat{\omega}_{n_{2j}+1}}$. $\quad\square$

Put $M_b = \sum_{k \in \mathbb{N}} k b(k)$. For the remainder of the proof we distinguish two cases: (I) $M_b < \infty$ and (II) $M_b = \infty$.

(I) Estimate, using Jensen's inequality and Lemma 11.6,

$$
\frac{1}{n} \mathbb{E}(\log Z_n^\omega) \geq \frac{1}{n} \mathbb{E}\left(E\left[\sum_{i=1}^n (\lambda \omega_i - h)\delta_i (1 - \widehat{\delta}_i) + \sum_{i=1}^n \widehat{\omega}_i \widehat{\delta}_i \right] \right)
$$
$$
= -h \frac{1}{n} \sum_{i=1}^n E(\delta_i) + h \frac{1}{n} \sum_{i=1}^n E(\delta_i \widehat{\delta}_i) + e(\lambda, h) \frac{1}{n} \sum_{i=1}^n E(\widehat{\delta}_i).
$$
(11.23)

By the standard renewal theorem, we have

$$
\lim_{n \to \infty} \frac{1}{n} \sum_{i=1}^n E(\delta_i) = \frac{1}{M_b}.
$$
(11.24)

Since

$$
E(\widehat{\delta}_{2i-1}) = P(S_{2i-2} = S_{2i} = 0) = E(\delta_{2i-2})\left[b(1)^2 + b(2) \right],
$$
$$
E(\delta_{2i-1}\widehat{\delta}_{2i-1}) = P(S_{2i-2} = S_{2i-1} = S_{2i} = 0) = E(\delta_{2i-2})\, b(1)^2,
$$
(11.25)

the statement in (11.24) implies that

$$
\lim_{n \to \infty} \frac{1}{n} \sum_{i=1}^n E(\widehat{\delta}_i) = \frac{b(1)^2 + b(2)}{2M_b}, \qquad \lim_{n \to \infty} \frac{1}{n} \sum_{i=1}^n E(\delta_i \widehat{\delta}_i) = \frac{b(1)^2}{2M_b}, \qquad (11.26)
$$

where the factor $\frac{1}{2}$ arises because $\widehat{\delta}_i = 0$ when i is even. Therefore, letting $n \to \infty$ in (11.23) and using (11.24–11.26), we obtain

$$
f(\lambda, h) = \lim_{n \to \infty} \frac{1}{n} \mathbb{E}(\log Z_n^\omega)
$$
$$
\geq \frac{1}{M_b} \left[e(\lambda, h) \frac{b(1)^2 + b(2)}{2} + h \left(\frac{b(1)^2}{2} - 1 \right) \right].
$$
(11.27)

This in turn yields

$$
e(\lambda, h) > h \frac{2 - b(1)^2}{b(1)^2 + b(2)} \quad \Longrightarrow \quad (\lambda, h) \in \mathcal{L}, \qquad (11.28)
$$

which proves the claim because of (11.17).

(II) The lower bound in (11.28) is useless when $M_b = \infty$. However, we may repeat the argument in (I) after tilting $b(\cdot)$. To that end, define, for $r > 0$ (compare with (7.18)),

$$b^r(k) = \frac{e^{-rk}b(k)}{B(r)}, \qquad k \in \mathbb{N}, \tag{11.29}$$

where $B(r) = \sum_{k \in \mathbb{N}} e^{-rk}b(k) < \infty$. Write P^r for the path measure corresponding to the excursion law $b^r(\cdot)$. We will exploit the inequality

$$\log \int e^X \, d\mu = \log \int e^{X - \log(d\nu/d\mu)} \, d\nu \geq \int X \, d\nu - R(\nu|\mu) \tag{11.30}$$

$$\forall \mu, \nu, X: \ \mu \ll \nu, \nu \ll \mu, \int |X| d\nu < \infty,$$

where $R(\nu|\mu) = \int \log(d\nu/d\mu) \, d\nu \in [0, \infty]$ denotes the relative entropy of μ w.r.t. ν. Using (11.30) with $\mu = P$, $\nu = P^r$ and X the r.h.s. of (11.18), we may again use Lemma 11.6 to estimate

$$\mathbb{E}(\log Z_n^\omega) \geq \mathbb{E}\left(E_r \left[\sum_{i=1}^{n} (\lambda \omega_i - h)\delta_i(1 - \widehat{\delta}_i) + \sum_{i=1}^{n} \widehat{\omega}_i \widehat{\delta}_i \right] \right) - R(P_n^r|P_n). \tag{11.31}$$

Therefore, letting $n \to \infty$, inserting the relation

$$\lim_{n \to \infty} \frac{1}{n} R(P_n^r|P_n) = R(r) \text{ with } R(r) = -r + \frac{[-\log B(r)]}{M_{b^r}} \tag{11.32}$$

(see Giacomin [116], Appendix B1; note that $M_{b^r} = \mathrm{d}[-\log B(r)]/dr$), and using the analogues of (11.24–11.26) for P^r, we obtain

$$f(\lambda, h) \geq \frac{1}{M_{b^r}} \left[e(\lambda, h) \frac{b^r(1)^2 + b^r(2)}{2} + h \left(\frac{b^r(1)^2}{2} - 1 \right) \right] - R(r). \tag{11.33}$$

Now let $r \downarrow 0$, note that $R(r) = O(r)$, and use that, by (11.5), $M_{b^r} = O(r^{-1+a})$, $a > 0$, up to a slowly varying correction. Then (11.33) once again yields (11.28). This completes the proof of Theorem 11.5 when $b(1), b(2) > 0$.

The condition $b(1), b(2) > 0$ is less restrictive than it seems. For instance, for SRW we have $\Lambda = \{k \in \mathbb{N}: b(k) > 0\} = 2\mathbb{N}$, and so the polymer does not see the disorder on the odd-numbered sites. The random pinning problem therefore is the same as for excursions with length distribution $\bar{b}(\cdot)$ given by $\bar{b}(k) = b(2k)$, $k \in \mathbb{N}$, for which $\bar{b}(1), \bar{b}(2) > 0$.

We finish by explaining what to do when $b(1) = 0$ and $b(2), b(3) > 0$. The sketch given below easily generalizes to arbitrary Λ.

Let $\lambda_1, \lambda_2, \lambda_3, h \geq 0$, and define

$$H_n^\omega(w) = -\sum_{i=1}^{n} \left(\lambda_{1+(i-1)(\mathrm{mod}\,3)}\omega_i - h \right) 1_{\{w_i=0\}}, \qquad w \in \mathcal{W}_n. \tag{11.34}$$

This Hamiltonian gives weight λ_j to the disorder at the sites in $j + 3\mathbb{N}_0$, $j = 1, 2, 3$. Let $f(\lambda_1, \lambda_2, \lambda_3, h)$ denote the associated quenched free energy. For every $h \geq 0$, this function is convex and symmetric in $\lambda_1, \lambda_2, \lambda_3$, and hence

$$f(\lambda_1, \lambda_2, \lambda_3, h) \geq f(0, 0, \lambda_3, h). \tag{11.35}$$

This in turn implies that $h_c(\lambda_1, \lambda_2, \lambda_3) \geq h_c(0, 0, \lambda_3)$ for the critical surface separating the localized phase from the delocalized phase (compare with Fig. 11.2). Picking $\lambda_1 = \lambda_2 = \lambda_3 = \lambda$ to recover the original Hamiltonian in (11.3), we find that $h_c(\lambda) \geq h_c(0, 0, \lambda)$, and so it suffices to prove that $h_c(0, 0, \lambda) > 0$ for $\lambda > 0$.

The Hamiltonian in (11.34) with $\lambda_1 = \lambda_2 = 0$ has disorder only at the sites in $3\mathbb{N}$. We can therefore repeat the argument given above for the case $b(1), b(2) > 0$, replacing the even sites by $6\mathbb{N}$, the odd sites by $3 + 6\mathbb{N}$, $b(1)^2$ by $b(3)^2$, and $b(2)$ by $b(2)^3 e^{-2h}$. Indeed, $b(2)^3 e^{-2h}$ is the contribution to the partition sum coming from 3 excursions of length 2 that hit the substrate at distances 2 and 4 but miss the substrate at distance 3. Consequently, in (11.16) we get $\chi = \chi(h) = b(3)^2/[b(3)^2 + b(2)^3 e^{-2h}]$, but (11.17) continues to hold, and the rest of the argument is the same. $\square$

The proof of Theorem 11.5 yields that (see (11.16–11.17) and (11.28))

$$h_c(\lambda) \geq c_1 \lambda^2, \qquad \lambda \in [0, c_2]. \tag{11.36}$$

Together with the upper bound in Theorem 11.4 this implies that

$$h_c(\lambda) \asymp \lambda^2, \qquad \lambda \downarrow 0. \tag{11.37}$$

Pétrélis [264] gives an alternative proof of Theorem 11.5 for the case of SRW by proving (11.36).

11.4.3 Weak Interaction Limit

Pétrélis [265] proves for SRW that

$$\lim_{a \downarrow 0} a^{-2} f(a\lambda, ah) = \widetilde{f}(h) \qquad \forall \lambda \geq 0, \tag{11.38}$$

where $\widetilde{f}(h)$ is the free energy of the space-time continuous Hamiltonian $H_t(B) = -h L_t(B)$, where $B = (B_s)_{s \geq 0}$ is standard Brownian motion. Thus, the disorder *vanishes* in the weak interaction limit, which is in sharp contrast with what we found in Section 9.4.3 for the copolymer near a selective interface. Since $\widetilde{f}(h) > 0$ for all $h > 0$ (which is the Brownian analogue of what we found in Theorem 7.2 for the homopolymer), it follows from (11.38) that $\lim_{\lambda \downarrow 0} h_c(\lambda)/\lambda = 0$, which is in agreement with (11.37). In Section 11.6 we will see that, actually, $\lim_{\lambda \downarrow 0} h_c(\lambda)/\lambda^2 = \lim_{\lambda \downarrow 0} h_c^{\mathrm{ann}}(\lambda)/\lambda^2 = \frac{1}{2}$.

11.5 Qualitative Properties of the Phases

Theorems 9.9–9.12 carry over to the random pinning model. We refer to Giacomin and Toninelli [120], [122], [123], [124] for proofs. Similarly as in in Chapter 9, the localization strategies that are used to obtain lower bounds on the free energies are those where the polymer hits the substrate at stretches carrying an atypically high density of +1's in ω. This allows for a comparison of the free energy at different value of the parameters, which is needed to carry out the proper perturbation arguments close to the critical curve.

The fact that the phase transition is at least of second order shows that the disorder has a tendency of *smoothing*. Indeed, for the homogeneous pinning model we saw in Section 7.1.3 that the order of the phase transition depends on the tail exponent a, namely, the order is ≥ 2 when $a > \frac{1}{2}$ and < 2 when $a < \frac{1}{2}$. The smoothing is in accordance with what is called the *Harris criterion* (Harris [149]):

- "Arbitrary weak disorder modifies the nature of a phase transition when the order of the phase transition in the non-disordered system is < 2".

11.6 Relevant versus Irrelevant Disorder

In this section we address the question:

- When is the upper bound in Theorem 11.4 sharp?

We will see that the answer depends on the excursion length distribution, i.e., on the exponent a and the slowly varying function $L(\cdot)$ in (11.5), and possibly also on λ.

- The disorder is said to be *relevant* when $h_c(\lambda) < h_c^{\mathrm{ann}}(\lambda)$ and *irrelevant* when $h_c(\lambda) = h_c^{\mathrm{ann}}(\lambda)$.

Relevant disorder means that the ability of the quenched polymer to mimic the behavior of the annealed polymer breaks down close to the annealed critical curve. Theorem 11.7 below summarizes results derived in Alexander [5], Toninelli [293], [294], Giacomin and Toninelli [125], Derrida, Giacomin, Lacoin and Toninelli [85], Alexander and Zygouras [7], [8], and Giacomin, Lacoin and Toninelli [119]. The exponent $a = \frac{1}{2}$ turns out to be critical, in accordance with the Harris criterion stated above (see Fig. 11.3), and the exponent $a = 0$ is included as well.

Theorem 11.7. *Subject to (11.5), the disorder is:*
- *relevant*
(1) *for all $\lambda > 0$ when $a \in (\frac{1}{2}, \infty)$;*
(2) *for all $\lambda > 0$ when $a = \frac{1}{2}$ and $L(\infty) = 0$ or $\lim_{K \to \infty}[L(K)]^{-1} \sum_{k=1}^{K} k^{-1}[L(k)]^{-2} = \infty$;*

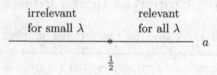

Fig. 11.3. Relevant vs. irrelevant disorder as summarized in Theorem 11.7. At the threshold value $a = \frac{1}{2}$ the behavior is delicate (marginally relevant vs. irrelevant disorder). At $a = 0$ the disorder is irrelevant for all λ.

- *irrelevant*
  (3) *for small $\lambda > 0$ when $a = \frac{1}{2}$ and $\sum_{k \in \mathbb{N}} k^{-1}[L(k)]^{-2} < \infty$;*
  (4) *for small $\lambda > 0$ when $a \in (0, \frac{1}{2})$;*
  (5) *for all $\lambda > 0$ when $a = 0$.*

Proof. A complete proof would take up far too much space. We give a brief sketch of the underlying heuristics, explaining why the case $a = \frac{1}{2}$ is critical. The following argument is taken from Toninelli [296]. For simplicity the disorder is taken to be standard Gaussian instead of binary. In that case, $f^{\mathrm{ann}}(\lambda, h) = f^{\mathrm{homo}}(\frac{1}{2}\lambda^2 - h)$, with $f^{\mathrm{homo}}(h)$ the free energy of the homopolymer with parameter h (i.e., (7.1–7.2) with $\zeta = h$), implying that $h_c^{\mathrm{ann}}(\lambda) = \frac{1}{2}\lambda^2$.

We begin by writing down the identity

$$f_n(\lambda, h) = \frac{1}{n}\,\mathbb{E}(\log Z_n^\omega) = f_n^{\mathrm{homo}}(\Delta) + R_n(\lambda, \Delta), \tag{11.39}$$

where $\Delta = \frac{1}{2}\lambda^2 - h$,

$$R_n(\lambda, \Delta) = \frac{1}{n}\log\mathbb{E}\left(E_n^\Delta\left(\exp\left[\sum_{i=1}^n (\lambda\omega_i - \tfrac{1}{2}\lambda^2)\,1_{\{S_i = 0\}}\right]\right)\right), \tag{11.40}$$

and E_n^Δ denotes expectation w.r.t. the path measure P_n^Δ of the homopolymer of length n with parameter Δ (i.e., (7.3) with $\zeta = \Delta$). Note that *irrelevance* of the disorder for small λ amounts to the second term in the r.h.s. of (11.39) being *much smaller* than the first term. Formally expanding the r.h.s. of (11.40) in λ for n, Δ fixed, we get

$$R_n(\lambda, \Delta) = -\frac{1}{2}\lambda^2 \frac{1}{n}\sum_{i=1}^n \left[P_n^\Delta(S_i = 0)\right]^2 + O_\Delta(\lambda^3), \qquad \lambda \downarrow 0, \tag{11.41}$$

where the notation O_Δ means that the error term may depend on Δ. Since

$$\lim_{n \to \infty} \frac{1}{n}\sum_{i=1}^n P_n^\Delta(S_i = 0) = \frac{\partial f^{\mathrm{homo}}(\Delta)}{\partial \Delta} \tag{11.42}$$

and $P_n^\Delta(S_i = 0)$ is essentially independent of i as long as $1 \ll i \ll n$, we obtain from (11.39–11.42) that

$$f\left(\lambda, \tfrac{1}{2}\lambda^2 - \Delta\right) = f^{\text{homo}}(\Delta) - \frac{1}{2}\lambda^2 \left[\frac{\partial f^{\text{homo}}(\Delta)}{\partial \Delta}\right]^2 + O_\Delta(\lambda^3), \quad \lambda \downarrow 0. \quad (11.43)$$

What this relation does is approximate the quenched free energy close to the annealed critical curve in terms of the free energy of the homopolymer. By Theorem 7.4, we have

$$f^{\text{homo}}(\Delta) \sim \Delta^{1 \vee (1/a)} L^*(1/\Delta), \quad \frac{\partial f^{\text{homo}}}{\partial \Delta}(\Delta) \sim \Delta^{0 \vee (1-a)/a} L^{**}(1/\Delta), \quad \Delta \downarrow 0, \quad (11.44)$$

with L^* and L^{**} slowly varying functions. Inserting this into (11.43), we see that the second term in (11.43) is much smaller than the first term if and only if $1/a < 2(1-a)/a$, i.e., $a \in (0, \tfrac{1}{2})$. Therefore, this is the range of a-values for which the disorder is irrelevant.

It is straightforward to extend the above argument to arbitrary disorder with finite moment generating function (see Section 11.9, Extension (3)). The l.h.s. of (11.43) then reads $f(\lambda, h_c^{\text{ann}}(\lambda) - \Delta)$. See [296] for details.

With the help of interpolation and replica coupling techniques, it is shown in Giacomin and Toninelli [125] that for cases (3) and (4) the expansion in (11.43) holds true, even in the sharper form

$$f\left(\lambda, h_c^{\text{ann}}(\lambda) - \Delta\right) = f^{\text{homo}}(\Delta) - \frac{1}{2}\lambda^2 \left[\frac{\partial f^{\text{homo}}(\Delta)}{\partial \Delta}\right]^2 [1 + O(1)], \quad \lambda, \Delta \downarrow 0. \quad (11.45)$$

From this it is deduced that $f(\lambda, h_c^{\text{ann}}(\lambda) - \Delta)$ and $f^{\text{homo}}(\Delta)$ are comparable when $\lambda, \Delta \downarrow 0$, implying that $h_c^{\text{ann}}(\lambda) = h_c(\lambda)$ for λ sufficiently small.

Cases (1) and (2) are dealt with in Derrida, Giacomin, Lacoin and Toninelli [85], and in Alexander and Zygouras [7]. The main idea in [85] is the following. Choose h close to $h_c(\lambda)$ such that the *annealed system is localized*, i.e., $h < h_c^{\text{ann}}(\lambda)$. Consider the quenched system with n of the order of the correlation length in the annealed system with $n = \infty$, and show that $\mathbb{E}([Z_n^\omega]^\gamma)$ is small for some appropriate $\gamma \in (0, 1)$ depending on a. After that show that, as n is increased, this fractional moment does not grow appreciably. The latter implies that the *quenched system is delocalized*, i.e., $h \geq h_c(\lambda)$, and so it follows that $h_c^{\text{ann}}(\lambda) > h_c(\lambda)$.

Case (5) is settled in Alexander and Zygouras [8]. □

Not surprisingly, the critical exponent $a = \tfrac{1}{2}$ is the most delicate. The physics literature contains conflicting statements. For instance, Forgacs, Luck, Nieuwenhuizen and Orland [104] claim irrelevance for small $\lambda > 0$ (for Gaussian disorder and ballot paths), Derrida, Hakim and Vannimenus [87] claim relevance for all $\lambda > 0$ (for arbitrary mean zero disorder and generalized ballot paths), while Gangardt and Nechaev [110] predict irrelevance for all $\lambda > 0$ (for binary disorder and ballot paths). Ballot paths and generalized ballot paths correspond to $a = \frac{1}{2}$ and $L(\infty) \in (0, \infty)$. Case (2) of Theorem 11.7 therefore shows that in all these examples the disorder is actually relevant.

For the cases of relevant disorder, bounds on the gap $h_c^{\mathrm{ann}}(\lambda) - h_c(\lambda)$ have been derived in the papers cited prior to Theorem 11.7. As $\lambda \downarrow 0$, this gap decays like

$$h_c^{\mathrm{ann}}(\lambda) - h_c(\lambda) \asymp \begin{cases} \lambda^2, & \text{if } a \in (1, \infty), \\ \lambda^2 \psi(1/\lambda), & \text{if } a = 1, \\ \lambda^{2a/(2a-1)}, & \text{if } a \in (\frac{1}{2}, 1), \end{cases} \qquad (11.46)$$

with $\psi(\cdot)$ slowly varying and vanishing at infinity when $L(\infty) \in (0, \infty)$.

Partial results are known for $a = \frac{1}{2}$. For instance, when $\sum_{k \in \mathbb{N}} k^{-1}[L(k)]^{-2} = \infty$ the gap decays faster than any polynomial, which implies that the disorder can at most be *marginally relevant*, a situation where standard perturbative arguments cannot work. When $L(\infty) \in (0, \infty)$, the gap lies between $\exp[-\lambda^{-4}]$ and $\exp[-\lambda^{-2}]$ for $\lambda > 0$ small enough, modulo constants in the exponent. When $L(k) = O([\log k]^{-\frac{1}{2}-\theta})$, $k \to \infty$, $\theta > 0$, the gap lies above $\exp[-\lambda^{-\theta'}]$ for all $\theta' < \theta$ and $\lambda > 0$ small enough. Both cases correspond to marginal relevance.

The more the stretches where the quenched polymer hits the interface look like the typical disorder, the closer $h_c(\lambda)/h_c^{\mathrm{ann}}(\lambda)$ is to 1. Thus, a moderate gap means that the depinning transition is fairly uniform until close to the critical curve, while a small gap means that it is fairly uniform all the way to the critical curve. The above mentioned results on the gap imply that $\lim_{\lambda \downarrow 0} h_c(\lambda)/h_c^{\mathrm{ann}}(\lambda) = 1$ when $a \in [0, 1]$ and < 1 when $a \in (1, \infty)$.

In the physics literature, relevant disorder is used both for a change in the critical curve *and* a change in the order of the phase transition. As we see from the above, the two come together, except possibly in the marginally relevant case.

11.7 A Polymer Near a Linear Impenetrable Random Substrate: Wetting

Essentially all the results stated in Sections 11.1–11.6 carry over from pinning to wetting. There is the same parallel here as between Sections 7.1 and 7.2. For details see e.g. the review paper by Toninelli [296]. Garel and Monthus [112] provides extensive numerics for the critical behavior in the wetting model.

DNA is a string of adenine-thymine and cytosine-guanine base pairs forming a double helix. A and T share two hydrogen bonds, C and G share three. If we think of the two strands as performing random walks in two-dimensional space subject to the restriction that they do not cross each other, then the distance between the two strands is a random walk in the presence of a wall. This representation of DNA is called the Poland-Sheraga model (see reference [268] and Fig. 11.4). The localized phase $\mathcal{L}$ corresponds to the bounded

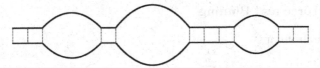

Fig. 11.4. Schematic representation of the two strands of DNA in the Poland-Sheraga model. The dotted lines are the interacting base pairs, the loops are the denaturated segments without interaction.

phase of DNA, where the two strands are attached, the delocalized phase $\mathcal{D}$ corresponds to the denaturated phase, where the two strands are detached. Since the order of the base pairs in DNA is irregular and their binding energies are different, we may think of DNA as a wetted copolymer with binary disorder. The order of the base pairs will of course not be i.i.d., but the comparison with the wetting model is reasonable for a qualitative description. If we want to allow for mismatches between the bases and assign them zero binding energy, then we have ternary disorder. Upon heating, the hydrogen bonds that keep the base pairs together can break and the two strands can separate, either partially or completely. This is called *denaturation*. See Cule and Hwa [79] and Kafri, Mukamel and Peliti [202] for background.

It is not realistic to presume that the two strands can be modeled as SRWs. However, the theory of wetting allows for a general excursion law, as long as it satisfies the regularity property in (11.5). Hence, we may attempt to pick an excursion law that approximates the true spatial behavior of DNA strands, one that takes into account for instance the self-avoidance *within* the denaturated segments. There is an extended literature on this subject (see e.g. Kafrie, Mukamel and Peliti [202], Richard and Guttmann [270]). A typical range of values for the tail exponent a of the excursion law used for DNA is $a \in [1.1, 1.2]$. In [202] it is claimed that the denaturation transition is first-order. This indicates that the Poland-Sheraga model has its limitations, because we know from Section 11.5 that the presence of disorder is smoothing the denaturation transition to second order or higher. Perhaps the DNA molecules that are used in experiments are not long enough ($n \leq 10^6$) for the asymptotic analysis ($n \to \infty$) to apply, or the self-avoidance *between* the different denaturated segments plays a role.

11.8 Pulling a Polymer off a Substrate by a Force

The results in Section 7.3, pertaining to what happens when a force is applied to a homopolymer that is pinning or wetting a substrate, can be extended to the disordered setting. What is described below is taken from Giacomin and Toninelli [121].

11.8.1 Force and Pinning

Our Hamiltonian is

$$H_n^\omega = -\lambda \sum_{i=1}^n \omega_i \, 1_{\{w_i=0\}} - \phi w_n, \qquad w \in \mathcal{W}_n, \tag{11.47}$$

with $\lambda, \phi > 0$ and $\mathcal{W}_n$ the set of paths in (11.1) (compare with (7.48)). It is not hard to prove that the quenched free energy

$$f(\lambda, \phi) = \lim_{n\to\infty} \frac{1}{n} \log Z_n^\omega \tag{11.48}$$

exists ω-a.s. and is self-averaging. The following is the analogue of Theorem 7.6, where we restricted ourselves to the special case where the reference random walk can only make steps of size ≤ 1, namely, $P(S_1 = -1) = P(S_1 = 1) = \frac{1}{2}p$ and $P(S_1 = 0) = 1 - p$ for some $p \in (0, 1)$, as in (7.50).

Theorem 11.8. *For every $\lambda, \phi > 0$, the free energy exists and is given by*

$$f(\lambda, \phi) = f(\lambda) \vee g(\phi), \tag{11.49}$$

with $f(\lambda)$ the free energy of the pinned copolymer without force, and

$$g(\phi) = \log \left[p \cosh(\phi) + (1 - p) \right]. \tag{11.50}$$

Proof. The proof is similar to that of Theorem 7.6, with $f(\lambda)$ taking over the role of the free energy of the homopolymer. For details we refer to [121].

The critical force is given by the formula

$$\phi_c(\lambda) = g^{-1}\big(f(\lambda)\big) \tag{11.51}$$

with g^{-1} the inverse of g. As for homopolymers in Section 7.3, we may put $\lambda = 1/T$, $F = \phi/T$, $F_c(T) = \phi_c(1/T)/T$ as in (7.62), and attempt to plot $T \mapsto F_c(T)$. Since we do not have a closed form expression for $f(\lambda)$, this cannot be done in full detail. However, the qualitative properties of the force-temperature diagram can be deduced by looking at the asymptotics of $f(\lambda)$ for $\lambda \downarrow 0$, respectively, $\lambda \to \infty$. Our main point of interest here is to see whether the force-diagram is *re-entrant*, i.e., whether $T \mapsto F_c(T)$ goes through a minimum or a maximum, at least for some values of p. Precisely when re-entrant behavior occurs depends on the fine details of the model, as we already saw in Section 7.3 for the homopolymer.

It is shown in [121] that

$$F_c(T) = \begin{cases} 0, & \text{as } T \to \infty, \\ \frac{1}{2} + T \left[\frac{1}{2} \sum_{k\in\mathbb{N}} 2^{-k} \log b_p(k) - \log(\frac{p}{2}) \right] + o(T), & \text{as } T \downarrow 0, \end{cases}$$

$$\tag{11.52}$$

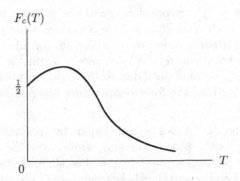

Fig. 11.5. Plot of $T \mapsto F_c(T)$ for p small.

with $b_p(\cdot)$ the excursion length distribution of the random walk with parameter p. Numerically this gives re-entrant behavior as soon as $p < 0.8006$ (see Fig. 11.5). The second line in (11.52) comes from the fact that, in the strong interaction limit, the free energy is dominated by paths that visit the substrate precisely at those i with $\omega_i = 1$ (compare with (11.11–11.13)). Note that the first line gives limit 0 instead of $1/p$ found for the homopolymer in (7.64). This is because in the weak interaction limit the free energy $f(\lambda)$ for the copolymer vanishes faster than the free energy $f^{\text{homo}}(\zeta)$ for the homopolymer, namely, $f(\lambda) = O(\lambda^4)$, $\lambda \downarrow 0$, compared to $f^{\text{homo}}(\zeta) \asymp \zeta^2$, $\zeta \downarrow 0$. Indeed, this follows from the annealed bound $f(\lambda) \leq f^{\text{homo}}(\log \cosh(\lambda))$ (recall (11.14)).

11.8.2 Force and Wetting

The extension to wetting is straightforward. The analogue of Theorem 7.7 can be guessed by looking at Theorem 11.8: simply replace $f(\lambda)$ by $f^+(\lambda)$. It is again possible to determine when the phase transition is re-entrant. In [121] it is shown that

$$F_c^+(T) = \begin{cases} 0, & \text{as } T \to \infty, \\ \frac{1}{2} + T \left[\frac{1}{2} \sum_{k \in \mathbb{N}} 2^{-k} \log b_p^+(k) - \log(\frac{p}{2}) \right] + o(T), & \text{as } T \downarrow 0, \end{cases}$$
(11.53)

with $b_p^+(1) = b_p(1)$ and $b_p^+(k) = \frac{1}{2}b_p(k)$, $k \in \mathbb{N}\setminus\{1\}$. Numerically this gives re-entrant behavior as soon as $p < 0.7162$.

Iliev, Orlandini and Whittington [181] compute the free energy of the wetted polymer with force in the first-order Morita approximation, i.e., by averaging Z_n^ω over ω conditioned on $n^{-1} \sum_{i=1}^n \omega_i = 0$. Even though this does not give the correct quenched free energy, it is argued that the force-temperature diagram is qualitatively similar. An argument is given why the quenched critical force is bounded from above by the critical force in the Morita approximation, which is explicitly calculable. Both ballot paths ($p = 1$) and generalized ballot paths ($p = \frac{2}{3}$) are considered, and the disorder is taken to be binary with $\mathbb{P}(\omega_1 = 1) = \rho$ and $\mathbb{P}(\omega_1 = 0) = 1 - \rho$ for some $\rho \in [0, 1]$. The force-temperature diagram is found to be re-entrant for all $\rho \in (0, 1)$. The reason is that at zero temperature the 1-monomers lie on the substrate, while the 0-monomers may or may not. Hence, the copolymer has a strictly positive entropy, implying that the force-temperature diagram has a strictly positive slope at $T = 0$.

The two strands of DNA can be pulled apart by applying a force to the end of one strand (e.g. with optical tweezers), while anchoring the end of the other strand to some physical support (e.g. a glass slide). In this way an *unzipping transition* can be observed. For background, see Lubensky and Nelson [229], Marenduzzo, Trovato and Maritan [234], who argue in favor of a re-entrant force-temperature phase diagram. Danilowicz, Kafri, Conroy, Coljee, Weeks and Prentiss [80] describe an experiment with double-stranded lambda-phage DNA at temperatures ranging from 15 to 50 degrees Celcius. Only in the range from 24 to 35 degrees Celcius do the measurements agree with the predictions of the theory based on the Poland-Sheraga model.

11.9 Extensions

(1) Similarly as noted in Section 7.6, Extension (1), for the homopolymer pinning model, all results for the random pinning model carry over when the reference random walk S is transient, by normalization of $b(\cdot)$. Alexander and Sidoravicius [6] show that the lower bound in Theorem 11.5 extends to an arbitrary excursion length distribution and an arbitrary disorder under only minor regularity conditions. The main result of [6] is that *disorder strictly enhances pinning*, i.e., $h_c(\lambda) > h_c^{\mathrm{homo}}(\lambda)$ for all $\lambda > 0$, the latter being the critical curve of the homopolymer whose Hamiltonian is obtained from (11.3) after replacing the disorder by its average value. Moreover, it is shown that $\lim_{\lambda \to \infty} h_c(\lambda)/\lambda = \infty$ when the disorder is unbounded, and that $h_c(\lambda) = \infty$ for all $\lambda > 0$ when the disorder does not have a finite exponential moment and the tail of the excursion length distribution is bounded from below by a power law with exponent < -1.

(2) Alexander [4] looks at the order of the phase transition when the disorder is standard Gaussian and the excursions are either recurrent or transient with a length distribution that has an exponential tail. It turns out that in the transient case the disorder is relevant for all $\lambda > 0$, and both the quenched and the annealed free energy have a phase transition of order < 2. In the recurrent case the behavior depends in a subtle way on the subexponential prefactors modulating the exponential decay, and various scenarios are possible. This, once more, shows that the issue of relevance vs. irrelevance is highly delicate.

(3) With the exception of the analogue of Theorem 9.11, i.e., the fact that the phase transition is second order or higher, all the results described in

Sections 11.1–11.8 extend to the situation where the ω_i's are $\mathbb{R}$-valued with a moment generating function that is finite on the positive halfline. For details we refer the reader to Giacomin [116], Chapter 5. In most of the papers cited prior to Theorem 11.7 proofs are written out for Gaussian disorder only, and it is stated "somewhat loosely" that proofs carry over to arbitrary disorder. In principle this is straightforward, although the critical exponent $a = \frac{1}{2}$ (marginally relevant vs. irrelevant disorder) requires attention. As in Section 9.6, Extension (1), the analogue of Theorem 9.11 has been proved for bounded disorder and for continuous disorder subject to a mild entropy condition (Giacomin and Toninelli [122], [123]).

It is shown in Toninelli [294] that any disorder that is *unbounded from above* is relevant for λ large enough, no matter what the choice of $a \in (0, \infty)$ and $L(\cdot)$ is. This supplements parts (1) and (2) of Theorem 11.7.

(4) Bolthausen, Caravenna and de Tilière [30] consider the limit when the disorder is strong but has a low density (referred to as a reduced wetting model). They take as Hamiltonian $H_n(w) = \lambda \sum_{i=1}^{n} \omega_i \, 1_{\{w_i=0\}}$, with the ω_i's i.i.d. $\{0, 1\}$-valued such that $\mathbb{P}(\omega_1 = 1) = e^{-c\lambda}$ and $\mathbb{P}(\omega_1 = 0) = 1 - e^{-c\lambda}$ for some $c \in (0, \infty)$, and let $\lambda \to \infty$. They find that for ballot paths (SRW with $a = \frac{1}{2}$) the value $c = \frac{2}{3}$ is the critical threshold for localization, in the sense that $f_c(\lambda)$, the quenched free energy, is > 0 for λ large when $c < \frac{2}{3}$ and $= 0$ for λ large when $c > \frac{2}{3}$. The proof is based on a renormalization argument, showing that the main contribution to the free energy comes from those paths that hit the substrate precisely at those i with $\omega_i = 1$. Toninelli [294] gives a simpler proof of the same result based on fractional moment estimates of the partition sum, and shows that the critical threshold is $c = 1/(1 + a)$ when the excursion length distribution has exponent $a \in (0, \infty)$.

(5) Toninelli [291] obtains finite-size estimates for the depinning transition. It is found that, for any $a \in (0, \infty)$ and $\lambda > 0$, if $h = h_n$ varies with the length n of the polymer such that $\lim_{n \to \infty} n^\chi [h_n - h_c(\lambda)] = c$ with $\chi \geq 0$ and $c \in \mathbb{R}$, then the number of visits of the polymer to the substrate is $O(n^{\frac{1}{3}} \log n)$ for $\chi \geq \frac{1}{3}$, $O(n^{2\chi} \log n)$ for $0 \leq \chi < \frac{1}{3}$ and $c > 0$, and $O(n^{1-\chi})$ for $0 \leq \chi < \frac{1}{3}$ and $c < 0$.

(6) Toninelli [291], [292] studies the correlation length in the pinning and wetting model in the localized phase. Let

$$\text{Cov}_l^\omega(k) = P_\infty^\omega(S_l = S_{l+k} = 0) - P_\infty^\omega(S_l = 0) \, P_\infty^\omega(S_{l+k} = 0), \qquad (11.54)$$

where P_∞^ω is the weak limit of the path measure P_n^ω as $n \to \infty$. For the wetting model with SRW, it is shown that, uniformly in $l \in \mathbb{N}$,

$$\lim_{k \to \infty} \frac{1}{k} \log \text{Cov}_l^\omega(k) = -f(\lambda, h),$$

$$\lim_{k \to \infty} \frac{1}{k} \log \mathbb{E}\big(\text{Cov}_l^\omega(k)\big) = -\mu(\lambda, h),$$

$$(11.55)$$

where

$$\mu(\lambda, h) = -\lim_{n \to \infty} \frac{1}{n} \log \mathbb{E}(1/Z_n^\omega) \qquad \omega - a.s. \qquad (11.56)$$

For wetting with a random walk different from SRW, and for pinning, only upper bounds are derived. What (11.55) says is that the *quenched* and the *average quenched* correlation length are controlled by the free energy, respectively, the "inverted free energy". It is shown in Giacomin and Toninelli [124] that $\mu(\lambda, h) < f(\lambda, h)$ throughout the localized phase, and in Giacomin and Toninelli [125] that $\mu(\lambda, h) \asymp f(\lambda, h)$ as $h \uparrow h_c(\lambda)$ for every $\lambda > 0$ when the disorder is irrelevant (recall cases (3) and (4) in Theorem 11.7).

The quantity $\mu(\lambda, h)$ plays a central role in the estimate of the largest gap between two successive hits of the substrate. Namely, it is shown in [124] that, in the localized phase,

$$\lim_{n \to \infty} P_n^\omega \left(\left| \frac{\mathrm{maxgap}_n(T)}{\log n} - \frac{1}{\mu(\lambda, h)} \right| > \epsilon \right) = 0 \quad \forall \epsilon > 0 \quad \text{in } \mathbb{P}\text{-probability}.$$

$$(11.57)$$

This is to be compared with what we found for the homopolymer in the proof of Theorem 7.3(a).

(7) Caravenna, Giacomin and Zambotti [57], [58] derive the scaling properties of the path measure for periodic disorder. The results are similar to those mentioned in Section 7.6, Extension (2).

(8) Cheliotis and den Hollander [65] derive a *variational formula* for the critical curve $\lambda \mapsto h_c(\lambda)$ in Theorem 11.3. The derivation is based on a quenched large deviation principle obtained by Birkner, Greven and den Hollander [17] for the empirical process of "words read off from a random letter sequence by an independent renewal process". The variational formula leads to a *necessary and sufficient* criterion for relevant disorder in terms of positivity of a certain relative entropy. From this criterion it is deduced that, for any choice of the disorder distribution and the excursion length distribution, there is a *critical* value $\lambda_c \in [0, \infty]$ such that the disorder is irrelevant for $\lambda \in [0, \lambda_c]$ and relevant for $\lambda \in (\lambda_c, \infty)$. Moreover, upper and lower bounds are derived for λ_c.

(9) Janvresse, de la Rue and Velenik [196] provide a necessary and sufficient condition for localization on a *single sample* of the environment. The Hamiltonian is (11.3) with $\lambda > 0$ and $h = 0$, and it is shown that

$$\liminf_{n \to \infty} E_n^\omega \left(\frac{1}{n} \sum_{i=1}^n \omega_i \, 1_{\{w_i=0\}} \right) > 0 \quad \Longleftrightarrow \quad \liminf_{n \to \infty} \frac{1}{n} \sum_{i=1}^n \omega_i > 0. \qquad (11.58)$$

(10) Pétrélis [264] looks at the model with Hamiltonian

$$H_n^\omega(w) = \lambda \sum_{i=1}^{n} (1 + s\eta_i) 1_{\{w_i = 0\}} + h \sum_{i=1}^{n} \text{sign}(w_{i-1}, w_i), \qquad (11.59)$$

with $\lambda, h, s \geq 0$ and $\eta = (\eta_i)_{i \in \mathbb{N}}$ an i.i.d. sequence such that η_1 has zero mean. This model describes the random pinning of a hydrophobic homopolymer at an interface between oil and water in which random droplets of a third solvent are present. For ballot paths (SRW) it is shown that the quenched free energy $f(\lambda, h, s)$ exists, and has a localized phase with $f(\lambda, h, s) > h$ and a delocalized phase with $f(\lambda, h, s) = h$. The critical curve $\lambda \mapsto h_c(\lambda, s)$ is finite, strictly increasing and convex on $[0, \lambda_c(s))$, for some $\lambda_c(s) \in (0, \infty)$, and is infinite on $(\lambda_c(s), \infty)$. It is not known whether $h_c(\lambda_c(s), s)$ is finite or not. Trivial bounds are

$$h_c(\lambda, 0) \leq h_c(\lambda, s) \leq h_c \left(\lambda + \log \mathbb{E}(e^{\lambda s \eta_1}), 0 \right), \qquad (11.60)$$

with $h_c(\lambda, 0) = -\frac{1}{4} \log[1 - 4(1 - e^{-\lambda})^2]$ the critical curve for the homopolymer (see Chapter 7). This in turn yields bounds on $\lambda_c(s)$, namely, $\bar{\lambda}_c(s) \leq \lambda_c(s) \leq \log 2$ with $\bar{\lambda}_c(s)$ the solution of the equation $\lambda + \log \mathbb{E}(e^{\lambda s \eta_1}) = \log 2$. In [264] it is shown that if η_1 has finite variance, then there exist $c_1, c_2 > 0$ such that

$$h_c(\lambda, s) \geq -\frac{1}{4} \log \left[1 - 4 \left(1 - e^{-\lambda - c_2 s^2 \lambda^2} \right)^2 \right], \qquad (11.61)$$

$$s \in [0, c_1], \ \lambda \in [0, \log 2 - c_2 s^2 \lambda^2).$$

Together with the upper bound in (11.60), this implies that $h_c(\lambda, s) \asymp \lambda^2$, $\lambda \downarrow 0$. The bounds in (11.60–11.61) are drawn in Fig. 11.6.

(11) Pétrélis [265] studies a combination of the models considered in Chapters 9 and 11 when the interface has a finite width, namely,

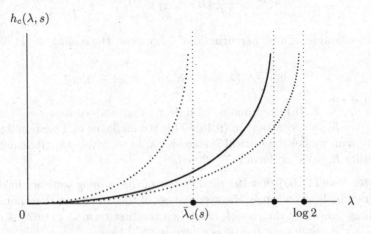

Fig. 11.6. The thick dotted lines are the upper and lower bound in (11.60). The thick drawn line is the lower bound in (11.61). The critical curve $\lambda \mapsto h_c(\lambda, s)$ lies somewhere in between the top two lines.

$$H_n^{\omega,\eta}(w) = -\lambda \sum_{i=1}^{n} (\omega_i + h) \,\mathrm{sign}(w_{i-1}, w_i) - \bar\lambda \sum_{i=1}^{n} \sum_{j=-M}^{M} (\eta[M]_i)_j \, 1_{\{w_i = j\}},$$

$$(11.62)$$

where $\lambda, \bar\lambda, h \geq 0$, $\omega = (\omega_i)_{i\in\mathbb{N}}$ is an i.i.d. sequence of random variables with a symmetric distribution whose variance equals 1, $M \in \mathbb{N}$, and $\eta[M] = (\eta[M]_i)_{i\in\mathbb{N}}$ is an i.i.d. sequence of random $(2M+1)$-vectors, independent of ω, whose components have a finite moment generating function. It is shown for ballot paths (SRW) that the quenched free energy $f(\lambda, \bar\lambda, h)$ exists, is non-analytic at a critical surface $(\lambda, \bar\lambda) \mapsto h_c(\lambda, \bar\lambda)$, and satisfies a weak scaling property similar to (9.46–9.48), namely,

$$\lim_{a\downarrow 0} a^{-2} f(a\lambda, a\bar\lambda, ah) = \widetilde{f}(\lambda, \bar\lambda\Sigma, h) \qquad (11.63)$$

with

$$\Sigma = \sum_{j=-M}^{M} \mathbb{E}\big((\eta[M]_1)_j\big) \qquad (11.64)$$

the average of the sum of the components of $\eta[M]_1$, $f(\lambda, \bar\lambda, h)$ the free energy of the space-time continuous Hamiltonian

$$H_t^b(B) = -\lambda \int_0^t (\mathrm{d}b_s + h \,\mathrm{d}s)\,\mathrm{sign}(B_s) - \bar\lambda L_t(B), \qquad (11.65)$$

$B = (B_s)_{s\geq 0}$ and $b = (b_s)_{s\geq 0}$ two independent standard Brownian motions, and $L_t(B)$ the local time of B at 0. The path measure is

$$\frac{\mathrm{d}P_t^b}{\mathrm{d}P}(B) = \frac{1}{Z_t^b}\, e^{-H_t^b(B)}. \qquad (11.66)$$

After a "stable against perturbations" argument, the scaling in (11.63) yields that

$$\lim_{\lambda\downarrow 0} \frac{1}{\lambda} h_c(\lambda, u\lambda) = K_c(u), \qquad u \in [0, \infty), \qquad (11.67)$$

with $u \mapsto K_c(u)$ continuous, non-decreasing and convex on $[0, \infty)$, and $K_c(0) = K_c$, the constant in (9.45). This the analogue of Theorem 9.8. In the continuum model the critical surface is $(\lambda, \bar\lambda) \mapsto \lambda K_c(\bar\lambda/\lambda)$. It is not known whether $K_c(u) < \infty$ for all $u \in [0, \infty)$.

We see from (11.65) that the randomness of the pinning vanishes in the weak interaction limit, whereas the randomness of the copolymer does not. For the pinning version of the model, i.e., when the first term in (11.62) is replaced by $-h \sum_{i=1} \mathrm{sign}(w_{i-1}, w_i)$, it is shown in [265] that

$$\widetilde{f}(0, \bar\lambda, h) = h \vee \tfrac{1}{2}\left(\lambda^2 + \tfrac{h^2}{\lambda^2}\right) \qquad (11.68)$$

and

$$\lim_{\bar{\lambda}\downarrow 0} \frac{1}{\bar{\lambda}^2} h_c(0,\bar{\lambda}) = \Sigma^2. \tag{11.69}$$

This describes the weak scaling limit of a homopolymer consisting of hydrophobic monomers near an oil-water interface of finite width with random pinning.

(12) Wüthrich [322] considers the modification of the pinning model in which the Hamiltonian is (compare with (11.3))

$$H_n^{\omega,\eta}(w) = -\lambda \sum_{i=1}^{n} (\omega_i + h)\, 1_{\{w_i=0\}}\, 1_{\{\eta_i=1\}}, \qquad w \in \mathcal{W}_n, \tag{11.70}$$

with $(\lambda, h) \in \text{QUA}$, $\omega = (\omega_i)_{i\in\mathbb{N}}$ as in (11.2), and $\eta = (\eta_i)_{i\in\mathbb{N}}$ a second i.i.d. sequence with $\mathbb{P}(\eta_1 = 1) = p$ and $\mathbb{P}(\eta_1 = -1) = 1 - p$, $p \in (0,1)$. This describes a copolymer consisting of hydrophobic and hydrophilic monomers (labeled by ω) in a medium consisting of water in the plane and droplets of oil located at random positions on the horizontal axis (labeled by η). The $(1 + d)$-version of the directed model is studied. It is proved that there is a critical line $\lambda \mapsto h_c^p(\lambda)$, separating a localized phase from a delocalized phase, which is continuous and non-decreasing on $(0, \infty)$. It is further shown that the distance between the endpoint of the copolymer and the interface is of order 1 in the localized phase and of order $\sqrt{n}$ deep inside the delocalized phase.

(13) Lacoin [225] considers pinning on a class of hierarchical lattices. Pick $b, s \in \mathbb{N}\backslash\{1\}$, and let D_0 be the graph with two vertices, A and B, connected by a single edge. Define D_n, $n \in \mathbb{N}_0$, iteratively as follows: to get D_{n+1}, replace each edge in D_n by b branches in parallel, each with s edges in series. For $n \in \mathbb{N}_0$, the path space $\mathcal{W}_n$ consists of all directed paths from A to B, of length s^n. The interface is chosen to be a reference path $w^* \in \mathcal{W}_n$, and the Hamiltonian is

$$H_n^{\omega}(w) = \sum_{i=1}^{s^n-1} (\lambda\omega_i - h)\, 1_{\{w_i=w_i^*\}}, \qquad w \in \mathcal{W}_n. \tag{11.71}$$

For the case $b \in (1, s)$ (the case $b \in (s, \infty)$ can be dealt with via duality), the following is shown: (1) $b < \sqrt{s}$: the disorder is relevant for all $\lambda > 0$, with gap

$$h_c^{\text{ann}}(\lambda) - h_c(\lambda) \asymp \lambda^{2a/(2a-1)}, \qquad \lambda \downarrow 0, \tag{11.72}$$

where $a = \log(s/b)/\log s \in (\frac{1}{2}, 1)$; (2) $b > \sqrt{s}$, the disorder is irrelevant for small $\lambda > 0$; (3) $b = \sqrt{s}$: the disorder is marginally relevant for small $\lambda > 0$, with a gap that lies between $\exp[-\lambda^{-4}]$ and $\exp[-\lambda^{-2}]$, modulo constants in the exponent. Giacomin, Lacoin and Toninelli [118], [119] consider a different class of hierarchical lattices, with bond rather than site disorder, and obtain similar results. Earlier work on hierarchical pinning appeared in Derrida and Griffiths [83], and Derrida, Hakim and Vannimenus [87].

(14) Cranston, Hryniv and Molchanov [78] consider a continuous-time SRW on the complete graph K_n (with n vertices and $\binom{n}{2}$ edges) jumping at rate 1. Pinning takes place at a single site, say 0. For the homopolymer, the Hamiltonian up to time t reads $H_t[K_n](w) = -\lambda \int_0^t ds\, 1_{\{w_s=0\}}$, and the free energy for $n, t \to \infty$ such that $t = o(n)$ equals $f(\lambda) = 0 \vee (\lambda - 1)$. For the copolymer, the Hamiltonian up to time t reads $H_t^b[K_n](w) = -\lambda \int_0^t (db_s + hds)\, 1_{\{w_s=0\}}$, with $(b_s)_{s \geq 0}$ a standard Brownian motion, and the quenched free energy (which is self-averaging) is shown to be explicitly computable with critical curve $h_c(\lambda) = 1/\lambda$.

(15) Velenik [305] contains an overview of pinning of higher-dimensional interfaces by disorder on a surface.

11.10 Challenges

(1) Improve the lower bound in Theorem 11.5. Is there a nice formula comparable with that in Theorem 9.7 for the copolymer near a selective interface?

(2) Complete the classification of relevant vs. irrelevant disorder in Section 11.6, i.e., for $a = \frac{1}{2}$ find the necessary and sufficient condition on the slowly varying function $L(\cdot)$ in (11.5) that separates relevant from irrelevant disorder.

(3) Find out whether the phase transition in the pinning model is second order or higher order when the disorder is relevant (recall Section 11.5). Numerical analysis seems to indicate that the order is higher. This is the analogue of Section 9.7, Challenge (1).

(4) Find out whether or not $(\lambda, h) \mapsto f(\lambda, h)$ is analytic on $\mathcal{L}$. Is $\lambda \mapsto h_c(\lambda)$ analytic for those cases where it differs from $h_c^{\mathrm{ann}}(\lambda)$ for all $\lambda \in (0, \infty)$? This is the analogue of Section 9.7, Challenge (2).

(5) Investigate to what extent the relations in (11.55) for wetting carry over when the reference random walk is not SRW, in particular, for $(\lambda, h) \in \mathcal{L}$ close to the quenched critical curve. It is shown in Giacomin [117] that they do not hold in general for pinning. Clarify the relation between $\mu(\lambda, h)$ and $f(\lambda, h)$ in the case of relevant disorder.

(6) For the model described in Section 11.9, Extension (11), with $M = 1$, is the critical surface $(\lambda, \bar{\lambda}) \mapsto h_c(\lambda, \bar{\lambda})$ different for $\bar{\lambda} \neq 0$ than for $\bar{\lambda} = 0$, i.e., does arbitrarily small disorder modify the critical behavior, as it does for the copolymer model studied in Chapter 9?

(7) Extend the analysis to disorder that is Markov. This is the analogue of Section 9.7, Challenge (5).

(8) Determine the fine details of the critical curve in the droplet model in Section 11.9, Extension (12), using the techniques from Chapter 9.

12

Polymers in a Random Potential

In Chapters 9–10 we studied a copolymer in the vicinity of a linear, respectively, a random selective interface between two solvents. The application we had in mind was a copolymer consisting of a random concatenation of hydrophobic and hydrophilic monomers, living in a medium consisting of oil and water located in two halfspaces, respectively, in *mesoscopic* droplets arranged in a percolation-type fashion. In Chapter 10 it was important that the droplets were large compared to the size of the monomers, so that a coarse-graining technique w.r.t. the random environment could be put to use. In the present chapter we look at what happens when the oil and water droplets are *microscopic*, i.e., have the same size as the monomers.

The model we will look at can be thought of as describing a homopolymer living on $\mathbb{N} \times \mathbb{Z}^d$, with $\mathbb{N}$ playing the role of time and $\mathbb{Z}^d$, $d \geq 1$, playing the role of space. The disorder is associated with the sites in $\mathbb{N} \times \mathbb{Z}^d$ (see Fig. 12.1). This model is commonly referred to as the *directed polymer in random environment*, but after what we saw in Chapters 9–10 this name is no longer distinctive.

In Section 12.1 we define the model and record a few basic properties, most notably, we identify an underlying *martingale* that plays a key role in the analysis. In Section 12.2 we state three theorems. The first theorem shows that there is a unique critical temperature separating a phase of *weak disorder* from a phase of *strong disorder*, in which the behavior of the polymer is diffusive, respectively, superdiffusive. The second and third theorem give sufficient conditions for the occurrence of the two phases in terms of the parameters in the model. In Sections 12.3–12.6 we give the proofs of these theorems. In Section 12.7 we address the question to what extent the sufficient condition for weak disorder can be sharpened. This leads to an interesting characterization of the critical temperature, and also reveals a link with the random pinning model studied in Chapter 11. Part of what is written below is drawn from the overview given in Comets, Shiga and Yoshida [70].

The one-dimensional version of the model made its first appearance in the physics literature, in work by Huse and Henley [176], Huse, Henley

F. den Hollander, *Random Polymers*,
Lecture Notes in Mathematics 1974, DOI: 10.1007/978-3-642-00333-2_12,
© Springer-Verlag Berlin Heidelberg 2009

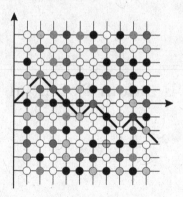

Fig. 12.1. A directed homopolymer in random environment. Different shades of white, grey and black represent different values of the disorder, e.g. microscopic droplets of oil and water of different sizes.

and Fisher [177], Kardar [205], Kardar and Nelson [206], and Kardar and Zhang [207], where it was used to describe roughening of domain walls in the two-dimensional Ising model with random impurities. Since then it has been used to describe a variety of different growth phenomena, including the formation of magnetic domains in spin-glasses, turbulence in viscous incompressible fluids, and the motion of fronts in forest fires and in bacterial colonies on Petri dishes. The zero-temperature limit of the model, where the polymer is controlled by the lowest energies it can find in the disorder, corresponds to oriented first-passage percolation. This is why the directed polymer in random environment is sometimes referred to as "oriented first-passage percolation at positive temperature".

12.1 A Homopolymer in a Micro-emulsion

The set of paths is

$$\mathcal{W}_n = \big\{ w = (w_i)_{i=0}^n \in (\mathbb{Z}^d)^{n+1} \colon w_0 = 0, \, \|w_{i+1} - w_i\| = 1 \; \forall \, 0 \le i < n \big\}. \tag{12.1}$$

The random environment

$$\omega = \{\omega(i,x) \colon i \in \mathbb{N}, \, x \in \mathbb{Z}^d\} \tag{12.2}$$

consists of an i.i.d. field of $\mathbb{R}$-valued non-degenerate random variables with cumulant generating function

$$c(\beta) = \log \mathbb{E}\big(e^{\beta \omega(1,0)}\big), \qquad \beta \in [0, \infty), \tag{12.3}$$

which is assumed to be finite. We write $\mathbb{P}$ to denote the law of ω. The Hamiltonian is

$$H_n^{\beta,\omega}(w) = -\beta \sum_{i=1}^{n} \omega(i, w_i), \qquad w \in \mathcal{W}_n, \tag{12.4}$$

where β plays the role of the *strength of the disorder*. The *quenched path measure* is

$$P_n^{\beta,\omega}(w) = \frac{1}{Z_n^{\beta,\omega}} e^{-H_n^{\beta,\omega}(w)} P_n(w), \qquad w \in \mathcal{W}_n, \tag{12.5}$$

where P_n is the projection onto $\mathcal{W}_n$ of the law P of SRW.

The interpretation of (12.1–12.5) is that of a directed polymer in a random potential that picks up random penalties or rewards depending on the sign and the value of the potential at the sites that it visits (see Fig. 12.1). For binary disorder the model can be used to describe a hydrophobic homopolymer in a random micro-emulsion of oil and water droplets located at the sites in $\mathbb{N} \times \mathbb{Z}^d$, with $\omega(i,x) = +1$ denoting the presence of an oil droplet at site (i,x) and $\omega(i,x) = -1$ the presence of a water droplet. Alternatively, we may think of a copolymer with hydrophobic and hydrophilic monomers, in which case $\omega(i,x) = +1$ stands for a match hydrophobic-oil or hydrophilic-water and $\omega(i,x) = -1$ for a mismatch hydrophilic-oil or hydrophobic-water. Yet another application is that of a polymer in a *gel* with different sizes of pockets. This is important because DNA molecules can be separated from one another by *electrophoresis* in a gel.

Note that $\beta \mapsto c(\beta)$ defined in (12.3) is strictly increasing and strictly convex on $[0, \infty)$ with $c(0) = 0$. Key examples of disorder are standard Gaussian with $c(\beta) = \frac{1}{2}\beta^2$ and p-Bernoulli with $c(\beta) = \log[pe^{\beta} + (1-p)e^{-\beta}]$, $p \in (0,1)$.

12.2 A Dichotomy: Weak and Strong Disorder

12.2.1 Key Martingale

The existence of the *quenched* free energy

$$f(\beta) = \lim_{n \to \infty} \frac{1}{n} \log Z_n^{\beta,\omega} \qquad \omega - a.s. \tag{12.6}$$

is proved in Carmona and Hu [61] (Gaussian disorder) and Comets, Shiga and Yoshida [69] (general disorder), by combining a subadditivity argument with a concentration inequality. This technique is by now standard, and we refer the reader to Chapters 9–11. The *annealed* free energy equals

$$f^{\mathrm{ann}}(\beta) = \lim_{n \to \infty} \frac{1}{n} \log \mathbb{E}(Z_n^{\beta,\omega}) = c(\beta), \tag{12.7}$$

since $\mathbb{E}(Z_n^{\beta,\omega}) = E(\mathbb{E}(\exp[\beta \sum_{i=1}^{n} \omega(i, S_i)])) = e^{nc(\beta)}$ for all $n \in \mathbb{N}_0$.

The key object in the analysis of the model defined in Section 12.1 is the following quantity:

$$M_n^{\beta,\omega} = \frac{Z_n^{\beta,\omega}}{\mathbb{E}(Z_n^{\beta,\omega})} = e^{-nc(\beta)} Z_n^{\beta,\omega}, \qquad n \in \mathbb{N}_0. \qquad (12.8)$$

The reason is that $(M_n^{\beta,\omega})_{n\in\mathbb{N}_0}$ is a *martingale* w.r.t. the filtration generated by ω, i.e., $(\mathcal{F}_{(0,n]}^{\omega})_{n\in\mathbb{N}_0}$ with $\mathcal{F}_{(0,n]}^{\omega} = \sigma[\omega(i,x)\colon 1 \leq i \leq n,\, x \in \mathbb{Z}^d]$ the sigma-algebra of the random environment up to time n. This fact is easily verified by writing

$$M_n^{\beta,\omega} = E\left(\prod_{i=1}^{n} e^{\beta\omega(i,S_i)-c(\beta)}\right), \qquad M_0^{\beta,\omega} = 1, \qquad (12.9)$$

and recalling (12.3). Note that $\mathbb{E}(M_n^{\beta,\omega}) = 1$ and $M_n^{\beta,\omega} > 0$ for all $n \in \mathbb{N}_0$.

By the martingale convergence theorem (Durrett [96], Section 4.2), we have

$$M^{\beta,\omega} = \lim_{n\to\infty} M_n^{\beta,\omega} \quad \text{exists } \omega - a.s. \qquad (12.10)$$

Moreover, since the event $\{\omega\colon M^{\beta,\omega} > 0\}$ is measurable w.r.t. the tail sigma-algebra $\cap_{n\in\mathbb{N}_0}\sigma[\omega(i,x)\colon i > n,\, x \in \mathbb{Z}^d]$, it follows from the Kolmogorov zero-one law (Durrett [96], Section 1.8) that the following dichotomy holds:

$$\begin{aligned}
(\text{WD})\colon &\quad \mathbb{P}(M^{\beta,\omega} > 0) = 1, \\
(\text{SD})\colon &\quad \mathbb{P}(M^{\beta,\omega} = 0) = 1.
\end{aligned} \qquad (12.11)$$

We will see in what follows that the first case characterizes *weak disorder*, for which the behavior of the polymer is diffusive in the $\mathbb{Z}^d$-direction, while the second case characterizes *strong disorder*, for which the behavior is superdiffusive (see Fig. 12.2 for an impression of the typical path behavior). In Sections 12.4–12.5 we will derive *sufficient conditions* on d, β and $c(\cdot)$ for each of the two cases. Note that if $\beta = 0$, then the polymer performs SRW in the $\mathbb{Z}^d$-direction, which fits with (WD) because $M_n^{0,\omega} = 1$ for all $n \in \mathbb{N}_0$.

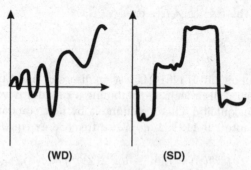

(WD)   **(SD)**

Fig. 12.2. Typical path behavior in the phases of weak disorder (WD) and strong disorder (SD).

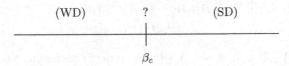

Fig. 12.3. Separation of the two phases.

12.2.2 Separation of the Two Phases

The following theorem is due to Comets and Yoshida [74], and shows that
(WD) and (SD) are separated by a unique critical value of the interaction
strength β (see Fig. 12.3).

Theorem 12.1. *For any choice of the disorder distribution, $\beta \mapsto \mathbb{E}(\sqrt{M^{\beta,\omega}})$
is non-increasing on $[0, \infty)$. Consequently, there exists a $\beta_c \in [0, \infty]$ such that*

$$\begin{aligned}
\beta \in [0, \beta_c) &\implies \text{(WD)}, \\
\beta \in (\beta_c, \infty) &\implies \text{(SD)}.
\end{aligned} \tag{12.12}$$

Theorem 12.1 will be proved in Section 12.3. It is not known what happens
at the critical value (see Section 12.9, Challenge (2)).
   It follows from (12.6–12.12) that

$$f(\beta) = f^{\text{ann}}(\beta) \qquad \forall \beta \in [0, \beta_c], \tag{12.13}$$

where the equality at $\beta = \beta_c$ can be added because both free energies are
convex and hence continuous in β. It is expected that $f(\beta) < f^{\text{ann}}(\beta)$ for
all $\beta \in (\beta_c, \infty)$, but this has not yet been proved in full generality. See
Section 12.8, Extension (2).

12.2.3 Characterization of the Two Phases

In this section we state two results for the path behavior in (WD), respectively,
(SD). These results are formulated in Theorems 12.2–12.3 below and will be
proved in Sections 12.5–12.6. They also yield a lower and an upper bound
on β_c.
   We need some more notation. Let

$$\pi_d = (P \times P')(\exists n \in \mathbb{N}: S_n = S'_n) \tag{12.14}$$

denote the collision probability of two independent copies of the random
walk S. Since S is SRW, $S - S'$ is SRW at twice the speed, and so we have
$\pi_d = F_d$ with $F_d = P(\exists n \in \mathbb{N}: S_n = 0)$ the probability of return to the
origin, for which we saw in Section 2.1 that $F_d = 1$ in $d = 1, 2$ and $F_d < 1$ in
$d \geq 3$. For $\beta \in [0, \infty)$, define

$$\Delta_1(\beta) = c(2\beta) - 2c(\beta),$$
$$\Delta_2(\beta) = \beta c'(\beta) - c(\beta).$$

$$(12.15)$$

Both $\beta \mapsto \Delta_1(\beta)$ and $\beta \mapsto \Delta_2(\beta)$ are strictly increasing on $[0, \infty)$ with $\Delta_1(0) = \Delta_2(0) = 0$. From the representations

$$\Delta_1(\beta) = \Delta_2(\beta) + \int_\beta^{2\beta} c''(v)(2\beta - v)\, dv,$$
$$\Delta_2(\beta) = \int_0^\beta c''(v) v\, dv,$$

$$(12.16)$$

we see that $\Delta_1 > \Delta_2$ on $(0, \infty)$. Furthermore, define

$$\max_n^{\beta,\omega} = \max_{x \in \mathbb{Z}^d} P_{n-1}^{\beta,\omega}(S_n = x), \qquad n \in \mathbb{N}, \qquad (12.17)$$

where the r.h.s. concerns the n-step polymer weighted according to the Hamiltonian up to time $n-1$ rather than time n, so that the last step is like a step of the random walk (this is done for later convenience). The quantity in (12.17) is a measure of how "localized" the polymer is at time n in the given random environment ω.

Theorem 12.2. *Suppose that*

(I) $d \geq 3$ *and* $\Delta_1(\beta) < \log(1/\pi_d)$.

Then

$$\lim_{n \to \infty} \frac{1}{n} E_n^{\beta,\omega}(\|S_n\|^2) = 1 \qquad \omega - a.s. \qquad (12.18)$$

and

$$\lim_{n \to \infty} \max_n^{\beta,\omega} = 0 \qquad \omega - a.s. \qquad (12.19)$$

Theorem 12.3. *Suppose that*

(II) $d = 1, 2$ *and* $\beta > 0$ *or* $d \geq 3$ *and* $\Delta_2(\beta) > \log(2d)$.

Then there exists a $c = c(d, \beta) > 0$ *such that*

$$\limsup_{n \to \infty} \max_n^{\beta,\omega} \geq c \qquad \omega - a.s. \qquad (12.20)$$

The first half of Theorem 12.2 is due to Imbrie and Spencer [183], Bolthausen [24], Sinai [275] and Song and Zhou [282] (see also Olsen and Song [250] and Kifer [213]). The second half of Theorem 12.2, as well as Theorem 12.3, are due to Carmona and Hu [61] and Comets, Shiga and Yoshida [69]. The proofs in [183] and [275] are based on expansion techniques, the proofs in [24] and [282] on martingale arguments.

12.2.4 Diffusive versus Non-diffusive Behavior

The statements in (12.18–12.19) are typical for *diffusive* behavior. What is noteworthy is that the diffusion constant is *not renormalized* by the disorder. This is markedly different from what we found e.g. for the soft polymer in Chapter 4, Theorem 4.1, where the self-repellence – no matter how weak – renormalizes the diffusion constant by a factor > 1.

The statement in (12.20), which indicates *non-diffusive* behavior, suggests that the endpoint of the polymer is concentrated in a small region around a "most favorable site" (which depends on ω). Possibly the whole polymer is concentrated inside a "most favorable corridor". It has been conjectured that, throughout (SD), ω-a.s. as $n \to \infty$

$$E_n^{\beta,\omega}(\|S_n\|^2) \approx n^{2\nu(d)}, \qquad \log Z_n^{\beta,\omega} - \mathbb{E}(\log Z_n^{\beta,\omega}) \approx n^{\chi(d)}, \qquad (12.21)$$

where the symbol $\approx$ is to be interpreted in the following sense:

$$\nu(d) = \inf\left\{\nu > 0 \colon \mathbb{E}\left(\lim_{n\to\infty} P_n^{\beta,\omega}\left(\max_{0\le i\le n}\|S_i\| \le n^\nu\right)\right) = 1\right\},$$
$$\chi(d) = \sup\left\{\chi > 0 \colon \liminf_{n\to\infty} \mathbb{E}\left([\log Z_n^{\beta,\omega} - \mathbb{E}(\log Z_n^{\beta,\omega})]^2\right) \ge n^{2\chi}\right\}. \qquad (12.22)$$

Both exponents are believed not to depend on β, throughout (SD), and to satisfy $\chi(d) = 2\nu(d) - 1$, with $\nu(1) = \frac{2}{3}$ and $\nu(2) \in (\frac{1}{2}, \frac{2}{3})$. Partial results towards (12.21) have been obtained by Piza [267] (for SRW and general disorder), Petermann [262] and Mejane [240] (for Gaussian random walk and Gaussian disorder), and Carmona and Hu [62] (for SRW and Gaussian disorder). See Section 12.7, Extension (7), for more details. There are deep links between (12.21) and results for fluctuations of surfaces in first-passage percolation and the Ising model, and fluctuations of spectra of random matrices. For an overview we refer the reader to Krug and Spohn [221].

12.2.5 Bounds on the Critical Temperature

The conditions in (I) and (II) are mutually exclusive because $\Delta_1 > \Delta_2$ on $(0, \infty)$ and $\pi_d > 1/2d$. In fact, there is a gap between (I) and (II), so that Theorems 12.2–12.3 do not cover the full parameter regime. Note that (I) holds for *all* $c(\cdot)$ when β is sufficiently small, while (II) holds for *many* $c(\cdot)$ when β is sufficiently large. Indeed, there are examples of disorder distributions for which (I) holds for all $\beta \in [0, \infty)$ and (II) is empty, for instance, for the p-Bernoulli distribution with $p \ge \pi_d$, because $\Delta_1(\infty) = \Delta_2(\infty) = \log(1/p)$. Only when $p < \pi_d$ is (II) non-empty. For the standard Gaussian distribution we have $\Delta_1(\beta) = \beta^2$ and $\Delta_2(\beta) = \frac{1}{2}\beta^2$, and (II) is non-empty.

In view of the above observations, it is natural to introduce two critical values,

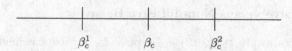

Fig. 12.4. Bounds on the critical temperature. Depending on the disorder distribution, either $0 < \beta_c^1 < \beta_c^2 < \infty$ or $0 < \beta_c^1 < \beta_c^2 = \infty$ or $\beta_c^1 = \beta_c^2 = \infty$.

$$\beta_c^1 = \sup \left\{ \beta \in [0, \infty): \; \Delta_1(\beta) < \log(1/\pi_d) \right\},$$
$$\beta_c^2 = \inf \left\{ \beta \in [0, \infty): \; \Delta_2(\beta) > \log(2d) \right\},$$
(12.23)

which satisfy

$$0 < \beta_c^1 \leq \beta_c^2 \leq \infty.$$
(12.24)

To ensure that $\beta_c^2 < \infty$, so that (II) is non-empty, it suffices that $\Delta_2(\infty) = \infty$, which by (12.16) amounts to $\int_0^\infty c''(v)v \, dv = \infty$. The latter holds, for instance, when the disorder is: (1) unbounded; (2) bounded with no mass at the supremum of its support and a regularly varying density near this supremum. Indeed, in case (1) we have $\infty = c'(\infty) = c'(0) + \int_0^\infty c''(v) \, dv$, while in case (2) an easy computation gives $c''(v) \asymp 1/v^2$ as $v \to \infty$. An example of (2) is the uniform distribution on $[0, 1]$, for which $\Delta_1(\beta) \sim \Delta_2(\beta) \sim \log \beta$ as $\beta \to \infty$.

Theorems 12.2 and 12.3 will be proved in Sections 12.5 and 12.6, respectively. The proofs will show that

$$\text{(I)} \Longrightarrow \text{(WD)}, \qquad \text{(II)} \Longrightarrow \text{(SD)}.$$
(12.25)

Recalling Theorem 12.1 and (12.23), we thus have (see Fig. 12.4)

$$\beta_c \in [\beta_c^1, \beta_c^2].$$
(12.26)

Note that, because $\Delta_1 > \Delta_2$ on $(0, \infty)$, if $\beta_c^2 < \infty$, then $\beta_c^1 < \beta_c^2$.

12.3 Proof of Uniqueness of the Critical Temperature

In this section we prove Theorem 12.1.

Proof. Abbreviate $f(x) = \sqrt{x}$, $x \in [0, \infty)$. We will show that, for all $n \in \mathbb{N}_0$ and $\beta \in \mathbb{R}$,

$$\mathbb{E}\left(f(M_n^{\beta, \omega}) \right) < \infty, \qquad \mathbb{E}\left(\frac{\partial}{\partial \beta} f(M_n^{\beta, \omega}) \right) < \infty,$$
(12.27)

and

$$\frac{\partial}{\partial \beta} \mathbb{E}(f(M_n^{\beta, \omega})) = \mathbb{E}\left(\frac{\partial}{\partial \beta} f(M_n^{\beta, \omega}) \right) \leq 0.$$
(12.28)

Since $\mathbb{E}(M_n^{\beta, \omega}) = 1$ for all $n \in \mathbb{N}_0$ and $\beta \in \mathbb{R}$, we have

$$\lim_{n\to\infty} \mathbb{E}\left(f(M_n^{\beta,\omega})\right) = \mathbb{E}\left(f(M^{\beta,\omega})\right). \tag{12.29}$$

Combining (12.28–12.29), we get the claim in Theorem 12.1.

To prove (12.27–12.28), we fix any $\beta_0 \in (0,\infty)$ and first show that, for all $n \in \mathbb{N}$ and $p \in [1,\infty)$,

$$\sup_{0\le\beta\le\beta_0} M_n^{\beta,\omega}, \quad \sup_{0\le\beta\le\beta_0} [M_n^{\beta,\omega}]^{-1}, \quad \sup_{0\le\beta\le\beta_0} \left|\frac{\partial}{\partial\beta} M_n^{\beta,\omega}\right| \in L^p(\mathbb{P}). \tag{12.30}$$

Indeed, recalling (12.4), abbreviating

$$e_n^{\beta,\omega}(S) = e^{\beta I_n^\omega(S) - nc(\beta)}, \qquad I_n^\omega(S) = \sum_{i=1}^n \omega(i, S_i), \tag{12.31}$$

and using Fubini's theorem, we have

$$\mathbb{E}\left([M_n^{\beta,\omega}]^p\right) = \mathbb{E}\left(\left[E\left(e_n^{\beta,\omega}(S)\right)\right]^p\right)$$
$$\le \mathbb{E}\left(E\left([e_n^{\beta,\omega}(S)]^p\right)\right) = \mathbb{E}\left(E\left(e^{p\beta I_n^\omega(S) - pnc(\beta)}\right)\right) = e^{n[c(p\beta) - pc(\beta)]} \tag{12.32}$$

and

$$\mathbb{E}\left([M_n^{\beta,\omega}]^{-p}\right) = \mathbb{E}\left(\left[E\left(e_n^{\beta,\omega}(S)\right)\right]^{-p}\right) \tag{12.33}$$
$$\le \mathbb{E}\left(E\left([e_n^{\beta,\omega}(S)]^{-p}\right)\right) = e^{n[c(-p\beta) + pc(\beta)]},$$

both of which are finite. A similar estimate can be done for the third quantity in (12.30), because

$$\frac{\partial}{\partial\beta} M_n^{\beta,\omega} = \frac{\partial}{\partial\beta} E\left(e_n^{\beta,\omega}(S)\right) = E\left([I_n^\omega(S) - nc'(\beta)] e_n^{\beta,\omega}(S)\right). \tag{12.34}$$

Next, since $f'(x) = \frac{1}{2\sqrt{x}} \le \frac{1}{2}(x + x^{-1})$, $x \in (0,\infty)$, we have

$$\left|\frac{\partial}{\partial\beta} f(M_n^{\beta,\omega})\right| = \left|f'(M_N^{\beta,\omega}) \frac{\partial}{\partial\beta} M_n^{\beta,\omega}\right| \le \left|\frac{1}{2}\left(M_n^{\beta,\omega} + [M_n^{\beta,\omega}]^{-1}\right) \frac{\partial}{\partial\beta} M_n^{\beta,\omega}\right|. \tag{12.35}$$

Together with (12.30) and Hölder's inequality this yields that, for all $n \in \mathbb{N}_0$,

$$\sup_{0\le\beta\le\beta_0} \left|\frac{\partial}{\partial\beta} f(M_n^{\beta,\omega})\right| \in L^1(\mathbb{P}). \tag{12.36}$$

We can now complete the proof (12.27–12.28). To that end we write

$$f(M_n^{\beta_0,\omega}) = f(M_n^{0,\omega}) + \int_0^{\beta_0} d\beta \frac{\partial}{\partial\beta} f(M_n^{\beta,\omega}). \tag{12.37}$$

The claim in (12.27) and the equality in (12.28) follow by combining (12.37) with (12.30) and using Fubini's theorem. To get the inequality in (12.28), we recall (12.34) to write

$$\mathbb{E}\left(\frac{\partial}{\partial \beta} f(M_n^{\beta,\omega})\right) = \mathbb{E}\left(f'(M_n^{\beta,\omega}) \frac{\partial}{\partial \beta} M_n^{\beta,\omega}\right)$$
$$= E\Big(\mathbb{E}\big(f'(M_n^{\beta,\omega}) \left[I_n^\omega(S) - nc'(\beta)\right] e_n^{\beta,\omega}(S)\big)\Big). \tag{12.38}$$

For fixed S, $e_n^{\beta,\omega}(S)\mathbb{P}(d\omega)$ is a product measure. Since $\omega \mapsto [I_n^\omega(S) - nc'(\beta)]$ is non-decreasing and $\omega \mapsto f'(M_n^{\beta,\omega})$ is non-increasing (because $f' \leq 0$ and $\omega \mapsto M_n^{\beta,\omega}$ is non-decreasing), it therefore follows from the FKG-inequality (Fortuin, Kasteleyn and Ginibre [106]) that

$$\mathbb{E}\Big(f'(M_n^{\beta,\omega}) \left[I_n^\omega(S) - nc'(\beta)\right] e_n^{\beta,\omega}(S)\Big)$$
$$\leq \mathbb{E}\left(f'(M_n^{\beta,\omega}) e_n^{\beta,\omega}(S)\right) \mathbb{E}\left(\left[I_n^\omega(S) - nc'(\beta)\right] e_n^{\beta,\omega}(S)\right). \tag{12.39}$$

However, for all S,

$$\mathbb{E}\left(\left[I_n^\omega(S) - nc'(\beta)\right] e_n^{\beta,\omega}(S)\right) = \mathbb{E}\left(\frac{\partial}{\partial \beta} e_n^{\beta,\omega}(S)\right)$$
$$= \frac{\partial}{\partial \beta} \mathbb{E}\left(e_n^{\beta,\omega}(S)\right) = \frac{\partial}{\partial \beta} 1 = 0, \tag{12.40}$$

and so the inequality in (12.28) follows by combining (12.38–12.40). $\square$

12.4 Martingale Estimates

We first note that, subject to (I),

$$\lim_{n\to\infty} M_n^{\beta,\omega} = M^{\beta,\omega} \quad \text{exists } \omega - a.s. \text{ and in } L^2(\mathbb{P}). \tag{12.41}$$

Indeed,

$$\mathbb{E}([M_n^{\beta,\omega}]^2) = (E \times E')\left(\prod_{i=1}^n \mathbb{E}\left(e^{\beta[\omega(i,S_i)+\omega(i,S_i')]-2c(\beta)}\right)\right)$$
$$= (E \times E')\left(\prod_{i=1}^n \left[1_{\{S_i \neq S_i'\}} + e^{c(2\beta)-2c(\beta)} 1_{\{S_i=S_i'\}}\right]\right) \tag{12.42}$$
$$= (E \times E')\left(e^{[c(2\beta)-2c(\beta)] V_n(S,S')}\right),$$

where S, S' are two independent copies of the random walk, and $V_n(S,S') = \sum_{i=1}^n 1_{\{S_i=S_i'\}}$ denotes their *collision local time* up to time n. Put $V(S,S') =$

$\lim_{n\to\infty} V_n(S,S')$. Since $(P \times P')(V(S,S') = k) = (1 - \pi_d)(\pi_d)^k$, $k \in \mathbb{N}$, we see that

$$(I) \quad \Longleftrightarrow \quad \sup_{n\in\mathbb{N}} \mathbb{E}([M_n^{\beta,\omega}]^2) < \infty. \tag{12.43}$$

The claim in (12.41) therefore follows from the martingale convergence theorem.

The proofs of Theorems 12.2–12.3 given in Sections 12.5–12.6 rely on two technical lemmas, which are extensions of (12.41). These lemmas, which are taken from Comets, Shiga and Yoshida [70], will be stated and proved in Sections 12.4.1–12.4.2 below. The first lemma is a generalization of an earlier result by Bolthausen [24] and Song and Zhou [282].

12.4.1 First Estimate

Let $\phi \colon \mathbb{N}_0 \times \mathbb{Z}^d \to \mathbb{R}$ be such that

(C1) $|\phi(n,x)| \leq C_1 + C_2 n^{q/2} + C_3 \|x\|^q$ for all $(n,x) \in \mathbb{N}_0 \times \mathbb{Z}^d$ and some $C_1, C_2, C_3 \in [0,\infty)$, $q \in [0,\infty)$,

(C2) $(\phi(n,S_n))_{n\in\mathbb{N}_0}$ is a martingale w.r.t. the filtration generated by S, i.e., $(\mathcal{F}_{(0,n]}^S)_{n\in\mathbb{N}_0}$ with $\mathcal{F}_{(0,n]}^S = \sigma[S_i \colon 0 \leq i \leq n]$ the sigma-algebra of SRW up to time n.

Define

$$M_n^{\beta,\omega}(\phi) = E\left(\phi(n,S_n) \prod_{i=1}^n e^{\beta\omega(i,S_i)-c(\beta)}\right), \qquad M_0^{\beta,\omega}(\phi) = \phi(0,0), \tag{12.44}$$

and note that, by (C2), $(M_n^{\beta,\omega}(\phi))_{n\in\mathbb{N}_0}$ is a martingale w.r.t. the filtration generated by ω (compare with (12.9)).

Lemma 12.4. *Assume* (I) *and* (C1–C2). *If $q > 0$, then*

$$\lim_{n\to\infty} n^{-q/2} \max_{0\leq i\leq n} M_i^{\beta,\omega}(\phi) = 0 \qquad \omega - a.s. \tag{12.45}$$

Moreover, if $0 \leq q < (d-2)/2$, then

$$\lim_{n\to\infty} M_n^{\beta,\omega}(\phi) = M^{\beta,\omega}(\phi) \qquad \text{exists } \omega - a.s. \text{ and in } L^2(\mathbb{P}). \tag{12.46}$$

Proof. We already proved (12.46) for $q = 0$ in (12.41), and so we may assume that $q > 0$. The proof will be based on the following additional estimate.

Lemma 12.5. *Assume* (I) *and* (C1–C2). *Then*

$$\mathbb{E}([M_n^{\beta,\omega}(\phi)]^2) = O(\psi_n) \qquad \text{as } n \to \infty, \tag{12.47}$$

with

$$\psi_n = \begin{cases} 1, & \text{if } \theta < 0, \\ \log n, & \text{if } \theta = 0, \\ n^\theta, & \text{if } \theta > 0, \end{cases} \tag{12.48}$$

where $\theta = q - (d-2)/2$.

Proof. First, following a computation similar as in (12.42), we write

$$\mathbb{E}\big([M_n^{\beta,\omega}(\phi)]^2\big) = [\phi(0, S_0)]^2$$
$$+ \sum_{k=1}^n \left(e^{\Delta_1(\beta)} - 1\right)^k \sum_{1 \le i_1 < \cdots < i_k \le n} (E \times E') \left([\phi(i_k, S_{i_k})]^2 \chi_{i_1 \ldots i_k}(S, S')\right), \tag{12.49}$$

where $i_0 = 0$ and we abbreviate

$$\chi_{i_1 \ldots i_k}(S, S') = 1_{\left\{S_{i_1} = S'_{i_1}, \ldots, S_{i_k} = S'_{i_k}\right\}}. \tag{12.50}$$

By (C1), the first term in the r.h.s. of (12.49) is bounded from above by C_1^2, while the summand in the second term is bounded from above by

$$3(C_3)^2 A_{i_1, \ldots, i_k} + [3(C_1)^2 + 3(C_2)^2 (i_k)^q] B_{i_1, \ldots, i_k}, \tag{12.51}$$

with

$$A_{i_1, \ldots, i_k} = (E \times E')(\|S_{i_k}\|^{2q} \chi_{i_1 \ldots i_k}(S, S')),$$
$$B_{i_1, \ldots, i_k} = (E \times E')(\chi_{i_1 \ldots i_k}(S, S')). \tag{12.52}$$

Next, the first quantity in (12.52) can be estimated as follows (C denotes a generic constant that may change from line to line):

$$A_{i_1, \ldots, i_k} \le k^{2q-1} \sum_{l=1}^k (E \times E')(\|S_{i_l} - S_{i_{l-1}}\|^{2q} \chi_{i_1 \ldots i_k}(S, S'))$$

$$= k^{2q-1} \sum_{l=1}^k \left\{ \prod_{m=1}^{l-1} (E \times E')(\chi_{i_m - i_{m-1}}(S, S')) \right\}$$
$$\times (E \times E')(\|S_{i_l - i_{l-1}}\|^{2q} \chi_{i_l - i_{l-1}}(S, S'))$$
$$\times \left\{ \prod_{m=l+1}^k (E \times E')(\chi_{i_m - i_{m-1}}(S, S')) \right\}$$

$$= k^{2q-1} \sum_{l=1}^k C(i_l - i_{l-1})^{q-d/2} \prod_{\substack{1 \le m \le k \\ m \ne l}} (E \times E')(\chi_{i_m - i_{m-1}}(S, S')). \tag{12.53}$$

Here, the first line uses the triangle inequality and Jensen's inequality (for $2q \ge 1$; see later for $2q < 1$), the second line uses the Markov property of S and S', while the third line uses the bound

$$(E \times E')\big(\|S_n\|^{2q}\chi_n\big) = \sum_{x \in \mathbb{Z}^d} E\big(\|S_n\|^{2q}1_{\{S_n = x\}}\big)\,P'(S'_n = x)$$

$$\leq C\,n^{-d/2}\,E(\|S_n\|^{2q}) \leq C\,n^{q-(d/2)}, \qquad n \in \mathbb{N}.$$
$$(12.54)$$

Summing (12.53), we obtain

$$\sum_{1 \leq i_1 < \cdots < i_k \leq n} A_{i_1,\dots,i_k} \leq C\,\psi_n\,k^{2q}\left(\frac{\pi_d}{1-\pi_d}\right)^{k-1}, \qquad (12.55)$$

where we use that $\sum_{j=1}^{n} j^{q-d/2} \leq C\psi_n$, $n \in \mathbb{N}$, and

$$\sum_{j=1}^{\infty}(E \times E')\big(\chi_j(S,S')\big) = \frac{\pi_d}{1-\pi_d}, \qquad (12.56)$$

noting that $\pi_d < 1$ when $d \geq 3$.

A similar bound holds for the second quantity in (12.52), namely,

$$\sum_{1 \leq i_1 < \cdots < i_k \leq n} (i_k)^q B_{i_1,\dots,i_k} \leq C\,\psi_n\,k^{2q}\left(\frac{\pi_d}{1-\pi_d}\right)^{k-1}, \qquad (12.57)$$

as is easily checked after replacing everywhere $\|S_i\|^{2q}$ by i^q, $i \in \mathbb{N}$. Inserting (12.55–12.57) into (12.51) and the latter into (12.49), we arrive at

$$\mathbb{E}\big([M_n^{\beta,\omega}(\phi)]^2\big) \leq C + C\,\psi_n \sum_{k=1}^{n} k^{2q}\,\big(e^{\Delta_1(\beta)} - 1\big)^k \left(\frac{\pi_d}{1-\pi_d}\right)^{k-1}, \qquad (12.58)$$

which gives the claim because $(e^{\Delta_1(\beta)} - 1)\frac{\pi_d}{1-\pi_d} < 1$ precisely when $\Delta_1(\beta) < \log(1/\pi_d)$.

It remains to deal with the case $2q < 1$. We can keep (12.53) provided we drop the factor k^{2q-1}. This is harmless, because the geometric factor in (12.58) absorbs the polynomial factor anyway. $\square$

We are finally ready to give the proof of (12.45–12.46). To get (12.46), note that $q < (d-2)/2$ corresponds to $\theta < 0$, and so Lemma 12.5 gives that $\sup_{n \in \mathbb{N}} \mathbb{E}([M_n^{\beta,\omega}(\phi)]^2) < \infty$. The claim therefore follows from the martingale convergence theorem. To get (12.45), abbreviate

$$\bar{M}_n^{\beta,\omega}(\phi) = \max_{0 \leq i \leq n} M_n^{\beta,\omega}(\phi). \qquad (12.59)$$

Pick $\delta > 0$, put $m(n) = \lceil n^{1/\delta}\rceil$, and estimate

$$\mathbb{E}\left(\bar{M}_{m(n)}^{\beta,\omega}(\phi) > m(n)^\delta\sqrt{\psi_{m(n)}}\right) \leq [m(n)^{2\delta}\psi_{m(n)}]^{-1}\,\mathbb{E}\left([\bar{M}_{m(n)}^{\beta,\omega}(\phi)]^2\right)$$

$$\leq [m(n)^{2\delta}\psi_{m(n)}]^{-1}\,4\,\mathbb{E}\left([M_{m(n)}^{\beta,\omega}(\phi)]^2\right)$$

$$\leq C\,m^{-2\delta} \leq C\,n^{-2},$$
$$(12.60)$$

where the second inequality follows from Doob's inequality for martingales (Durrett [96], Section 4.4), and the third inequality uses Lemma 12.5. Therefore, by the Borel-Cantelli lemma, the events

$$\left\{ \bar{M}^{\beta,\omega}_{m(n)}(\phi) > m(n)^\delta \sqrt{\psi_{m(n)}} \right\}, \qquad n \in \mathbb{N}, \tag{12.61}$$

occur finitely often ω-a.s. Since $n \mapsto \bar{M}^{\beta,\omega}_n(\phi)$ is non-decreasing and $n \mapsto m(n)$ grows polynomially, the latter implies that also the events $\{\bar{M}^{\beta,\omega}_n(\phi) > n^\delta \sqrt{\psi_n}\}$, $n \in \mathbb{N}$, occur finitely often ω-a.s. However, by (12.48), we have $n^\delta \sqrt{\psi_n} = o(n^{q/2})$ for δ small enough, because $q > 0$ and $\theta < q$ (the latter because (I) requires that $d \geq 3$). This proves the claim. $\square$

12.4.2 Second Estimate

Let

$$I^{\beta,\omega}_n = \sum_{x \in \mathbb{Z}^d} [P^{\beta,\omega}_{n-1}(S_n = x)]^2, \qquad n \in \mathbb{N}. \tag{12.62}$$

Lemma 12.6. *If $\beta \neq 0$, then $\{M^{\beta,\omega} = 0\} = \{\sum_{n \in \mathbb{N}} I^{\beta,\omega}_n = \infty\}$ ω-a.s. If, moreover, $\mathbb{P}(M^{\beta,\omega} = 0) = 1$, then there exist $c_1, c_2 \in (0, \infty)$ such that*

$$\exp\left[-c_1 \sum_{i=1}^n I^{\beta,\omega}_i \right] \leq M^{\beta,\omega}_n \leq \exp\left[-c_2 \sum_{i=1}^n I^{\beta,\omega}_i \right] \qquad \omega - a.s. \tag{12.63}$$

Proof. It suffices to prove that the following implications hold ω-a.s.:

$$M^{\beta,\omega} = 0 \implies \sum_{n \in \mathbb{N}} I^{\beta,\omega}_n = \infty \implies (12.63). \tag{12.64}$$

The proof of (12.64) is based on Doob's decomposition theorem (Durrett [96], Section 4.2). Let $X_n = -\log M^{\beta,\omega}_n$, $n \in \mathbb{N}_0$ ($X_0 = 0$), where for ease of notation we suppress β, ω from the notation. Split X_n into a martingale part and an associated compensator part,

$$X_n = A_n + B_n, \qquad n \in \mathbb{N}_0 \ (A_0 = B_0 = 0), \tag{12.65}$$

defined by

$$\Delta A_n = A_n - A_{n-1} = -\log(1 + \delta_n) + \mathbb{E}\left(\log(1 + \delta_n) \mid \mathcal{F}^\omega_{(0,n-1]} \right),$$
$$\Delta B_n = B_n - B_{n-1} = -\mathbb{E}\left(\log(1 + \delta_n) \mid \mathcal{F}^\omega_{(0,n-1]} \right), \tag{12.66}$$

where

$$\delta_n = E^{\beta,\omega}_{n-1}\left(e^{\beta\omega(n,S_n) - c(\beta)} \right) - 1, \tag{12.67}$$

and $\mathcal{F}^\omega_{(0,n-1]}$ is the sigma-algebra generated by ω up to time $n-1$ (recall Section 12.2.1). The split in (12.65–12.67) arises from $\Delta X_n = X_n - X_{n-1} = \Delta A_n + \Delta B_n$ and the relation

$$\frac{M_n^{\beta,\omega}}{M_{n-1}^{\beta,\omega}} = 1 + \delta_n, \tag{12.68}$$

the latter being immediate from (12.4–12.5) and (12.8–12.9). Note that $n \mapsto A_n$ is a martingale w.r.t. the filtration generated by ω, and that (12.66) determines A_n and B_n recursively since $A_0 = B_0 = 0$.

With the help of the relations

$$\delta_n = \sum_{x \in \mathbb{Z}^d} P_{n-1}^{\beta,\omega}(S_n = x)\left[e^{\beta\omega(n,x)-c(\beta)} - 1\right],$$

$$\mathbb{E}\left(\delta_n \mid \mathcal{F}^\omega_{(0,n-1]}\right) = 0, \tag{12.69}$$

it can be easily shown (see [70] for details) that there is a $c \in (0, \infty)$ such that

$$\Delta\langle A\rangle_n \le c I_n^{\beta,\omega}, \qquad c^{-1} I_n^{\beta,\omega} \le \Delta B_n \le c I_n^{\beta,\omega}. \tag{12.70}$$

Here, $n \mapsto \langle A\rangle_n$ is the compensator of $n \mapsto (X_n)^2$, determined recursively from

$$\Delta\langle A\rangle_n = \mathbb{E}\left([\Delta A_n]^2 \mid \mathcal{F}^\omega_{(0,n-1]}\right), \qquad n \in \mathbb{N} \ (\langle A\rangle_0 = 0). \tag{12.71}$$

It follows from (12.70) that ω-a.s.

$$\sum_{n\in\mathbb{N}} I_n^{\beta,\omega} < \infty \implies \langle A\rangle \in \mathbb{R}, B \in \mathbb{R} \implies A \in \mathbb{R}, B \in \mathbb{R}$$

$$\implies X \in \mathbb{R} \implies M^{\beta,\omega} > 0, \tag{12.72}$$

where $X, A, B, \langle A\rangle$ denote the limits of $X_n, A_n, B_n, \langle A\rangle_n$ as $n \to \infty$. Note that $n \mapsto \langle A\rangle_n$ and $n \mapsto B_n$ are non-decreasing (the latter because of Jensen's inequality in combination with the second lines in (12.66) and (12.69)), and that $-\log M^{\beta,\omega} = X = A + B$. The second implication in (12.72) uses that A exists and is finite a.s. on the set $\{\langle A\rangle < \infty\}$ (Durrett [96], Section 4.4). This settles the first implication in (12.64).

To get the second implication in (12.64), we argue that

$$B = \infty \implies \lim_{n\to\infty} \frac{-\log M_n^{\beta,\omega}}{B_n} = \lim_{n\to\infty} \frac{X_n}{B_n} = 1. \tag{12.73}$$

Indeed, either $\langle A\rangle < \infty$, in which case $A \in \mathbb{R}$ and $\lim_{n\to\infty} A_n/B_n = A/B = 0$, or $\langle A\rangle = \infty$, in which case we write $A_n/B_n = (A_n/\langle A\rangle_n)(\langle A\rangle_n/B_n)$ and use that the first factor tends to zero while the second factor is bounded away from infinity by (12.70), to again get $\lim_{n\to\infty} A_n/B_n = 0$. Combining (12.73) and (12.70), we arrive at the second implication in (12.64). $\quad\square$

12.5 The Weak Disorder Phase

In this section we prove Theorem 12.2. The proof relies on Lemmas 12.4 and 12.6.

- Proof of the first implication in (12.25):

Proof. Pick $\phi \equiv 1$ ($q = 0$) in (12.44). Then $M_n^{\beta,\omega}(\phi) = M_n^{\beta,\omega}$, and so (12.41) yields that $\mathbb{E}(M^{\beta,\omega}) = \lim_{n\to\infty} \mathbb{E}(M_n^{\beta,\omega}) = 1$. By the zero-one law for the event $\{M^{\beta,\omega} > 0\}$, we therefore have (WD). $\square$

- Proof of (12.18):

Proof. Pick $\phi(n,x) = \|x\|^2 - n$ ($q = 2$) in (12.44), and write

$$E_n^{\beta,\omega}(\|S_n\|^2) - n = M_n^{\beta,\omega}(\phi)/M_n^{\beta,\omega}. \tag{12.74}$$

By (12.45) in Lemma 12.4, we have $\lim_{n\to\infty} n^{-1} M_n^{\beta,\omega}(\phi) = 0$ ω-a.s., while we know from (WD) proved above that $M_n^{\beta,\omega}$ is ω-a.s. bounded away from 0 uniformly in n. $\square$

- Proof of (12.19):

Proof. Recalling (12.17) and (12.62), we have the sandwich

$$[\max_n^{\beta,\omega}]^2 \le I_n^{\beta,\omega} \le \max_n^{\beta,\omega}. \tag{12.75}$$

It follows from Lemma 12.6 and (WD) proved above that $\sum_{n\in\mathbb{N}} I_n^{\beta,\omega} < \infty$ ω-a.s. Therefore

$$\sum_{n\in\mathbb{N}} [\max_n^{\beta,\omega}]^2 < \infty \qquad \omega - a.s., \tag{12.76}$$

which implies the claim. $\square$

12.6 The Strong Disorder Phase

In this section we prove Theorem 12.3. The proof relies on Lemma 12.6.

- Proof of the second implication in (12.25):

Proof. Pick $\gamma \in (0,1)$. Recalling (12.9) and using the estimate $(u + v)^\gamma \le u^\gamma + v^\gamma$, $u, v \ge 0$, we have

$$[M_n^{\beta,\omega}]^\gamma = \left[E\left(\prod_{i=1}^n e^{\beta\omega(i,S_i)-c(\beta)} \right) \right]^\gamma$$

$$= \left[\sum_{\substack{x\in\mathbb{Z}^d \\ \|x\|=1}} \tfrac{1}{2d} e^{\beta\omega(1,x)-c(\beta)} E\left(\prod_{j=1}^{n-1} e^{\beta\omega(j+1,x+S_j)-c(\beta)} \right) \right]^\gamma \tag{12.77}$$

$$\le \sum_{\substack{x\in\mathbb{Z}^d \\ \|x\|=1}} [\tfrac{1}{2d} e^{\beta\omega(1,x)-c(\beta)}]^\gamma \left[E\left(\prod_{j=1}^{n-1} e^{\beta\omega(j+1,x+S_j)-c(\beta)} \right) \right]^\gamma.$$

Since the last expectation has the same law as $M_{n-1}^{\beta,\omega}$, we get

$$\mathbb{E}([M_n^{\beta,\omega}]^\gamma) \le r(\gamma)\,\mathbb{E}([M_{n-1}^{\beta,\omega}]^\gamma), \tag{12.78}$$

where

$$r(\gamma) = (2d)^{1-\gamma}\,\mathbb{E}\left(e^{\beta\omega(1,x)-c(\beta)}\right)^\gamma. \tag{12.79}$$

Next, $\gamma \mapsto \log r(\gamma)$ is convex and infinitely differentiable, with $r(0) = 2d$ and $r(1) = 1$. The inequality $\Delta_2(\beta) > \log(2d)$ in condition (II) is equivalent to the statement that $[\log r]'(1) > 0$, and hence guarantees that $r(\gamma) < 1$ for some $\gamma \in (0,1)$. Choosing this γ in the above estimates, we conclude from (12.78) that there exists a $c \in (0, \infty)$ such that

$$\limsup_{n\to\infty} n^{-1}\log M_n^{\beta,\omega} \le -c \qquad \omega - a.s. \tag{12.80}$$

Hence we have proved (SD) under the second half of condition (II). In a similar manner it can be shown that for every $\beta > 0$ there exists a $c \in (0, \infty)$ such that

$$d = 1: \quad \limsup_{n\to\infty} n^{-\frac{1}{3}}\log M_n^{\beta,\omega} \le -c \qquad \omega - a.s.,$$
$$d = 2: \quad \limsup_{n\to\infty} (\log n)^{-\frac{1}{2}}\log M_n^{\beta,\omega} \le -c \qquad \omega - a.s. \tag{12.81}$$

This implies that (SD) also holds under the first half of condition (II). For details we refer to [69]. □

• Proof of (12.20):

Proof. With the help of (12.62), (12.80) and Lemma 12.6, we estimate

$$\limsup_{n\to\infty} \max_{x\in\mathbb{Z}^d} P_{n-1}^{\beta,\omega}(S_n = x) \ge \limsup_{n\to\infty} \frac{1}{n}\sum_{i=1}^{n} I_i^{\beta,\omega}$$
$$\ge -\liminf_{n\to\infty}\frac{1}{c_1 n}\log M_n^{\beta,\omega} \ge c/c_1 > 0, \tag{12.82}$$

which proves the claim. □

12.7 Beyond Second Moments

Recall (12.43), which says that (I) holds if and only if $\sup_{n\in\mathbb{N}}\mathbb{E}([M_n^{\beta,\omega}]^2) < \infty$, implying (WD) as stated in (12.25). Higher moments are of no use to improve (I). Indeed, Coyle [77] shows that $\lim_{n\to\infty}\mathbb{E}([M_n^{\beta,\omega}]^\alpha)$ diverges for all $\alpha \ge \alpha_0$ for some $\alpha_0 = \alpha_0(d, \beta) \in (0, \infty)$ when $d \ge 3$ and $\beta \in (0, \infty)$. The way forward is to consider lower moments. Clearly, to get (WD) it is sufficient that

$$\sup_{n\in\mathbb{N}} \mathbb{E}([M_n^{\beta,\omega}]^\alpha) < \infty \quad \text{for some } \alpha \in (1,2], \tag{12.83}$$

and so we are looking for a condition on d and β under which the latter is true. Monthus and Garel [245] conjecture that the second moment estimate is sharp, i.e., $\beta_c^1 = \beta_c$ in (12.26). However, we shall see that this is wrong.

12.7.1 Fractional Moment Estimates

Evans and Derrida [97], using convexity arguments, derive a sufficient condition for (12.83), namely,

$$c(\alpha\beta) - \alpha c(\beta) < \log[1/\rho(\alpha)] \qquad (12.84)$$

with

$$\rho(\alpha) = \sum_{n \in \mathbb{N}} \sum_{x \in \mathbb{Z}^d} \left[(P \times P')(S_i \neq S_i' \text{ for } 0 < i < n, \, S_n = S_n' = x) \right]^{\frac{1}{2}\alpha}. \qquad (12.85)$$

Together with the definitions

$$\beta_c(\alpha) = \sup\left\{ \beta \in [0, \infty) : \, c(\alpha\beta) - \alpha c(\beta) < \log[1/\rho(\alpha)] \right\}, \qquad \alpha \in (1, 2], \qquad (12.86)$$

and

$$\bar{\beta}_c = \sup_{\alpha \in (1,2]} \beta_c(\alpha), \qquad (12.87)$$

this yields

$$\beta < \bar{\beta}_c \quad \Longrightarrow \quad \text{(WD)}. \qquad (12.88)$$

Note that $\rho(2) = \pi_d$, so that for $\alpha = 2$ the criterion in (12.84) reduces to (I), i.e., from (12.23) we have

$$\beta_c(2) = \beta_c^1. \qquad (12.89)$$

Consequently, $\bar{\beta}_c \geq \beta_c^1$. Thus, the task is to compute $\bar{\beta}_c$ for different choices of d and the disorder distribution and to see in which cases $\bar{\beta}_c > \beta_c(2)$. In [97] this task is taken up numerically for standard Gaussian disorder, and it is shown that $\bar{\beta}_c > \beta_c^1$ for d sufficiently large.

Camanes and Carmona [48] carry out a systematic study of $\rho(\alpha)$ when $d \geq 3$, showing that there is an $\alpha_0 \in [1, 2)$ such that $\alpha \mapsto \rho(\alpha)$ is infinite on $(1, \alpha_0]$, is continuous and non-increasing on $(\alpha_0, 2]$, and satisfies $\lim_{\alpha \downarrow \alpha_0} \rho(\alpha) = \infty$. Since $\rho(2) = \pi_d < 1$, it follows that there is an $\alpha_1 \in (\alpha_0, 2)$ such that $\rho(\alpha_1) = 1$, so that $\alpha \mapsto \beta_c(\alpha)$ is defined on $[\alpha_1, 2]$ and satisfies $\beta_c(\alpha_1) = 0$. They then show that, subject to a certain entropy condition on the random walk and the disorder, $\alpha \mapsto \beta_c(\alpha)$ goes through a maximum at some $\bar{\alpha} \in (\alpha_1, 2)$, which therefore is the optimal value for the estimate. Based on this analysis, they show that $\bar{\beta}_c = \beta_c(\bar{\alpha}) > \beta_c^1$ for standard Gaussian disorder in $d \geq 5$, binomial disorder in $d \geq 4$ with small mean, and Poisson disorder in $d \geq 3$ with small mean (see Fig. 12.5).

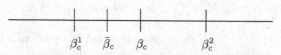

Fig. 12.5. Bounds on the critical temperature.

12.7.2 Size-biasing

Birkner [16] carries the analysis a crucial step further by proving the following result.

Theorem 12.7. *Let*

$$\theta^* = \sup \left\{ \theta \geq 1 \colon E\big(\theta^{V(S,S')}\big) < \infty \ S' - a.s. \right\} \tag{12.90}$$

with $V(S, S') = \sum_{n \in \mathbb{N}} 1_{\{S_n = S_n'\}}$ *the collision local time of two independent SRWs, both starting at 0. (Note that* θ^* *is non-random because the tail sigma-algebra of* S' *is trivial.) Define*

$$\beta_c^* = \sup \left\{ \beta \in [0, \infty) \colon c(2\beta) - 2c(\beta) < \log \theta^* \right\}. \tag{12.91}$$

Then

$$\beta < \beta_c^* \implies \text{(WD)}. \tag{12.92}$$

Proof. The proof is based on a size-biasing argument, which runs as follows. Define (suppress from β from the notation)

$$e = \{e(i, x)\}_{i \in \mathbb{N}, x \in \mathbb{Z}^d}, \qquad e(i, x) = e^{\beta \omega(i, x) - c(\beta)}, \tag{12.93}$$

which is an i.i.d. $(0, \infty)$-valued random field. Define a size-biased version of this random field, written

$$\hat{e} = \{\hat{e}(i, x)\}_{i \in \mathbb{N}, x \in \mathbb{Z}^d} \tag{12.94}$$

and independent of e, such that

$$\mathbb{P}(\hat{e}(1, 0) \in \cdot) = \mathbb{E}\big(e(1, 0) \, 1_{\{e(1,0) \in \cdot\}}\big) \tag{12.95}$$

(no normalization is needed because $\mathbb{E}(e(1, 0)) = 1$). Given S', put

$$\hat{e}_{S'} = \{\hat{e}_{S'}(i, x)\}_{i \in \mathbb{N}, x \in \mathbb{Z}^d}, \qquad \hat{e}_{S'}(i, x) = 1_{\{S_i' \neq x\}} \, e(i, x) + 1_{\{S_i' = x\}} \, \hat{e}(i, x), \tag{12.96}$$

and define

$$\hat{M}_n^{e, \hat{e}, S'} = E \left(\prod_{i=1}^n \hat{e}_{S'}(i, S_i) \right). \tag{12.97}$$

This is a size-biased version of the martingale defined in (12.9), which in the present notation is

$$M_n^e = E \left(\prod_{i=1}^n e(i, S_i) \right). \tag{12.98}$$

The point of (12.97) is that for any bounded function $f \colon [0, \infty) \to \mathbb{R}$,

$$\mathbb{E}\big(M_n^e \, f(M_n^e)\big) = (\mathbb{E} \times \hat{\mathbb{E}} \times E')\Big(f\big(\hat{M}_n^{e, \hat{e}, S'}\big)\Big), \tag{12.99}$$

where $\mathbb{E}, \hat{\mathbb{E}}, E, E'$ denote expectation w.r.t. $e, \hat{e}, S, S'$, respectively. Indeed, this identity follows from the following computation:

$$
\begin{aligned}
\mathbb{E}\big(M_n^e f(M_n^e)\big) &= \mathbb{E}\left[E'\left(\prod_{i=1}^n e(i,S_i')\right) f\left(E\left(\prod_{i=1}^n e(i,S_i)\right)\right)\right]\\
&= E'\left(\mathbb{E}\left[\left(\prod_{i=1}^n e(i,S_i')\right) f\left(E\left(\prod_{i=1}^n e(i,S_i)\right)\right)\right]\right)\\
&= E'\left((\mathbb{E}\times\hat{\mathbb{E}})\left[f\left(E\left(\prod_{i=1}^n \hat{e}_{S'}(i,S_i)\right)\right)\right]\right)\\
&= (\mathbb{E}\times\hat{\mathbb{E}}\times E')\big(f(\hat{M}_n^{e,\hat{e},S'})\big),
\end{aligned}
\tag{12.100}
$$

where the third equality uses (12.95–12.96).

Next, it follows from (12.99) that

$$
(M_n^e)_{n\in\mathbb{N}_0} \text{ is uniformly integrable} \quad \Longleftrightarrow \quad (\hat{M}_n^{e,\hat{e},S'})_{n\in\mathbb{N}_0} \text{ is tight.} \tag{12.101}
$$

However,

$$
(\mathbb{E}\times\hat{\mathbb{E}})\big(\hat{M}_n^{e,\hat{e},S'}\big) = E\left(\theta^{\sum_{i=1}^n 1_{\{S_i=S_i'\}}}\right) \le E\big(\theta^{V(S,S')}\big), \qquad \theta = e^{c(2\beta)-2c(\beta)},
\tag{12.102}
$$

and hence $E(\theta^{V(S,S')}) < \infty$ S'-a.s. is enough to ensure the r.h.s. of (12.101). This completes the proof because the l.h.s. of (12.101) is equivalent to (WD). □

Since $(E\times E')(\theta^{V(S,S')}) < \infty$ when $\theta < 1/\pi_d$, we see from (12.90) that $\theta^* \ge 1/\pi_d$, so that $\beta_c^* \ge \beta_c^1$. It was proved in Birkner, Greven and den Hollander [18] that $\theta^* > \pi_d$ in $d \ge 5$, implying that $\beta_c^* > \beta_c^1$ (see Fig. 12.6). Note that this gap holds *irrespective* of the distribution of the disorder, in contrast with the gap that was obtained by Camanes and Carmona [48] with the help of fractional moment estimates (see Section 12.7.1). In [18] it is conjectured that the gap persists in $d = 3, 4$. For $d = 4$ this conjecture is settled in Birkner and Sun [19].

Theorem 12.7 is optimal, in the sense that the size-biasing is a fractional moment estimate of order α with α "infinitesimally close" to 1. The only point where something is lost in the argument is (12.102), where an expectation is

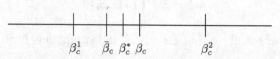

Fig. 12.6. Bounds on the critical temperature.

taken w.r.t. to the disorder. Presumably there exists a $\beta_c^{**} > \beta_c^{*}$ such that the r.h.s. of (12.101) holds for $\beta < \beta_c^{**}$. If so, then $\beta_c^{**} = \beta_c$, the critical threshold between (WD) and (SD) found in Theorem 12.1.

12.7.3 Relation with Random Pinning

As noted in Birkner and Sun [19], there is an interesting connection between the gap $\beta_c^{*} > \beta_c(2)$ obtained in Section 12.7.2 and the question of relevant vs. irrelevant disorder in the random pinning model discussed in Section 11.6.

Return to (12.96–12.97). Given S', put

$$\check{e}_{S'} = \{\check{e}_{S'}(i,x)\}_{i\in\mathbb{N}, x\in\mathbb{Z}^d}, \qquad \check{e}_{S'}(i,x) = 1_{\{S_i' \neq x\}}\, e(i,x) + 1_{\{S_i'=x\}}\, \hat{e}(i,0),$$
$$(12.103)$$

and define

$$\check{M}_n^{e,\check{e},S'} = E\left(\prod_{i=1}^n \check{e}_{S'}(i,S_i)\right). \tag{12.104}$$

Then, because the disorder is i.i.d., the distribution of $\check{M}_n^{e,\check{e},S'}$ as a function of $e, \check{e}, S'$ is the same as the distribution of $M_n^{e,\hat{e},S'}$ as a function of $e, \hat{e}, S'$. Therefore, tightness of $(M_n^{e,\hat{e},S'})_{n\in\mathbb{N}_0}$ is equivalent to tightness of $(\check{M}_n^{e,\check{e},S'})_{n\in\mathbb{N}_0}$. Consequently, recalling (12.101), we have

$$\sup_{n\in\mathbb{N}_0} (\mathbb{E} \times E')\left(\check{M}_n^{e,\check{e},S'}\right) < \infty \ \ a.s. \quad \Longrightarrow \quad \text{(WD)}. \tag{12.105}$$

Now, an easy computation gives

$$(\mathbb{E} \times E')\left(\check{M}_n^{e,\check{e},S'}\right) = (E \times E')\left(\exp\left[\sum_{i=1}^n (\beta\check{\omega}_i - c(\beta))\, 1_{\{S_i=S_i'\}}\right]\right)$$
$$= \tilde{E}\left(\exp\left[\sum_{i=1}^n (\beta\check{\omega}_i - c(\beta))\, 1_{\{\tilde{S}_i=0\}}\right]\right), \tag{12.106}$$

where $\check{\omega} = (\check{\omega}_i)_{i\in\mathbb{N}}$ is i.i.d. with moment generating function

$$\mathbb{\check{E}}\left(e^{\beta\check{\omega}_1}\right) = e^{c(2\beta)-c(\beta)}, \tag{12.107}$$

and $\tilde{S} = (\tilde{S}_i)_{i\in\mathbb{N}_0}$ is the random walk whose excursion lengths away the origin have the same distribution as the first collision time of two independent copies of the random walk $S = (S_i)_{i\in\mathbb{N}_0}$. Recalling (11.3–11.4), we see that the r.h.s. of (12.106) is the partition sum of the random pinning model with parameters $(\lambda, h) = (\beta, c(\beta))$, disorder $\check{\omega}$ and reference random walk $\tilde{S}$.

Let $\tilde{f}(\beta)$ denote the quenched free energy associated with the above random pinning model. The annealed free energy $\tilde{f}^{\text{ann}}(\beta)$ is the free energy of the homopolymer with parameter $\zeta = c(2\beta) - 2c(\beta)$, which is > 0 if and only if $c(2\beta) - 2c(\beta) > \log(1/\pi_d)$ (recall Section 7.6, Extension (1)), i.e., if and only if $\beta > \beta_c(2)$. It follows from the results in Section 11.6 that, for all $\tilde{S}$ whose excursion length distribution is such that the disorder is relevant in the random pinning model (which includes SRW), there exists a $\beta > \beta_c(2)$, sufficiently close to $\beta_c(2)$, such that $\tilde{f}(\beta) = 0 < \tilde{f}^{\mathrm{ann}}(\beta)$. This in turn implies that $\beta_c > \beta_c(2)$.

What the above considerations show is that the directed polymer in random environment with disorder ω and reference random walk S has a gap $\beta_c > \beta_c(2)$ as soon as the random pinning model with *biased disorder* $\breve{\omega}$ and reference random walk given by the *difference random walk* $\tilde{S}$ has relevant disorder. Note that if ω has finite moment generating function, then so has $\tilde{\omega}$. Thus, we can use the results in Section 11.6, which only depend on the type of random walk $\tilde{S}$.

12.8 Extensions

(1) Essentially all the results described in Sections 12.1–12.7 carry over from SRW to an arbitrary reference random walk, with π_d in (12.14) as the key quantity for the weak disorder regime. The diffusive scaling in (12.18) only holds when S has zero mean and finite variance, but can be adapted accordingly. It is natural to conjecture that $\beta_c > \beta_c^1$ for all transient $\tilde{S}$ with excursion length exponent $a \in (0, \infty)$. Indeed, presumably something is lost by taking the expectation w.r.t. the disorder ω in (12.105), so that the results for the random pinning model do not give the sharp criterion for the gap.

(2) It is proved in Comets and Yoshida [74] that for arbitrary disorder with finite moment generating function there exists a $\tilde{\beta}_c \in [0, \infty]$ such that

$$f(\beta) \begin{cases} = f^{\mathrm{ann}}(\beta), & \text{if } \beta \in [0, \tilde{\beta}_c], \\ < f^{\mathrm{ann}}(\beta), & \text{if } \beta \in (\tilde{\beta}_c, \infty). \end{cases} \tag{12.108}$$

This implies that $\tilde{\beta}_c$ is a point of non-analyticity of the quenched free energy, satisfying $\tilde{\beta}_c \geq \beta_c$ because of (12.13). It is proved in Comets and Vargas [71] that $\tilde{\beta}_c = 0$ for arbitrary disorder and $d = 1$, and in Lacoin [226] that $\tilde{\beta}_c = 0$ for Gaussian disorder and $d = 2$. We saw in Theorem 12.3 that $\beta_c = 0$ in $d = 1, 2$. It is expected that $\tilde{\beta}_c = \beta_c$ in general.

Note that, by (12.6–12.8),

$$f^{\mathrm{ann}}(\beta) - f(\beta) = -\lim_{n\to\infty} \frac{1}{n} \log M_n^{\beta,\omega} \qquad \omega - a.s. \tag{12.109}$$

Thus, the second half of Lemma 12.6 shows that, for all $\beta \in (\tilde{\beta}_c, \infty)$, $\liminf_{n\to\infty} \frac{1}{n} \sum_{i=1}^n I_i^{\beta,\omega} \geq c > 0$ ω-a.s. In combination with the sandwich in (12.75), the latter implies the localization property in (12.20), which therefore holds beyond condition (II) in Theorem 12.3.

(3) As shown in Imbrie and Spencer [183], Bolthausen [24], Sinai [275], Song and Zhou [282], under condition (I) in Theorem 12.2 the CLT holds for the endpoint of the polymer. Comets and Yoshida [74] prove that, in fact, $d \geq 3$ and (WD) imply the CLT. Conlon and Olsen [75] and Coyle [76] explain that the reason why the diffusion constant is not renormalized is that the disorder is *uncorrelated in time*. As a result, under the annealed path measure the polymer evolves as SRW. As soon as time-correlation is added, the diffusion constant in general becomes renormalized, similarly as for random walk in random environment (see the Saint-Flour lectures by Zeitouni [323]).

(4) Song and Zhou [282] and Albeverio and Zhou [2] derive a rate of convergence result for (12.18). In particular, they show that, subject to (I) and for any $\delta > 0$,

$$\mathbb{E}\big(E_n^{\beta,\omega}(\|S_n\|^2 - n)\big) \sim \begin{cases} O\big(n^{\frac{6-d}{4}+\delta}\big), & \text{if } d = 3,4,5, \\ O\big([\log n]^{1+\delta}\big), & \text{if } d = 6, \\ O(1), & \text{if } d \geq 7. \end{cases} \qquad (12.110)$$

Partial results had been obtained earlier in [183], [24].

(5) Carmona and Hu [61], [62] (Gaussian disorder) and Comets, Shiga and Yoshida [69] (general disorder) derive large deviation bounds for the endpoint of the polymer.

(6) Essentially all results in Sections 12.1–12.5 carry over from SRW to an arbitrary irreducible random walk. The gap $\beta_c^* > \beta_c^1$ is proved in [18] for any symmetric strongly transient random walk with a regularly varying tail at infinity, and is conjectured to be present for any symmetric transient random walk.

(7) It is observed numerically in Huse and Henley [176] that in $d = 1$

$$E_n^{\beta,\omega}(\|S_n\|^2) \asymp n^{\frac{4}{3}} \quad \text{as } n \to \infty \quad \omega - a.s. \qquad (12.111)$$

An intuitive explanation of this superdiffusive scaling is given in Huse, Henley and Fisher [177], Kardar and Nelson [206], and Hwa and Fisher [180], based on the idea that in low dimension even for small β the disorder is "effectively strong at large times and large distances" (see also Kardar [205] and Kardar and Zhang [207]). There still is no proof available. For a Gaussian version of the model, in which SRW is replaced by a random walk in $\mathbb{R}$ with Gaussian steps and the disorder is a Gaussian random field in $\mathbb{N} \times \mathbb{R}^d$ whose covariance function is a two-sided exponential, Petermann [262] shows that the exponent $\nu(1)$ in (12.21) satisfies $\nu(1) \geq \frac{3}{5}$. Mejane [240] complements this result by showing that $\nu(d) \leq \frac{3}{4}$ for $d \geq 1$ as soon as the covariance function is bounded and integrable, extending earlier work by Piza [267] for SRW and bounded disorder.

(8) Comets and Yoshida [72], [73], [74] study a continuous space-time version of the model in which the reference process is Brownian motion and the

random environment is a *Poisson point process* with constant intensity. The Hamiltonian at time t is chosen to be $-\beta$ times the number of points in the 1-neighborhood of the path up to time t. Essentially all the results described in Sections 12.2–12.4 carry over, including the bound $\nu(d) \le \frac{3}{4}$ for $d \ge 1$ mentioned in Extension (7). Stochastic analysis techniques allow for more elegant and flexible proofs. This model is interesting both for $\beta > 0$ (attractive points) and $\beta < 0$ (repulsive points), and has a close link with the work on Brownian motion in a random environment consisting of finite balls whose centers are located on a Poisson point process, as studied in the monograph by Sznitman [288] and in Wüthrich [318], [319], [320], [321].

(9) Sinai [275] proves a local central limit theorem for the discrete model, showing that, under condition (I) and conditionally on the event $\{S_n = x\}$ with $\|x\| = o(\sqrt{n})$, the polymer feels the disorder only near 0 and x, and in between behaves like a SRW conditioned to hit x at time n. Vargas [304] extends this result to the continuous model with Poisson point process disorder mentioned in Extension (8).

(10) Moriarty and O'Connell [246] consider a continuous version of the directed polymer in random environment given by the partition sum

$$Z_N^{\beta,N}(B) = \int dt_1 \cdots \int dt_{N-1} \, 1_{\{0 \le t_1 < \cdots < t_{N-1} \le N\}}$$
$$\times \exp\left[\beta \sum_{i=1}^{N} [B_i(t_i) - B_i(t_{i-1})]\right], \tag{12.112}$$

where $N \in \mathbb{N}$, $\beta \in \mathbb{R}$, $t_0 = 0$, $t_N = N$, and $B = (B_m)_{m \in \mathbb{N}}$ is a sequence of independent standard Brownian motions on $\mathbb{R}$. This models a directed polymer of length N visiting N layers successively, running along each layer for some time and picking up a Brownian increment. It is shown that, for every $\beta \ne 0$,

$$f(\beta) = \lim_{N \to \infty} \frac{1}{N} \log Z_N^{\beta,N}(B) = -\mathcal{L}[-\psi](-\beta^2) - \log(\beta^2) \quad B\text{-}a.s., \tag{12.113}$$

where $\psi = \Gamma'/\Gamma$ with Γ the Gamma-function, and $\mathcal{L}[-\psi]$ is the Legendre transform of $-\psi$. This settles a conjecture put forward in O'Connell and Yor [249]. It is further shown that $\beta \mapsto f(\beta)$ is analytic and strictly convex on $\mathbb{R}$ (with $f(0) = 1$). Hence, there is no phase transition as a function of β. The reason for the latter is that the system is in the strong disorder phase for every $\beta \ne 0$. Indeed, the annealed free energy is $f^{\mathrm{ann}}(\beta) = \frac{1}{2}\beta^2 + 1$, because

$$\mathbb{E}(Z_N^{\beta,N}(B)) = \int dt_1 \cdots \int dt_{N-1} \, 1_{\{0 \le t_1 < \cdots < t_{N-1} \le N\}}$$
$$\times \mathbb{E}\left(e^{\beta[B_1(N) - B_1(0)]}\right) = \frac{N^{N-1}}{(N-1)!} e^{\frac{1}{2}\beta^2 N}, \tag{12.114}$$

and it is easy to check that $f(\beta) < f^{\mathrm{ann}}(\beta)$ for all $\beta \ne 0$.

It is immediate from (12.112) that

$$\lim_{\beta \to \infty} \frac{1}{\beta} \log Z_N^{\beta,N}(B) = \sup_{0 \leq t_1 < \cdots < t_{N-1} \leq N} \sum_{i=1}^{N} [B_i(t_i) - B_i(t_{i-1})]. \quad (12.115)$$

Let us abbreviate the r.h.s. by $M_N^N(B)$. Then it follows from the work of Baryshnikov [13] and Gravner, Tracy and Widom [128] on spectra of random matrices and related growth models that $N^{-1/3}[M_N^N(B) - 2N]$ converges in distribution as $N \to \infty$ to the Tracy-Widom law (see Tracy and Widom [297]). This fits well with the conjecture that the fluctuation exponent for the partition sum of the directed polymer in random environment equals $\chi(1) = \frac{1}{3}$ (recall Section 12.2.4).

(11) Buffet, Patrick and Pulé [45] consider the directed polymer in random environment on a binary tree, extending earlier work by Derrida and Spohn [84]. Paths can only move forward in the tree, and the reference measure on path space is taken to be the counting measure. Since $\mathbb{E}(Z_n^{\beta,\omega}) = [2e^{c(\beta)}]^n$, the annealed free energy equals $f^{\text{ann}}(\beta) = c(\beta) + \log 2$. The quenched free energy is self-averaging and satisfies

$$f(\beta) = \begin{cases} f^{\text{ann}}(\beta), & \text{if } \beta \in [0, \beta_c], \\ f^{\text{ann}}(\beta_c), & \text{if } \beta \in (\beta_c, \infty), \end{cases} \quad (12.116)$$

where $\beta_c \in (0, \infty]$ is the critical threshold separating (WD) and (SD). This threshold is shown to solve the equation $g'(\beta) = 0$ with $g(\beta) = \beta^{-1} f^{\text{ann}}(\beta)$. The function $\beta \mapsto g(\beta)$ is strictly decreasing on $(0, \beta_c)$ and strictly increasing on (β_c, ∞). It is plotted in Fig. 12.7 for the case where the disorder distribution is exponential on the negative axis with mean $-\lambda^{-1}$, $\lambda > 0$, in which case $c(\beta) = \log[\lambda/(\beta+\lambda)]$ and β_c solves the equation $\log[2\lambda/(\beta+\lambda)] = -\beta/(\beta+\lambda)$. For standard Gaussian disorder $c(\beta) = \frac{1}{2}\beta^2$ and $\beta_c = \sqrt{2 \log 2}$.

On the tree, (12.83) holds whenever $g'(\beta) < 0$, i.e., for $\beta \in (0, \beta_c)$. The computations are easy because after two paths separate they never meet again.

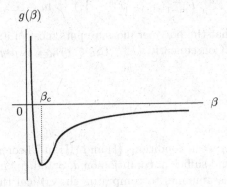

Fig. 12.7. Plot of $\beta \mapsto g(\beta)$ on the binary tree for a disorder distribution that is exponential on the negative axis.

(12) Ioffe and Velenik [185] study what happens for *undirected* paths. They consider the following partition sum:

$$Z^{\lambda,\beta,\omega}(N) = \sum_{w \in \mathcal{W}(N)} e^{-\lambda|w|-\beta\sum_{i=1}^{|w|}\omega(w_i)}, \qquad \lambda > \log(2d),\ \beta > 0, \quad (12.117)$$

where $\omega = \{\omega(x)\colon x \in \mathbb{Z}^{d+1}\}$ are i.i.d. $[0,\infty)$-valued random variables whose support contains 0, $\mathcal{W}(N)$ is the set of nearest-neighbor paths on $\mathbb{Z}^{d+1}$ of arbitrary length starting at 0 and ending at the hyperplane at distance N from 0 perpendicular to the 1-st direction, $|w|$ is the length of $w \in \mathcal{W}(N)$, and w_i is the site that w visits at time i. They show that if $d \geq 3$, then for every $\lambda > \log(2d)$ and $\beta \in [0,\beta_0)$ for some $\beta_0 = \beta_0(d,\lambda) \in (0,\infty)$,

$$\lim_{N\to\infty} M^{\lambda,\beta,\omega}(N) = \lim_{N\to\infty} \frac{Z^{\lambda,\beta,\omega}(N)}{\mathbb{E}(Z^{\lambda,\beta,\omega}(N))} \qquad (12.118)$$

exists and is strictly positive ω-a.s., and the path obeys diffusive scaling with a *renormalized* diffusion constant. This is precisely the analogue of Theorem 12.2, but now for undirected paths. The proofs are hard, because the martingale property no longer is available, and are based on an elaborate coarse-graining argument.

(13) Vargas [303] extends (SD) to disorder without finite moment generating function, where the martingale method breaks down. Vargas [304] proves that if β and the disorder distribution are such that $c(\beta) = \infty$, then for every $\delta \in (0,1)$ there exists an $\epsilon(\delta) > 0$ such that

$$\liminf_{n\to\infty} \mathbb{E}\left(\frac{1}{n}\sum_{m=1}^{n} P_{m-1}^{\beta,\omega}\left(S_m \in \mathcal{A}_m^{\beta,\omega,\epsilon(\delta)}\right)\right) \geq \delta, \qquad (12.119)$$

where

$$\mathcal{A}_m^{\beta,\omega,\epsilon} = \left\{x \in \mathbb{Z}^d \colon P_{m-1}^{\beta,\omega}(S_m = x) > \epsilon\right\}, \qquad m \in \mathbb{N},\ \epsilon > 0. \quad (12.120)$$

What this says is that the polymer measure puts most of its mass on ϵ-atoms for ϵ small enough. Consequently, $(P_{n-1}^{\beta,\omega}(S_n \in \cdot))_{n\in\mathbb{N}}$ is *asymptotically purely atomic*.

12.9 Challenges

(1) Close the gap between conditions (I) and (II) in Theorems 12.2–12.3, i.e., find the necessary and sufficient condition on d, β and $c(\cdot)$ under which (WD) and (SD) hold. This amounts to computing the critical threshold $\beta_c^{**} = \beta_c$ mentioned at the end of Section 12.6.

(2) Prove that $\tilde{\beta}_c = \beta_c$ in general, as already indicated in Section 12.8, Extension (2). Determine what happens at the critical threshold β_c. Is this part of (WD) or (SD)?

(3) Prove that $\beta_c > \beta_c^1$ as soon as the random walk S is such that the difference random walk $\tilde{S}$ is transient with excursion length exponent $a \in (0, \infty)$.

(4) Clarify the relation between $\bar{\beta}_c$ in (12.87) and β_c^* in (12.91). Is it the case that $\beta_c^* > \bar{\beta}_c$?

(5) Prove (12.111) and compute $\nu(1)$ and $\nu(2)$. Is there a scaling limit for the path that is sample-independent and, if so, then what is this scaling limit?

(6) Prove that in (SD) the path is confined to a "most favorable corridor" (recall the remarks made below Theorems 12.2–12.3). Show that in $d = 1$ the width of this corridor is of order 1 and its typical fluctuations are of order $n^{2/3}$ (see Section 12.2.4).

(7) Investigate what happens when the polymer lives on a randomly branching tree or on a supercritical percolation cluster, both carrying an i.i.d. random environment.

References

1. S. Albeverio and X.Y. Zhou, Free energy and some sample path properties of a random walk with random potential, J. Stat. Phys. 83 (1996) 573–622.
2. S. Albeverio and X.Y. Zhou, A martingale approach to directed polymers in a random environment, J. Theoret. Probab. 9 (1996) 171–189.
3. D. Aldous, Self-intersections of 1-dimensional random walks, Probab. Theory Relat. Fields 72 (1986) 559–587.
4. K.S. Alexander, Ivy on the ceiling: first-order polymer depinning transitions with quenched disorder, Markov Proc. Relat. Fields 13 (2007) 663–680.
5. K.S. Alexander, The effect of disorder on polymer depinning transitions, Commun. Math. Phys. 279 (2008) 117–146.
6. K.S. Alexander and V. Sidoravicius, Pinning of polymers and interfaces by random potentials, Ann. Appl. Probab. 16 (2006) 636–669.
7. K.S. Alexander and N. Zygouras, Quenched and annealed critical points in polymer pinning models, preprint 2008.
8. K.S. Alexander and N. Zygouras, Equality of critical points for polymer depinning transitions with loop exponent one, preprint 2008.
9. S.E. Alm and S. Janson, Random self-avoiding walks on one-dimensional lattices, Commun. Statist. Stochastic Models 6 (1990) 169–212.
10. S. Asmussen, *Applied Probability and Queues* (2nd. ed.), Applications of Mathematics, Vol. 51, Springer, New York, 2003.
11. A. Asselah, Annealed large deviation estimates for the energy of a polymer, work in progress.
12. M.N. Barber and B.W. Ninham, *Random and Restricted Walks: Theory and Applications*, Gordon and Breach, New York, 1970.
13. Yu. Baryshnikov, GUEs ans queues, Probab. Theory Relat. Fields 119 (2001) 256–274.
14. P. Billingsley, *Probability and Measure*, John Wiley & Sons, New York, 1968.
15. N.H. Bingham, C.M. Goldie and J.L. Teugels, *Regular Variation*, Cambridge University Press, Cambridge, 1987.
16. M. Birkner, A condition for weak disorder for directed polymers in random environment, Elect. Comm. Prob. 9 (2004) 22–25.
17. M. Birkner, A. Greven and F. den Hollander, Quenched large deviation principle for words in a letter sequence, EURANDOM Report 2008–034, to appear in Probab. Theory Relat. Fields.

18. M. Birkner, A. Greven and F. den Hollander, Collision local time of transient random walks and intermediate phases in interacting stochastic systems, preprint 2008.

19. M. Birkner and R. Sun, Annealed vs quenched critical points for a random walk pinning model, preprint 2008.

20. M. Biskup and F. den Hollander, A heteropolymer near a linear interface, Ann. Appl. Probab. 9 (1999) 668–687.

21. M. Biskup and W. König, Long-time tails in the parabolic Anderson model with bounded potential, Ann. Probab. 29 (2001) 636–682.

22. T. Bodineau and G. Giacomin, On the localization transition of random copolymers near selective interfaces, J. Stat. Phys. 117 (2004) 801–818.

23. T. Bodineau, G. Giacomin, H. Lacoin and F.L. Toninelli, Copolymers at selective interfaces: new bounds on the phase diagram, J. Stat. Phys. 132 (2008) 603–626.

24. E. Bolthausen, A note on diffusion of directed polymers in a random environment, Commun. Math. Phys. 123 (1989) 529–534.

25. E. Bolthausen, On self-repellent one-dimensional random walks, Probab. Theory Relat. Fields 86 (1990) 423–441.

26. E. Bolthausen, On the construction of the three-dimensional polymer measure, Probab. Theory Relat. Fields 97 (1993) 81–101.

27. E. Bolthausen, Localization of a two-dimensional random walk with an attractive path interaction, Ann. Probab. 22 (1994) 875–918.

28. E. Bolthausen, *Large Deviations and Interacting Random Walks*, Lecture Notes in Mathematics 1781, Springer, Berlin, 2002, pp. 1–124.

29. E. Bolthausen and G. Giacomin, Periodic copolymers at selective interfaces: a large deviations approach, Ann. Appl. Probab. 15 (2005) 963–983.

30. E. Bolthausen, F. Caravenna and B. de Tilière, The quenched critical point of a diluted polymer model, to appear in Stoch. Proc. Appl.

31. E. Bolthausen and F. den Hollander, Localization transition for a polymer near an interface, Ann. Probab. 25 (1997) 1334–1366.

32. E. Bolthausen and C. Ritzmann, A central limit theorem for convolution equations and weakly self-avoiding walks, preprint 2007, unpublished.

33. E. Bolthausen and U. Schmock, On self-attracting random walks, in: *Stochastic Analysis* (eds. M.C. Cranston and M.A. Pinsky), Proc. Sympos. Pure Math., Vol. 57, American Mathematical Society, Providence, RI, 1995, pp. 23–44.

34. E. Bolthausen and U. Schmock, On self-attracting d-dimensional random walks, Ann. Probab. 25 (1997) 531–572.

35. R. Brak, J.W. Essam and A.L. Owczarek, New results for directed vesicles and chains near an attractive wall, J. Stat. Phys. 93 (1998) 155–193.

36. R. Brak, A.J. Guttmann and S.G. Whittington, A collapse transition in a directed walk model, J. Phys. A: Math. Gen. 25 (1992) 2437–2446.

37. R. Brak, G.K. Iliev, A. Rechnitzer and S.G. Whittington, Motzkin path models of long chain polymers in slits, J. Phys. A: Math. Gen. 40 (2007) 4415–4437.

38. R. Brak, A.L. Owczarek and T. Prellberg, A scaling theory of the collapse transition in geometric cluster models of polymers and vesicles, J. Phys. A: Math. Gen. 26 (1993) 4565–4579.

39. R. Brak, A.L. Owczarek, A. Rechnitzer and S.G. Whittington, A directed model of a long chain polymer in a slit with attractive walls, J. Phys. A: Math. Gen. 38 (2005) 4309–4325.

40. V.A. Brazhnyi and S. Stepanow, Adsorption of a random heteropolymer with random self-interactions onto an interface, Eur. Phys. J. B27 (2002) 355–362.

41. D.C. Brydges and J.Z. Imbrie, End-to-end distance from the Green's function for a hierarchical self-avoiding walk in four dimensions, Commun. Math. Phys. 239 (2003) 523–547.

42. D.C. Brydges and G. Slade, A collapse transition for self-attracting walks, Resenhas do Instituto da Matemática e Estatística da Universidade de São Paulo, 1 (1994) 363–372.

43. D.C. Brydges and G. Slade, The diffusive phase of a model of self-interacting walks, Probab. Theory Relat. Fields 103 (1995) 285–315.

44. D.C. Brydges and T. Spencer, Self-avoiding walk in 5 or more dimensions. Commun. Math. Phys. 97 (1985) 125–148.

45. E. Buffet, A. Patrick and J.V. Pulé, Directed polymers on trees: a martingale approach, J. Phys. A: Math. Gen. 26 (1993) 1823–1834.

46. E. Buffet and J.V. Pulé, A model of continuous polymers with random charges, J. Math. Phys. 38 (1997) 5143–5152.

47. T.W. Burkhardt, Localization-delocalization transition in a solid-on-solid model with a pinning potential, J. Phys. A: Math. Gen. 14 (1981) L63–L68.

48. A. Camanes and P. Carmona, Directed polymers, critical temperature and uniform integrability, preprint 2007, submitted to Markov Proc. Relat. Fields.

49. P. Caputo, F. Martinelli and F.L. Toninelli, On the approach to equilibrium for a polymer with adsorption and repulsion, Elect. J. Probab. 13 (2008) 213–258.

50. S. Caracciolo, G. Parisi and A. Pelissetto, Random walks with short-range interaction and mean-field behavior, J. Stat. Phys. 77 (1994) 519–543.

51. F. Caravenna, *Random Walk Models and Probabilistic Techniques for Inhomogeneous Polymer Chains*, Ph.D. Thesis, University of Milan Bicocca and University of Paris 7, October 21, 2005.

52. F. Caravenna and J.-D. Deuschel, Pinning and wetting transition for (1+1)-dimensional fields with Laplacian interaction, to appear in Ann. Probab.

53. F. Caravenna and J.-D. Deuschel, Scaling limits of (1+1)-dimensional pinning models with Laplacian interaction, preprint 2008.

54. F. Caravenna and G. Giacomin, On constrained annealed bounds for pinning and wetting models, Elect. Comm. Probab. 10 (2005) 179–189.

55. F. Caravenna, G. Giacomin and M. Gubinelli, A numerical approach to copolymers at selective interfaces, J. Stat. Phys. 122 (2006) 799–832.

56. F. Caravenna, G. Giacomin and L. Zambotti, Sharp asymptotic behavior for wetting models in (1+1)-dimension, Elect. J. Probab. 11 (2006) 345–362.

57. F. Caravenna, G. Giacomin and L. Zambotti, A renewal theory approach to periodic copolymers with adsorption, Ann. Appl. Probab. 17 (2007) 1362–1398.

58. F. Caravenna, G. Giacomin and L. Zambotti, Infinite volume limits of polymer chains with periodic charges, Markov Proc. Relat. Fields 13 (2007) 679–730.

59. F. Caravenna and N. Pétrélis, A polymer in a multi-interface medium, preprint 2008, submitted to Ann. Appl. Prob.

60. F. Caravenna and N. Pétrélis, Depinning of a polymer at an infinity of interfaces, preprint 2008.

61. P. Carmona and Y. Hu, On the partition function of a directed polymer in a Gaussian random environment, Probab. Theory Relat. Fields 124 (2002) 431–457.

62. P. Carmona and Y. Hu, Fluctuation exponents and large deviations for directed polymers in a random environment, Stoch. Proc. Appl. 112 (2004) 285–308.

63. M.C.T.P. Carvalho and V. Privman, Directed walk models of polymers at interfaces, J. Phys. A: Math. Gen. 21 (1988) L1033–L1037.

64. M.S. Causo and S.G. Whittington, A Monte Carlo investigation of the localization transition in random copolymers at an interface, J. Phys. A: Math. Gen. 36 (2003) L189–L195.

65. D. Cheliotis and F. den Hollander, Variational characterization of the critical curve for pinning of random polymers, work in progress.

66. X. Chen, Limit law for the energy of a charged polymer, preprint 2007, to appear in Ann. I. H. Poincaré P & S.

67. X. Chen, *Random Walk Intersections: Large Deviations and Some Related Topics*, to appear in Mathematical Surveys and Monographs, American Mathematical Society.

68. N. Clisby, R. Liang and G. Slade, Self-avoiding walk enumeration via the lace expansion. J. Phys. A: Math. Theor. 40 (2007) 10973–11017.

69. F. Comets, T. Shiga and N. Yoshida, Directed polymers in random environment: Path localization and strong disorder, Bernoulli 9 (2003) 705–723.

70. F. Comets, T. Shiga and N. Yoshida, Probabilistic analysis of directed polymers in a random environment: a review, in: *Stochastic Analysis on Large Scale Systems*, Adv. Stud. Pure Math. 39 (2004), pp. 115–142.

71. F. Comets and V. Vargas, Majorizing multiplicative cascades for directed polymers in random media, Alea 2 (2006) 267–277.

72. F. Comets and N. Yoshida, Brownian directed polymers in random environment, Commun. Math. Phys. 254 (2004) 257–287.

73. F. Comets and N. Yoshida, Some new results on Brownian directed polymers in random environment, RIMS Kokyuroku 1386 (2004) 50–66.

74. F. Comets and N. Yoshida, Directed polymers in random environment are diffusive at weak disorder, Ann. Probab. 34 (2006) 1746–1770.

75. J.G. Conlon and P.A. Olsen, A Brownian motion version of the directed polymer problem, J. Stat. Phys. 84 (1996) 415–454.

76. L.N. Coyle, A continuous time version of random walks in a random potential, Stoch. Proc. Appl. 64 (1996) 209–235.

77. L.N. Coyle, Infinite moments of the partition function for random walks in a random potential, J. Math. Phys. 39 (1998) 2019–2034.

78. M. Cranston, O. Hryniv and S. Molchanov, Homo- and hetero-polymers in the mean-field approximation, preprint 2008.

79. D. Cule and T. Hwa, Denaturation of heterogenous DNA, Phys. Rev. Lett. 79 (1997) 2375–2378.

80. C. Danilowicz, Y. Kafri, R.S. Conroy, V.W. Coljee, J. Weeks and M. Prentiss, Measurement of the phase diagram of DNA unzipping in the temperature-force plane, Phys. Rev. Lett. 93 (2004) 078101.

81. K. De'Bell and T. Lookman, Surface phase transitions in polymer systems, Rev. Mod. Phys. 65 (1993) 87–113.

82. A. Dembo and O. Zeitouni, *Large Deviations Techniques and Applications* (2nd. ed.), Springer, New York, 1998.

83. B. Derrida and R.B. Griffiths, Directed polymers on disordered hierarchical lattices, Europhys. Lett. 8 (1989) 111–116.

84. B. Derrida and H. Spohn, Polymers on disordered trees, spin glasses, and traveling waves, J. Stat. Phys. 51 (1988) 817–840.

85. B. Derrida, G. Giacomin, H. Lacoin and F.L. Toninelli, Fractional moment bounds and disorder relevance for pinning models, preprint 2007, to appear in Commun. Math. Phys.

86. B. Derrida, R.B. Griffiths and P.G. Higgs, A model of directed random walks with random self-interactions, Europhys. Lett. 18 (1992) 361–366.

87. B. Derrida, V. Hakim and J. Vannimenus, Effect of disorder on two-dimensional wetting, J. Stat. Phys. 66 (1992) 1189–1213.

88. B. Derrida and P.G. Higgs, Low-temperature properties of directed random walks with random self-intersections, J. Phys. A: Math. Gen. 27 (1994) 5485–5493.

89. J.-D. Deuschel, G. Giacomin and L. Zambotti, Scaling limits of equilibrium wetting models in (1+1)-dimension, Probab. Theory Relat. Fields 132 (2005) 471–500.

90. J.-D. Deuschel and W. Stroock, *Large Deviations*, Academic Press, London, 1989.

91. E.A. DiMarzio and R.J. Rubin, Adsorption of a chain polymer between two plates, J. Chem. Phys. 55 (1971) 4318–4336.

92. M. Doi and S.F. Edwards, *The Theory of Polymer Dynamics*, International Series of Monographs on Physics, Vol. 73, Clarendon Press, Oxford, 1986.

93. K. Dušek (ed.), *Polymer Networks*, Advances in Polymer Sciences 44, Springer, Berlin, 1982.

94. N. Dunford and J.T. Schwartz, *Linear Operators, Part II: Spectral Theory*, Interscience Publishers, New York, 1963.

95. B. Duplantier and H. Saleur, Exact tricritical exponents for polymers at the FTHETA point in two dimensions, Phys. Rev. Lett. 59 (1987) 539–542.

96. R. Durrett, *Probability: Theory and Examples* (2nd. ed.), Duxbury Press, Belmont, 1996.

97. M.R. Evans and B. Derrida, Improved bounds for the transition temperature of directed polymers in a finite-dimensional random medium, J. Stat. Phys. 69 (1992) 427–437.

98. M.E. Fisher, Walks, walls, wetting, and melting, J. Stat. Phys. 34 (1984) 667–729.

99. G.J. Fleer, M.A. Cohen Stuart, T. Cosgrove and J.M.H.M. Scheutjens and B. Vincent, *Polymers at Interfaces*, Chapmann and Hall, London, 1993.

100. P.J. Flory, *Principles of Polymer Chemistry*, Cornell University Press, Ithaca, 1949.

101. P.J. Flory, The configuration of a real polymer chain, J. Chem. Phys. 17 (1949) 303–310.

102. P. Flory, *The Statistical Mechanics of Chain Molecules*, Interscience Publishers, New York, 1969.

103. P.J. Flory, Spatial configuration of macromolecular chains, Nobel Lecture, 1974. (http://nobelprize.org/nobel_prizes/chemistry/laureates/1974/).

104. G. Forgacs, J.M. Luck, Th.M. Nieuwenhuizen and H. Orland, Wetting of a disordered substrate: Exact critical behavior in two dimensions, Phys. Rev. Lett. 57 (1986) 2184–2187.

105. G. Forgacs, V. Privman and H.L. Frisch, Adsorption-desorption transition of polymer chains interacting with surfaces, J. Chem. Phys. 90 (1989) 3339–3345.

106. C.M. Fortuin, P.W. Kasteleyn and J. Ginibre, Correlation inequalities on some partially ordered sets, Commun. Math. Phys. 22 (1971) 89–103.

107. D.P. Foster, Exact evaluation of the collapse phase boundary for two-dimensional directed polymers, J. Phys. A: Math. Gen. 23 (1990) L1135–L1138.

108. D.P. Foster, E. Orlandini and M.C. Tesi, Surface critical exponents for models of polymer collapse and adsorption: the universality of the θ and θ' points, J. Phys. A: Math. Gen. 25 (1992) L1211–L1217.

109. D.P. Foster and J. Yeomans, Competition between self-attraction and adsorption in directed self-avoiding polymers, Physica A 177 (1991) 443–452.

110. D.M. Gangardt and S.K. Nechaev, Wetting transition on a one-dimensional disorder, preprint 2007.

111. T. Garel, D.A. Huse, S. Leibler and H. Orland, Localization transition of random chains at interfaces, Europhys. Lett. 8 (1989) 9–13.

112. T. Garel and C. Monthus, Two-dimensional wetting with binary disorder: a numerical study of the loop statistics, Eur. Phys. J. B46 (2005) 117–125.

113. J. Gärtner and W. König, The parabolic Anderson model. In: *Interacting Stochastic Systems* (J.-D. Deuschel and A. Greven, eds.). Springer, Berlin, 2005, pp. 153–179.

114. P.-G. de Gennes, *Scaling Concepts in Polymer Physics*, Cornell University Press, Ithaca, 1979.

115. P.-G. de Gennes, Soft matter, (http://nobelprize.org/nobel_prizes/physics/laureates/1991/).

116. G. Giacomin, *Random Polymer Models*, Imperial College Press, London, 2007.

117. G. Giacomin, Renewal convergence rates and correlation decay for homogeneous pinning models, Elect. J. Probab. 13 (2008) 513–529.

118. G. Giacomin, H. Lacoin and F.L. Toninelli, Hierarchical pinning models, quadratic maps and quenched disorder, preprint 2007.

119. G. Giacomin, H. Lacoin and F.L. Toninelli, Marginal relevance of disorder for pinning models, preprint 2008.

120. G. Giacomin and F.L. Toninelli, Estimates on path delocalization for copolymers at selective interfaces, Probab. Theory Relat. Fields 133 (2005) 464–482.

121. G. Giacomin and F.L. Toninelli, Force-induced depinning of directed polymers, J. Phys. A: Math. Gen. 40 (2007) 5261–5275.

122. G. Giacomin and F.L. Toninelli, Smoothing of depinning transitions for directed polymers with quenched disorder, Phys. Rev. Lett. 96 (2006) 070602.

123. G. Giacomin and F.L. Toninelli, Smoothing effect of quenched disorder on polymer depinning transitions, Commun. Math. Phys. 266 (2006) 1–16.

124. G. Giacomin and F.L. Toninelli, The localized phase of disordered copolymers with adsorption, Alea 1 (2006) 149–180.

125. G. Giacomin and F.L. Toninelli, On the irrelevant disorder regime of pinning models, preprint 2007, submitted to Ann. Probab.

126. I. Golding and Y. Kantor, Two-dimensional polymers with random short-range interactions, Phys. Rev. E 56 (1997) R1318–R1321.

127. P. Grassberger and R. Hegger, On the collapse of random copolymers, Europhys. Lett. 31 (1995) 351–356.

128. J. Gravner, C.A. Tracy and H. Widom, Limit theorems for height fluctuations in a class of discrete space and time growth models, J. Stat. Phys. 102 (2001) 1085–1132.

129. C.T. Greenwood and E.A. Milne, *Natural High Polymers*, Contemporary Science Paperbacks 18, Oliver & Boyd, Edinburgh, 1968.

130. A. Greven and F. den Hollander, Variational characterization of the speed of a one-dimensional self-repellent random walk, Ann. Appl. Probab. 3 (1993) 1067–1099.

131. P. Griffin, Accelerating beyond the third dimension: Returning to the origin in simple random walk, Math. Scientist 15 (1990) 24–35.

132. R.B. Griffiths, Nonanalytic behavior above critical point in a random Ising ferromagnet, Phys. Rev. Lett. 23 (1969) 17–19.

133. A. Grosberg, S. Izrailev and S. Nechaev, Phase transition in a heteropolymer chain at a selective interface, Phys. Rev. E 50 (1994) 1912–1921.

134. H. Guida and J. Zinn-Justin, Critical exponents of the N-vector model, J. Phys. A: Math. Gen. 31 (1998) 8103–8121.

135. N. Gunari, A.C. Balazs and G.C. Walker, Force-induced globule-coil transition in single polystyrene chains in water, J. Am. Chem. Soc. 129 (2007) 10046–10047.

136. N. Gunari and G.C. Walker, Nanomechanical fingerprints of individual blocks of a diblock copolymer chain, Langmuir 24 (2008) 5197–5201.

137. A.J. Guttmann, Asymptotic analysis of power-series expansions, in: *Phase Transitions and Critical Phenomena* (eds. C. Domb and J.L. Lebowitz), Vol. 11, Academic Press, New York, 1989.

138. A.J. Guttmann (ed.), *Polygons, Polyominoes and Polyhedra*, Springer, Berlin, in press.

139. A.J. Guttman and S.G. Whittington, Self-avoiding walks in a slab of finite thickness: A model of steric stabilisation, J. Phys. A: Math. Gen. 11 (1978) L107–L110.

140. N. Habibzadah, G.K. Iliev, A. Saguia and S.G. Whittington, Some Motzkin path models of random and periodic copolymers, J. Phys.: Conf. Ser. 42 (2006) 111–123.

141. J.M. Hammersley, On the rate of convergence to the connective constant of the hypercubical lattice, Quart. J. Math. Oxford 12 (1961) 250–256.

142. J.M. Hammersley, G.M. Torrie and S.G. Whittington, Self-avoiding walks interacting with a surface, J. Phys. A: Math. Gen. 15 (1982) 539–571.

143. J.M. Hammersley and D.J.A. Welsh, Further results on the rate of convergence to the connective constant of the hypercubical lattice, Quart. J. Math. Oxford 13 (1962) 108–110.

144. J.M. Hammersley and S.G. Whittington, Self-avoiding walks in wedges, J. Phys. A: Math. Gen. 18 (1985) 101–111.

145. T. Hara, Decay of correlations in nearest-neighbour self-avoiding walk, percolation, lattice trees and animals, Ann. Probab. 36 (2008) 530–593.

146. T. Hara, R. van der Hofstad and G. Slade, Critical two-point functions and the lace expansion for spread-out high-dimensional percolation and related models, Ann. Probab. 31 (2003) 349–408.

147. T. Hara and G. Slade, Self-avoiding walk in five or more dimensions, I. The critical behaviour, Commun. Math. Phys. 147 (1992) 101–136.

148. T. Hara and G. Slade, The lace expansion for self-avoiding walk in five or more dimensions, Rev. Math. Phys. 4 (1992) 235–327.

149. A.B. Harris, Effect of random defects on the critical behaviour of Ising models, J. Phys. C 7 (1974) 1671–1692.

150. B.J. Haupt, T.J. Senden and E.M. Sevick, AFM evidence of Rayleigh instability in single polymer chains, Langmuir 18 (2002) 2174–2182.

151. M. Heydenreich, Long-range self-avoiding walk converges to α-stable processes, EURANDOM Report 2008–038.

152. M. Heydenreich, R. van der Hofstad and A. Sakai, Mean-field behavior for long- and finite-range Ising model, percolation and self-avoiding walk, J. Stat. Phys. 132 (2008) 1001–1049.

153. R. van der Hofstad, The constants in the central limit theorem for the one-dimensional Edwards model, J. Stat. Phys. 90 (1998) 1295–1310.

154. R. van der Hofstad, One-Dimensional Random Polymers, CWI Tract 123, Stichting Mathematisch Centrum, Amsterdam, 1998.

155. R. van der Hofstad, The lace expansion approach to ballistic behaviour for one-dimensional weakly self-avoiding walk, Probab. Theory Relat. Fields 119 (2001) 311–349.

156. R. van der Hofstad, Spread-out oriented percolation and related models above the upper critical dimension: induction and superprocesses, Ensaios Mathemáticos, Vol. 9, Sociedade Brasileira de Matemática, 2005, pp. 91–181.

157. R. van der Hofstad and F. den Hollander, Scaling for a random polymer, Commun. Math. Phys. 169 (1995) 397–440.

158. R. van der Hofstad, F. den Hollander and W. König, Central limit theorem for a weakly interacting random polymer, Markov Proc. Relat. Fields 3 (1997) 1–63.

159. R. van der Hofstad, F. den Hollander and W. König, Central limit theorem for the Edwards model, Ann. Probab. 25 (1997) 573–597.

160. R. van der Hofstad, F. den Hollander and W. König, Large deviations for the one-dimensional Edwards model, Ann. Probab. 31 (2003) 2003–2039.

161. R. van der Hofstad, F. den Hollander and W. König, Weak interaction limits for one-dimensional random polymers, Probab. Theory Relat. Fields 125 (2003) 483–521.

162. R. van der Hofstad, F. den Hollander and G. Slade, A new inductive approach to the lace expansion, Probab. Theory Relat. Fields 111 (1998) 253–286.

163. R. van der Hofstad and A. Klenke, Self-attractive random polymers, Ann. Appl. Prob. 11 (2001) 1079–1115.

164. R. van der Hofstad, A. Klenke and W. König, The critical attractive random polymer in dimension one, J. Stat. Phys. 106 (2002) 477–520.

165. R. van der Hofstad and W. König, A survey of one-dimensional random polymers, J. Stat. Phys. 103 (2001) 915–944.

166. R. van der Hofstad and G. Slade, A generalised inductive approach to the lace expansion, Probab. Theory Relat. Fields 122 (2002) 389–430.

167. R. van der Hofstad and G. Slade, The lace expansion on a tree with applications to networks of self-avoiding walks, Adv. Appl. Math. 30 (2003) 471–528.

168. F. den Hollander, Large Deviations, American Mathematical Society, Fields Institute Monographs 14, Providence, RI, 2000.

169. F. den Hollander and N. Pétrélis, A mathematical model for a copolymer in an emulsion, EURANDOM Report 2007–032, to appear in J. Math. Chem.

170. F. den Hollander and N. Pétrélis, On the localized phase of a copolymer in an emulsion: supercritical percolation regime, EURANDOM Report 2007–048, to appear in Commun. Math. Phys.

171. F. den Hollander and N. Pétrélis, On the localized phase of a copolymer in an emulsion: subcritical percolation regime, EURANDOM Report 2008–031, to appear in J. Stat. Phys.

172. F. den Hollander and S.G. Whittington, Localization transition for a copolymer in an emulsion, Theor. Prob. Appl. 51 (2006) 193–240.

173. F. den Hollander and M. Wüthrich, Diffusion of a heteropolymer in a multi-interface medium, J. Stat. Phys. 114 (2004) 849–889.

174. M. Holmes, A.A. Járai, A. Sakai and G. Slade, High-dimensional graphical networks of self-avoiding walks, Canad. J. Math. 56 (2004) 77–114.

175. B.D. Hughes, *Random Walks and Random Environments*, Vols. 1 and 2, Oxford Science Publications, Clarendon Press, Oxford, 1996.

176. D.A. Huse and C.L. Henley, Pinning and roughening of domain walls in Ising systems due to random impurities, Phys. Rev. Lett. 54 (1985) 2708–2711.

177. D.A. Huse, C.L. Henley and D. Fisher, [Roughening by impurities at finite temperatures] Huse, Henley, and Fisher respond, Phys. Rev. Lett. 55 (1985) 2094.

178. T. Hwa, Disorder-induced depinning transition, Phys. Rev. B 51 (1995) 455–469.

179. T. Hwa and D. Cule, Polymer adsorption on disordered substrates, Phys. Rev. Lett. 79 (1997) 4930.

180. T. Hwa and D.S. Fisher, Anomalous fluctuations of directed polymers in random media, Phys. Rev. B 49 (1994) 3136–3154.

181. G. Iliev, E. Orlandini and S.G. Whittington, Adsorption and localization of random copolymers subject to a force: The Morita apprximation, Eur. Phys. J. B 40 (2004) 63–71.

182. G. Iliev, A. Rechnitzer and S.G. Whittington, Random copolymer localization in the Morita approximation, J. Phys. A: Math. Gen. 38 (2005) 1209–1223.

183. J.Z. Imbrie and T. Spencer, Diffusion of directed polymers in a random environment, J. Stat. Phys. 52 (1988) 609–626.

184. D. Ioffe and Y. Velenik, Ballistic phase for self-interacting random walks, in: *Analysis and Stochastics of Growth Processes and Interface Models* (eds. P. Mörters, R. Moser, H. Schwetlick, M. Penrose and J. Zimmer), Oxford University Press, Oxford, 2008.

185. D. Ioffe and Y. Velenik, Diffusivity of semi-directed polymers at weak disorder, manuscript in preparation.

186. Y. Isozaki and N. Yoshida, Weakly pinned random walk on the wall: pathwise descriptions of the phase transition, Stoch. Proc. Appl. 96 (2001) 261–284.

187. E.J. Janse van Rensburg, Collapsing and adsorbing polygons, J. Phys. A: Math. Gen. 31 (1998) 8295–8306.

188. E.J. Janse van Rensburg, *The Statistical Mechanics of Interacting Walks, Polygons, Animals and Vesicles*, Oxford University Press, Oxford, 2000.

189. E.J. Janse van Rensburg, Adsorbing bargraph in a q-wedge, J. Phys. A: Math. Gen. 38 (2005) 8505–8525 (corrigendum: J. Phys. A: Math. Gen. 38 (2006) 8505–8525).

190. E.J. Janse van Rensburg, Forces in Motzkin paths in a wedge, J. Phys. A: Math. Gen. 39 (2006) 1581–1608.

191. E.J. Janse van Rensburg, Moments of directed paths in a wedge, J. Phys.: Conf. Ser. 42 (2006) 147–162.

192. E.J. Janse van Rensburg, E. Orlandini, A.L. Owczarek, A. Rechnitzer and S.G. Whittington, Self-avoiding walks in a slab with attractive walls, J. Phys. A: Math. Gen. 38 (2005) L823–L828.

193. E.J. Janse van Rensburg, E. Orlandini, M.C. Tesi and S.G. Whittington, Self-averaging in random self-attracting polygons, J. Phys. A: Math. Gen. 34 (2001) L37–L44.

194. E.J. Janse van Rensburg, E. Orlandini and S.G. Whittington, Self-avoiding walks in a slab: rigorous results, J. Phys. A: Math. Gen. 39 (2006) 13869–13902.

195. E.J. Janse van Rensburg and A. Rechnitzer, Multiple Markov chain Monte Carlo study of adsorbing self-avoiding walks in two and three dimensions, J. Phys. A: Math. Gen. 37 (2004) 6875–6898.

196. E.J. Janvresse, T. de la Rue and Y. Velenik, Pinning by a sparse potential, Stoch. Proc. Appl. 115 (2005) 1323–1331.

197. E.J. Janse van Rensburg and L. Ye, Forces in square lattice directed paths in wedges, J. Phys. A: Math. Gen. 38 (2005) 8493–8503.

198. E.W. James, C.E. Soteros and S.G. Whittington, Localization of a random copolymer at an interface: an exact enumeration study, J. Phys. A: Math. Gen. 36 (2003) 11575–11584.

199. E.W. James, C.E. Soteros and S.G. Whittington, Localization of a random copolymer at an interface: an untethrered self-avoiding walk model, J. Phys. A: Math. Gen. 36 (2003) 1–14.

200. I. Jensen, Enumeration of self-avoiding walks on the square lattice, J. Phys. A: Math. Gen. 37 (2004) 5503–5524.

201. I. Jensen, homepage (www.ms.unimelb.edu.au/~iwan).

202. Y. Kafri, D. Mukamel and L. Peliti, Why is the DNA denaturation transition first order?, Phys. Rev. Lett. 85 (2000) 4988-4991.

203. Y. Kantor and M. Kardar, Polymers with random self-interactions, Europhys. Lett. 14 (1991) 421–426.

204. Y. Kantor and M. Kardar, Collapse of randomly self-interacting polymers, Europhys. Lett. 28 (1994) 169–174.

205. M. Kardar, Roughening by impurities at finite temperatures, Phys. Rev. Lett. 55 (1985) 2923.

206. M. Kardar and D.R. Nelson, Commensurate-incommensurate transitions with quenched random impurities, Phys. Rev. Lett. 55 (1985) 1157–1160.

207. M. Kardar and Y.C. Zhang, Scaling of directed polymers in random media, Phys. Rev. Lett. 58 (1987) 2087–2090.

208. T. Kennedy, Ballistic behavior in a 1D weakly self-avoiding walk with decaying energy penalty, J. Stat. Phys. 77 (1994) 565–579.

209. T. Kennedy, Monte Carlo tests of Stochastic Loewner Evolution predictions for the 2D self-avoiding walk, Phys. Rev. Lett. 88 (2002) 130601.

210. H. Kesten, Ratio theorems for random walk II, J. d'Analyse Math. 11 (1963) 323–379.

211. H. Kesten, On the number of self-avoiding walks, J. Math. Phys. 4 (1963) 960–969.

212. H. Kesten and F. Spitzer, Ratio theorems for random walks, J. d'Analyse Math. 11 (1963) 285–322.

213. Y. Kifer, The Burgers equation with a random force and a general model for directed polymers in random environments, Probab. Theory Relat. Fields 108 (1997) 29–65.

214. J.F.C. Kingman, Subadditive ergodic theory, Ann. Probab. 6 (1973) 883–909.

215. F.B. Knight, Random walks and a sojourn density process of Brownian motion, Transactions of the AMS 109 (1963) 56–86.

216. W. König, The drift of a one-dimensional self-avoiding random walk, Probab. Theory Relat. Fields 96 (1993) 521–543.

217. W. König, The drift of a one-dimensional self-repellent random walk with bounded increments, Probab. Theory Relat. Fields 100 (1994) 513–544.

218. W. König, A central limit theorem for a one-dimensional polymer measure, Ann. Probab. 24 (1996) 1012–1035.

219. J. Krawczyk, T. Prellberg, A.L. Owczarek and A. Rechnitzer, Stretching of a chain polymer adsorbed at a surface, J. Stat. Mech. Theor. Exp. (2004) P10004.

220. J. Krawczyk, A.L. Owczarek, T. Prellberg and A. Rechnitzer, Pulling absorbing and collapsing polymers from a surface, J. Stat. Mech. Theor. Exp. (2005) P05008.

221. H. Krug and H. Spohn, Kinetic roughening of growing surfaces, in: *Solids far from Equilibrium* (ed. C. Godrèche), Cambridge University Press, Cambridge, 1991.

222. S. Kumar, I. Jensen, J. Jacobsen and A.J. Guttmann, Role of conformational entropy in force-induced biopolymer unfolding, Phys. Rev. Lett. 98 (2007) 128101.

223. S. Kusuoka, On the path property of Edwards' model for long polymer chains in three dimensions, Proc. Bielefeld Conf. on *Infinite Dimensional Analysis and Stochastic Processes* (ed. S. Albeverio), Res. Notes Math. 124 (Pitman Advanced Publishing Program IX), Boston, Pitman, 1985, pp. 48–65.

224. S. Kusuoka, Asymptotics of polymer measures in one dimension, Proc. Bielefeld Conf. on *Infinite Dimensional Analysis and Stochastic Processes* (ed. S. Albeverio), Res. Notes Math. 124 (Pitman Advanced Publishing Program IX), Boston, Pitman, 1985, pp. 66–82.

225. H. Lacoin, Hierarchical pinning model with site disorder: Disorder is marginally relevant, preprint 2008.

226. H. Lacoin, New bounds for the free energy of directed polymer in dimension $1 + 1$ and $1 + 2$, preprint 2008.

227. G.F. Lawler, O. Schramm and W. Werner, On the scaling limit of planar self-avoiding walks, Proc. Sympos. Pure Math., Vol. 72, Part 2, American Mathematical Society, Providence, RI, 2002, pp. 339–364.

228. J.M.J. van Leeuwen and H.J. Hilhorst, Pinning of a rough interface by an external potential, Phys. A 107 (1981) 319–329.

229. D.K. Lubensky and D.R. Nelson, Pulling pinned polymers and unzipping DNA, Phys. Rev. Lett. 85 (2000) 1572–1575.

230. N. Madras and G. Slade, *The Self-Avoiding Walk*, Birkhäuser, Boston, 1993.

231. N. Madras and A.D. Sokal, The pivot algorithm: A highly efficient Monte Carlo method for the self-avoiding walk, J. Stat. Phys. 50 (1988) 109–186.

232. N. Madras and S.G. Whittington, Self-averaging in finite random copolymers, J. Phys. A: Math. Gen. 35 (2002) L427–L431.

233. N. Madras and S.G. Whittington, Localization of a random copolymer at an interface, J. Phys. A: Math. Gen. 36 (2003) 923–938.

234. D. Marenduzzo, A. Trovato and A. Maritan, Phase diagram of force-induced DNA unzipping in exactly solvable models, Phys. Rev. E 64 (2001) 031901.

235. A. Maritan, M.P. Riva and A. Trovato, Heteropolymers in a solvent at an interface, J. Phys. A: Math. Gen. 32 (1999) L275–L280.

236. R. Martin, M.S. Causo and S.G. Whittington, Localization transition for a randomly coloured self-avoiding walk at an interface, J. Phys. A: Math. Gen. 33 (2000) 7903–7918.

237. R. Martin, E. Orlandini, A.L. Owczarek, A. Rechnitzer and S.G. Whittington, Exact enumeration and Monte Carlo results for self-avoiding walks in a slab, J. Phys. A: Math. Gen. 40 (2007) 7509–7521.

238. S. Martinez and D. Petritis, Thermodynamics of a Brownian bridge polymer model in a random environment, J. Phys. A29 (1996) 1267–1279.

239. B. McCoy, Incompleteness of the critical exponent description for ferromagnetic systems containing random impurities, Phys. Rev. Lett. 23 (1969) 383–386.

240. O. Mejane, Upper bound of a volume exponent for directed polymers in a random environment, Ann. I. H. Poincaré P & R 40 (2004) 299–308.

241. M. S. Moghaddam, T. Vrbová and S.G. Whittington, Adsorption of periodic copolymers at a planar interface, J. Phys. A: Math. Gen. 33 (2000) 4573–4584.

242. P. Monari and A.L. Stella, θ-point universality of random polyampholytes with screened interactions, Phys. Rev. E 59 (1999) 1887–1892.

243. C. Monthus, On the localization of random heteropolymers at the interface between two selective solvents, Eur. Phys. J. B13 (2000) 111–130.

244. C. Monthus and T. Garel, Delocalization transition of the selective interface model: distribution of pseudo-critical temperatures, J. Stat. Mech. (2005) P12011.

245. C. Monthus and T. Garel, Freezing transition of the directed polymer in a 1+d random medium: Location of the critical temperature and unusual critical properties, Phys. Rev. E 74 (2006) 011101.

246. J. Moriarty and N. O'Connell, On the free energy of a directed polymer in a Brownian environment, Markov Proc. Relat. Fields 13 (2007) 251–266.

247. D.H. Napper, *Polymeric Stabilization of Colloidal Dispersions*, Academic Press, London, 1983.

248. B. Nienhuis, Exact critical exponents of the $O(n)$ models in two dimensions, Phys. Rev. Lett. 49 (1982) 1062–1065.

249. N. O'Connell and M. Yor, Brownian analogues of Burke's theorem, Stoch. Proc. Appl. 96 (2001) 285–304.

250. R. Olsen and R. Song, Diffusion of directed polymers in a strong random environment, J. Stat. Phys. 83 (1996) 727–738.

251. Y. Oono, On the divergence of the perturbation series for the excluded-volume problem in polymers, J. Phys. Soc. Japan 39 (1975) 25–29.

252. Y. Oono, On the divergence of the perturbation series for the excluded-volume problem in polymers, II. Collapse of a single chain in poor solvents, J. Phys. Soc. Japan 41 (1976) 787–793.

253. E. Orlandini, A. Rechnitzer and S.G. Whittington, Random copolymers and the Morita approximation: polymer adsorption and polymer localization, J. Phys. A: Math. Gen. 35 (2002) 7729–7751.

254. E. Orlandini, M.C. Tesi and S.G. Whittington, A self-avoiding walk model of random copolymer adsorption, J. Phys. A: Math. Gen. 32 (1999) 469–477.

255. E. Orlandini, M.C. Tesi and S.G. Whittington, Self-averaging in models of random copolymer collapse, J. Phys. A: Math. Gen. 33 (2000) 259–266.

256. E. Orlandini, M.C. Tesi and S.G. Whittington, Adsorption of a directed polymer subject to an elongational force, J. Phys. A: Math. Gen. 37 (2004) 1535–1543.

257. E. Orlandini and S.G. Whittington, Statistical topology of closed curves: Some applications in polymer physics, Rev. Mod. Phys. 79 (2007) 611–642.

258. A.L. Owczarek and T. Prellberg, Exact solution of semi-flexible and super-flexible interacting partially directed walks, J. Stat. Mech. (2007) P11010.

259. A.L. Owczarek, T. Prellberg and R. Brak, The tricritical behavior of self-interacting partially directed walks, J. Stat. Phys. 72 (1993) 737–772.

260. A.L. Owczarek, T. Prellberg and A. Rechnitzer, Finite-size scaling functions for directed polymers confined between attracting walls, J. Phys. A: Math. Theor. 41 (2008) 035002.

261. A.L. Owczarek and S.G. Whittington, Interacting lattice polygons, in: *Polygons, Polynominoes and Polyhedra* (ed. A.J. Guttmann), Springer, Berlin, in press, pp. 305–319.

262. M. Petermann, Superdiffusivity of directed polymers in random environment, Ph.D. thesis, University of Zürich, 2000.

263. N. Pétrélis, *Localisation d'un Polymère en Interaction avec une Interface*, Ph.D. Thesis, University of Rouen, France, February 2, 2006.

264. N. Pétrélis, Polymer pinning at an interface, Stoch. Proc. Appl. 116 (2006) 1600–1621.

265. N. Pétrélis, Polymer pinning: weak coupling limits, to appear in Ann. I. H. Poincaré.

266. J.S. Phipps, R.M. Richardson, T. Cosgrove and A. Eaglesham, Neutron reflection studies of copolymers at the hexane/water interface, Langmuir 9 (1993) 3530–3537.

267. M.S.T. Piza, Directed polymers in a random environment: some results on fluctuations, J. Stat. Phys. 89 (1997) 581–603.

268. D. Poland and H.A. Sheraga (eds.), *Theory of Helix-Coil Transitions in Biopolymers*, Academic Press, New York, 1970.

269. V. Privman, G. Forgacs and H.L. Frisch, New solvable model of polymer chain adsorption at a surface, Phys. Rev. B 37 (1988) 9897–9900.

270. C. Richard and A.J. Guttmann, Poland-Sheraga models and the DNA denaturation transition, J. Stat. Phys. 115 (2004) 943–965.

271. B. Roynette, P. Vallois and M. Yor, Limiting laws associated with Brownian motion perturbed by normalized exponential weights, C. R. Math. Acad. Sci. Paris 337 (2003) 667–673.

272. R.J. Rubin, Random-walk model of chain-polymer adsorption at a surface, J. Chem. Phys. 43 (1965) 2392–2407.

273. F. Seno and A.L. Stella, θ point of a linear polymer in 2 dimensions: a renormalization group analysis of Monte Carlo enumerations, J. Physique 49 (1988) 739–748.

274. Ya.G. Sinai, A random walk with random potential, Theor. Prob. Appl. 38 (1993) 382–385.

275. Ya.G. Sinai, A remark concerning random walks with random potentials, Fund. Math. 147 (1995) 173–180.

276. Ya.G. Sinai and H. Spohn, Remarks on the delocalization transition for heteropolymers, in: *Topics in Statistical and Theoretical Physics* (eds. R.L. Dobrushin et al.), AMS Transl. 177, Providence, RI, 1996, pp. 219–223.

277. G. Slade, The diffusion of self-avoiding walk in high dimensions, Commun. Math. Phys. 110 (1987) 661–683.

278. G. Slade, Convergence of self-avoiding walk to Brownian motion in high dimensions, J. Phys. A 21 (1988) L417–L420.

279. G. Slade, The scaling limit of self-avoiding random walk in high dimensions, Ann. Proab. 17 (1989) 91–107.

280. G. Slade, *The Lace Expansion and its Applications*, Lecture Notes in Mathematics 1879, Springer, Berlin, 2006.

281. J. Sohier, Finite size scaling for homogeneous pinning models, preprint 2008.

282. R. Song and X.Y. Zhou, A remark on diffusion of directed polymers in random environments, J. Stat. Phys. 85 (1996) 277–289.

283. C.E. Soteros and S.G. Whittington, The statistical mechanics of random copolymers, J. Phys. A: Math. Gen. 37 (2004) R279–R325.

284. F. Spitzer, *Principles of Random Walk* (2nd. ed.), Springer, New York, 1976.

285. S. Stepanow, J.-U. Sommer and I.Ya. Erukhimovich, Localization transition of random copolymers at interfaces, Phys. Rev. Lett. 81 (1998) 4412–4415.

286. S.F. Sun, C.-C. Chou and R.A. Nash, Viscosity study of the collapsed state of polystyrene, J. Chem. Phys. 93 (1990) 7508–7509.

287. S.-T. Sun, I. Nishio, G. Swislow and T. Tanaka, The coil-globule transition: radius of gyration of polystyrene in cyclohexane, J. Chem. Phys. 73 (1980) 5971–5975.

288. A.-S. Sznitman, *Brownian Motion, Obstacles and Random Media*, Springer, Berlin, 1998.

289. M.C. Tesi, E.J. Janse van Rensburg, E. Orlandini and S.G. Whittington, Monte Carlo study of the interacting self-avoiding walk model in three dimensions, J. Stat. Phys. 82 (1996) 155–181.

290. M.C. Tesi, E.J. Janse van Rensburg, E. Orlandini and S.G. Whittington, Interacting self-avoiding walks and polygons in three dimensions, J. Phys. A: Math. Gen. 29 (1996) 2451–2463.

291. F.L. Toninelli, Critical properties and finite-size estimates for the depinning transition of directed random polymers, J. Stat. Phys. 126 (2007) 1025–1044.

292. F.L. Toninelli, Correlation lengths for random polymer models and for some renewal sequences, Elect. J. Probab. 12 (2007) 613–636.

293. F.L. Toninelli, A replica-coupling approach to disordered pinning models, Commun. Math. Phys. 280 (2008) 389–401.

294. F.L. Toninelli, Disordered pinning models and copolymers: beyond annealed bounds, Ann. Appl. Probab. 18 (2008) 1569–1587.

295. F.L. Toninelli, Coarse graining, fractional moments and the critical slope of random polymers, preprint 2008, to appear in Elect. J. Probab.

296. F.L. Toninelli, Localization transition in disordered pinning models. Effect of randomness on the critical properties, in: *Proceedings of the 5th Prague Summer School on Mathematical Statistical Mechanics* (September 2006) (ed. R. Kotecký), Springer, Berlin, to appear.

297. C.A. Tracy and H. Widom, Level-spacing distribution and the Airy kernel, Commun. Math. Phys. 159 (1994) 151–174.

298. A. Trovato and A. Maritan, A variational approach to the localization transition of heteropolymers at interfaces, Europhys. Lett. 46 (1999) 301–306.

299. D. Ueltschi, A self-avoiding walk with attractive interactions, Probab. Theory Relat. Fields 124 (2002) 189–203.

300. C. Vanderzande, *Lattice Models of Polymers*, Cambridge Lecture Notes in Physics 11, Cambridge University Press, 1998.

301. S.R.S. Varadhan, Appendix to K. Symanzik, Euclidean quantum field theory, in: *Local Quantum Theory* (ed. R. Jost), Academic Press, 1969.

302. V. Vargas, *Polymères dirigés en milieu aléatoire et champs multifractaux*, Ph.D. Thesis, University of Paris 7, France, November 23, 2006.

303. V. Vargas, A local limit theorem for directed polymers in random media: the continuous and the discrete case, Ann. Inst. H. Poincaré. Probab. Statist. 42 (2006) 521–534.

304. V. Vargas, Strong localization and macroscopic atoms for directed polymers, Probab. Theory Relat. Fields 138 (2007) 391–410.

305. Y. Velenik, Localization and delocalization of random interfaces, Probability Surveys 3 (2006) 112–169.

306. T. Vrbová and S.G. Whittington, Adsorption and collapse of self-avoiding walks and polygons in three dimensions, J. Phys. A: Math. Gen. 29 (1996) 6253–6264.

307. T. Vrbová and S.G. Whittington, Adsorption and collapse of self-avoiding walks in three dimensions: A Monte Carlo study, J. Phys. A: Math. Gen. 31 (1998) 3989–3998.

308. T. Vrbová and S.G. Whittington, Adsorption and collapse of self-avoiding walks at a defect line, J. Phys. A: Math. Gen. 31 (1998) 7031–7041.

309. W. Werner, *Random Planar Curves and Schramm-Loewner Evolutions*, Lecture Notes in Mathematics 1840, Springer, Berlin, 2004, pp. 107–195.

310. J. Westwater, On Edwards' model for polymer chains, Commun. Math. Phys. 72 (1980) 131–174.

311. J. Westwater, On Edwards' model for polymer chains. III. Borel summability, Comm. Math. Phys. 84 (1982) 459–470.

312. J. Westwater, On Edwards' model for polymer chains, in: *Trends and Developments in the Eighties* (S. Albeverio and P. Blanchard, eds.), Bielefeld Encounters in Math. Phys. 4/5, World Scientific, Singapore, 1984.

313. S.G. Whittington, Self-avoiding walks terminally attached to an interface, J. Chem. Phys. 63 (1975) 779–785.

314. S.G. Whittington, Self-avoiding walks with geometrical constraints, J. Stat. Phys. 30 (1983) 449–456.

315. S.G. Whittington, A directed walk model of copolymer adsorption, J. Phys. A: Math. Gen. 31 (1998) 8797–8803.

316. S.G. Whittington, Random copolymers, Physica A 314 (2002) 214–219.

317. S.G. Whittington and C.E. Soteros, Polymers in slabs, slits and pores, Isr. J. Chem. 31 (1991) 127–133.

318. M.V. Wüthrich, Scaling identity for crossing Brownian motion in a Poissonian potential, Probab. Theory Relat. Fields 112 (1998) 299–319.

319. M.V. Wüthrich, Superdiffusive behavior of two-dimensional Brownian motion in a Poissonian potential, Ann. Probab. 26 (1998) 1000–1015.

320. M.V. Wüthrich, Fluctuation results for Brownian motion in a Poissonian potential, Ann. Inst. H. Poincaré Probab. Statist. 34 (1998) 279–308.

321. M.V. Wüthrich, Numerical bounds for critical exponents of crossing Brownian motion, Proc. Amer. Math. Soc. 130 (2002) 217–225.

322. M.V. Wüthrich, A heteropolymer in a medium with random droplets, Ann. Appl. Probab. 3 (2006) 1653–1670.

323. O. Zeitouni, *Random Walks in Random Environment*, Lecture Notes in Mathematics 1837, Springer, Berlin, 2004, pp. 189–312.

Index

List of Participants

37th Probability Summer School, Saint-Flour, France

July 8-21, 2007

Lecturers

Jérôme BUZZI	École Polytechnique, Palaiseau, France
Frank den HOLLANDER	Leiden Univ., The Netherlands
Jonathan C. MATTINGLY	Duke Univ., Durham, USA

Participants

Jean-Baptiste BARDET	Univ. Rennes 1, F
Anne-Laure BASDEVANT	Univ. Pierre et Marie Curie, Paris, F
Mireia BESALU	Univ. Barcelona, Spain
Michel BONNEFONT	Univ. Paul Sabatier, Toulouse, F
Pavel BUBAK	Univ. Warwick, Coventry, UK
Alain CAMANES	Univ. Nantes, F
Francesco CARAVENNA	Univ. Padova, Italy
Pavel CHIGANSKY	Univ. du Maine, Le Mans, F
Alina CRUDU	Univ. Rennes 1, F
Sébastien DARSES	Univ. Pierre et Marie Curie, Paris, F
Omar de la CRUZ	Univ. Chicago, USA
Latifa DEBBI	Univ. Setif, Algérie
François DELARUE	Univ. Denis Diderot, Paris, F
Maxime DELEBASSEE	Univ. Paul Sabatier, Toulouse, F
Hacène DJELLOUT	Univ. Blaise Pascal, Clermont-Ferrand, F

Leif DOERING	TU Berlin, Germany
Pierre FOUGERES	Univ. Paris 10, F
Antoine GERBAUD	Institut Fourier, Grenoble, F
Ricardo GOMEZ	Univ. Nacional Autónoma México
Dan GOREAC	Univ. Brest, F
Ludovic GOUDENEGE	Univ. Rennes 1, F
Erika HAUSENBLAS	Univ. Salzburg, Austria
Martin HUTZENTHALER	Goethe Univ., Frankfurt, Germany
Barbara JASIULIS	Univ. Wroclaw, Poland
Wouter KAGER	EURANDOM, Eindhoven, The Netherlands
Tamas KOI	Budapest Univ. Technology, Hungary
Maxime LAGOUGE	Univ. Denis Diderot, Paris, F
Matthieu LERASLE	INSA Toulouse, F
Shuyan LIU	Univ. Lille 1, F
Eva LOECHERBACH	Univ. Paris 12, F
Grégory MAILLARD	École Polytechnique Fédérale Lausanne, CH
Florent MALRIEU	Univ. Rennes 1, F
Charles MANSON	Univ. Warwick, Coventry, UK
Scott McKINLEY	Duke Univ., Durham, USA
Oana MOCIOALCA	Kent State Univ., USA
Peter NANDORI	Budapest Univ. Technology, Hungary
Francesca R. NARDI	Eindhoven, The Netherlands
Eulalia NUALART	Univ. Paris 13, F
Alberto OHASHI	Univ. Campinas, Brazil
Nicolas PETRELIS	EURANDOM, Eindhoven, The Netherlands
Jean PICARD	Univ. Blaise Pascal, Clermont-Ferrand, F
Mikael ROGER	Univ. Rennes 1, F
Marco ROMITO	Univ. Firenze, Italy
Jean-Pierre ROZELOT	Observatoire Côte d'Azur, F
E. SAINT LOUBERT BIE	Univ. Blaise Pascal, Clermont-Ferrand, F
Christian SELINGER	Univ. Luxembourg
Laurent SERLET	Univ. Blaise Pascal, Clermont-Ferrand, F
Arvind SINGH	Univ. Pierre et Marie Curie, Paris, F
Julien SOHIER	Univ. Denis Diderot, Paris, F
Cristian SPITONI	Leiden Univ., The Netherlands
Mitja STADJE	ORFE, Princeton, USA

Ramon Van HANDEL	California Inst. Techn., Pasadena, USA
Andrea WATKINS	Duke Univ., Durham, USA
Hendrik WEBER	Univ. Bonn, Germany
Radoslaw WIECZOREK	Polish Academy Sciences, Katowice, Poland
Lorenzo ZAMBOTTI	Univ. Pierre et Marie Curie, Paris, F

Programme of the School

Main Lectures

Jérôme Buzzi	Hyperbolicity through entropies
Frank den Hollander	Random polymers
Jonathan Mattingly	Ergodicity of dissipative SPDEs

Short Lectures

Pavel Bubak	Asymptotic strong Feller property for degenerate diffusions
Alain Camanes	Directed polymers in random environment and uniform integrability
Francesco Caravenna	The quenched critical point of a diluted disordered polymer model
Sébastien Darses	Stochastic derivatives
Latifa Debbi	On deterministic and stochastic fractional partial differential equations
François Delarue	Stochastic analysis of a numerical scheme for transport
Maxime Delebassee	Uniqueness in graph percolation
Pierre Fougères	Markovian type semilinear problems associated with some (infinite dimensional) functional inequalities
Ricardo Gómez	Almost isomorphisms of Markov shifts
Erika Hausenblas	SPDEs driven by Poisson random measures
Wouter Kager	Patterns and ratio limit theorems for Markovian random fields

Shuyan Liu	Estimation of parameters of stable distributions on convex cones
Grégory Maillard	Intermittency on catalysts
Oana Mocioalca	Space regularity of stochastic heat equations driven by irregular Gaussian processes
Francesca Romana Nardi	Ideal gas approximation for a two-dimensional rarefied gas under Kawasaki dynamics
Eulalia Nualart	Hitting probabilities for systems of fractional stochastic heat equations on the circle
Alberto Ohashi	Finite dimensional invariant manifolds for SPDEs driven by fractional Brownian motion
Nicolas Pétrélis	Copolymer in an emulsion
Mikaël Roger	A central limit theorem for some (random) compositions of automorphisms of the torus
Marco Romito	Markov solutions for the stochastic Navier-Stokes equations in dimension three
Jean-Pierre Rozelot	A story of the sun oblateness: from Princeton, 1996 to Pic du Midi, 2006
Laurent Serlet	Can super-Brownian excursion become old, big or heavy?
Arvind Singh	The speed of a cookie random walk
Julien Sohier	The homogeneous polymer pinning model near criticality
Cristian Spitoni	Metastability for reversible probabilistic cellular automata with self-interaction
Ramon Van Handel	Robustness and approximations in nonlinear filtering
Radoslaw Wieczorek	Fragmentation-coagulation processes as limits of individual based models

Saint-Flour Probability Summer Schools

In order to facilitate research concerning previous schools we give here the names of the authors, the series*, and the number of the volume where their lectures can be found:

Summer School	Authors	Series*	Vol. Nr.
1971	Bretagnolle; Chatterji; Meyer	LNM	307
1973	Meyer; Priouret; Spitzer	LNM	390
1974	Fernique; Conze; Gani	LNM	480
1975	Badrikian; Kingman; Kuelbs	LNM	539
1976	Hoffmann-Jörgensen; Liggett; Neveu	LNM	598
1977	Dacunha-Castelle; Heyer; Roynette	LNM	678
1978	Azencott; Guivarc'h; Gundy	LNM	774
1979	Bickel; El Karoui; Yor	LNM	876
1980	Bismut; Gross; Krickeberg	LNM	929
1981	Fernique; Millar; Stroock; Weber	LNM	976
1982	Dudley; Kunita; Ledrappier	LNM	1097
1983	Aldous; Ibragimov; Jacod	LNM	1117
1984	Carmona; Kesten; Walsh	LNM	1180
1985/86/87	Diaconis; Elworthy; Föllmer; Nelson; Papanicolaou; Varadhan	LNM	1362
1986	Barndorff-Nielsen	LNS	50
1988	Ancona; Geman; Ikeda	LNM	1427
1989	Burkholder; Pardoux; Sznitman	LNM	1464
1990	Freidlin; Le Gall	LNM	1527
1991	Dawson; Maisonneuve; Spencer	LNM	1541
1992	Bakry; Gill; Molchanov	LNM	1581
1993	Biane; Durrett;	LNM	1608
1994	Dobrushin; Groeneboom; Ledoux	LNM	1648
1995	Barlow; Nualart	LNM	1690
1996	Giné; Grimmett; Saloff-Coste	LNM	1665
1997	Bertoin; Martinelli; Peres	LNM	1717
1998	Emery; Nemirovski; Voiculescu	LNM	1738
1999	Bolthausen; Perkins; van der Vaart	LNM	1781
2000	Albeverio; Schachermayer; Talagrand	LNM	1816
2001	Tavaré; Zeitouni	LNM	1837
	Catoni	LNM	1851
2002	Tsirelson; Werner	LNM	1840
	Pitman	LNM	1875
2003	Dembo; Funaki	LNM	1869
	Massart	LNM	1896
2004	Cerf	LNM	1878
	Slade	LNM	1879
	Lyons T.J., Caruana, Lévy	LNM	1908
2005	Doney	LNM	1897
	Evans	LNM	1920
	Villani	GL	338
2006	Bramson	LNM	1950
	Guionnet	LNM	1957
	Lauritzen	Forthcoming	
2007	den Hollander	LNM	1974
	Buzzi	Forthcoming	
	Mattingly	Forthcoming	
2008	Kenyon	Forthcoming	
	Koltchinskii	Forthcoming	
	Le Jan	Forthcoming	

*Lecture Notes in Mathematics (LNM), Lecture Notes in Statistics (LNS), Grundlehren der mathematischen Wissenschaften (GL)

Lecture Notes in Mathematics

For information about earlier volumes
please contact your bookseller or Springer
LNM Online archive: springerlink.com

Vol. 1830: M. I. Gil', Operator Functions and Localization of Spectra. XIV, 256 p, 2003.

Vol. 1831: A. Connes, J. Cuntz, E. Guentner, N. Higson, J. E. Kaminker, Noncommutative Geometry, Martina Franca, Italy 2002. Editors: S. Doplicher, L. Longo (2004)

Vol. 1832: J. Azéma, M. Émery, M. Ledoux, M. Yor (Eds.), Séminaire de Probabilités XXXVII (2003)

Vol. 1833: D.-Q. Jiang, M. Qian, M.-P. Qian, Mathematical Theory of Nonequilibrium Steady States. On the Frontier of Probability and Dynamical Systems. IX, 280 p, 2004.

Vol. 1834: Yo. Yomdin, G. Comte, Tame Geometry with Application in Smooth Analysis. VIII, 186 p, 2004.

Vol. 1835: O.T. Izhboldin, B. Kahn, N.A. Karpenko, A. Vishik, Geometric Methods in the Algebraic Theory of Quadratic Forms. Summer School, Lens, 2000. Editor: J.-P. Tignol (2004)

Vol. 1836: C. Năstăsescu, F. Van Oystaeyen, Methods of Graded Rings. XIII, 304 p, 2004.

Vol. 1837: S. Tavaré, O. Zeitouni, Lectures on Probability Theory and Statistics. Ecole d'Eté de Probabilités de Saint-Flour XXXI-2001. Editor: J. Picard (2004)

Vol. 1838: A.J. Ganesh, N.W. O'Connell, D.J. Wischik, Big Queues. XII, 254 p, 2004.

Vol. 1839: R. Gohm, Noncommutative Stationary Processes. VIII, 170 p, 2004.

Vol. 1840: B. Tsirelson, W. Werner, Lectures on Probability Theory and Statistics. Ecole d'Eté de Probabilités de Saint-Flour XXXII-2002. Editor: J. Picard (2004)

Vol. 1841: W. Reichel, Uniqueness Theorems for Variational Problems by the Method of Transformation Groups (2004)

Vol. 1842: T. Johnsen, A. L. Knutsen, K_3 Projective Models in Scrolls (2004)

Vol. 1843: B. Jefferies, Spectral Properties of Noncommuting Operators (2004)

Vol. 1844: K.F. Siburg, The Principle of Least Action in Geometry and Dynamics (2004)

Vol. 1845: Min Ho Lee, Mixed Automorphic Forms, Torus Bundles, and Jacobi Forms (2004)

Vol. 1846: H. Ammari, H. Kang, Reconstruction of Small Inhomogeneities from Boundary Measurements (2004)

Vol. 1847: T.R. Bielecki, T. Björk, M. Jeanblanc, M. Rutkowski, J.A. Scheinkman, W. Xiong, Paris-Princeton Lectures on Mathematical Finance 2003 (2004)

Vol. 1848: M. Abate, J. E. Fornaess, X. Huang, J. P. Rosay, A. Tumanov, Real Methods in Complex and CR Geometry, Martina Franca, Italy 2002. Editors: D. Zaitsev, G. Zampieri (2004)

Vol. 1849: Martin L. Brown, Heegner Modules and Elliptic Curves (2004)

Vol. 1850: V. D. Milman, G. Schechtman (Eds.), Geometric Aspects of Functional Analysis. Israel Seminar 2002-2003 (2004)

Vol. 1851: O. Catoni, Statistical Learning Theory and Stochastic Optimization (2004)

Vol. 1852: A.S. Kechris, B.D. Miller, Topics in Orbit Equivalence (2004)

Vol. 1853: Ch. Favre, M. Jonsson, The Valuative Tree (2004)

Vol. 1854: O. Saeki, Topology of Singular Fibers of Differential Maps (2004)

Vol. 1855: G. Da Prato, P.C. Kunstmann, I. Lasiecka, A. Lunardi, R. Schnaubelt, L. Weis, Functional Analytic Methods for Evolution Equations. Editors: M. Iannelli, R. Nagel, S. Piazzera (2004)

Vol. 1856: K. Back, T.R. Bielecki, C. Hipp, S. Peng, W. Schachermayer, Stochastic Methods in Finance, Bressanone/Brixen, Italy, 2003. Editors: M. Fritelli, W. Runggaldier (2004)

Vol. 1857: M. Émery, M. Ledoux, M. Yor (Eds.), Séminaire de Probabilités XXXVIII (2005)

Vol. 1858: A.S. Cherny, H.-J. Engelbert, Singular Stochastic Differential Equations (2005)

Vol. 1859: E. Letellier, Fourier Transforms of Invariant Functions on Finite Reductive Lie Algebras (2005)

Vol. 1860: A. Borisyuk, G.B. Ermentrout, A. Friedman, D. Terman, Tutorials in Mathematical Biosciences I. Mathematical Neurosciences (2005)

Vol. 1861: G. Benettin, J. Henrard, S. Kuksin, Hamiltonian Dynamics – Theory and Applications, Cetraro, Italy, 1999. Editor: A. Giorgilli (2005)

Vol. 1862: B. Helffer, F. Nier, Hypoelliptic Estimates and Spectral Theory for Fokker-Planck Operators and Witten Laplacians (2005)

Vol. 1863: H. Führ, Abstract Harmonic Analysis of Continuous Wavelet Transforms (2005)

Vol. 1864: K. Efstathiou, Metamorphoses of Hamiltonian Systems with Symmetries (2005)

Vol. 1865: D. Applebaum, B. V. R. Bhat, J. Kustermans, J. M. Lindsay, Quantum Independent Increment Processes I. From Classical Probability to Quantum Stochastic Calculus. Editors: M. Schürmann, U. Franz (2005)

Vol. 1866: O.E. Barndorff-Nielsen, U. Franz, R. Gohm, B. Kümmerer, S. Thorbjønsen, Quantum Independent Increment Processes II. Structure of Quantum Lévy Processes, Classical Probability, and Physics. Editors: M. Schürmann, U. Franz, (2005)

Vol. 1867: J. Sneyd (Ed.), Tutorials in Mathematical Biosciences II. Mathematical Modeling of Calcium Dynamics and Signal Transduction. (2005)

Vol. 1868: J. Jorgenson, S. Lang, $Pos_n(R)$ and Eisenstein Series. (2005)

Vol. 1869: A. Dembo, T. Funaki, Lectures on Probability Theory and Statistics. Ecole d'Eté de Probabilités de Saint-Flour XXXIII-2003. Editor: J. Picard (2005)

Vol. 1870: V.I. Gurariy, W. Lusky, Geometry of Müntz Spaces and Related Questions. (2005)

Vol. 1871: P. Constantin, G. Gallavotti, A.V. Kazhikhov, Y. Meyer, S. Ukai, Mathematical Foundation of Turbulent Viscous Flows, Martina Franca, Italy, 2003. Editors: M. Cannone, T. Miyakawa (2006)

Vol. 1872: A. Friedman (Ed.), Tutorials in Mathematical Biosciences III. Cell Cycle, Proliferation, and Cancer (2006)

Vol. 1873: R. Mansuy, M. Yor, Random Times and Enlargements of Filtrations in a Brownian Setting (2006)

Vol. 1874: M. Yor, M. Émery (Eds.), In Memoriam Paul-André Meyer - Séminaire de Probabilités XXXIX (2006)

Vol. 1875: J. Pitman, Combinatorial Stochastic Processes. Ecole d'Eté de Probabilités de Saint-Flour XXXII-2002. Editor: J. Picard (2006)

Vol. 1876: H. Herrlich, Axiom of Choice (2006)

Vol. 1877: J. Steuding, Value Distributions of L-Functions (2007)

Vol. 1878: R. Cerf, The Wulff Crystal in Ising and Percolation Models, Ecole d'Eté de Probabilités de Saint-Flour XXXIV-2004. Editor: Jean Picard (2006)

Vol. 1879: G. Slade, The Lace Expansion and its Applications, Ecole d'Eté de Probabilités de Saint-Flour XXXIV-2004. Editor: Jean Picard (2006)

Vol. 1880: S. Attal, A. Joye, C.-A. Pillet, Open Quantum Systems I, The Hamiltonian Approach (2006)

Vol. 1881: S. Attal, A. Joye, C.-A. Pillet, Open Quantum Systems II, The Markovian Approach (2006)

Vol. 1882: S. Attal, A. Joye, C.-A. Pillet, Open Quantum Systems III, Recent Developments (2006)

Vol. 1883: W. Van Assche, F. Marcellàn (Eds.), Orthogonal Polynomials and Special Functions, Computation and Application (2006)

Vol. 1884: N. Hayashi, E.I. Kaikina, P.I. Naumkin, I.A. Shishmarev, Asymptotics for Dissipative Nonlinear Equations (2006)

Vol. 1885: A. Telcs, The Art of Random Walks (2006)

Vol. 1886: S. Takamura, Splitting Deformations of Degenerations of Complex Curves (2006)

Vol. 1887: K. Habermann, L. Habermann, Introduction to Symplectic Dirac Operators (2006)

Vol. 1888: J. van der Hoeven, Transseries and Real Differential Algebra (2006)

Vol. 1889: G. Osipenko, Dynamical Systems, Graphs, and Algorithms (2006)

Vol. 1890: M. Bunge, J. Funk, Singular Coverings of Toposes (2006)

Vol. 1891: J.B. Friedlander, D.R. Heath-Brown, H. Iwaniec, J. Kaczorowski, Analytic Number Theory, Cetraro, Italy, 2002. Editors: A. Perelli, C. Viola (2006)

Vol. 1892: A. Baddeley, I. Bárány, R. Schneider, W. Weil, Stochastic Geometry, Martina Franca, Italy, 2004. Editor: W. Weil (2007)

Vol. 1893: H. Hanßmann, Local and Semi-Local Bifurcations in Hamiltonian Dynamical Systems, Results and Examples (2007)

Vol. 1894: C.W. Groetsch, Stable Approximate Evaluation of Unbounded Operators (2007)

Vol. 1895: L. Molnár, Selected Preserver Problems on Algebraic Structures of Linear Operators and on Function Spaces (2007)

Vol. 1896: P. Massart, Concentration Inequalities and Model Selection, Ecole d'Été de Probabilités de Saint-Flour XXXIII-2003. Editor: J. Picard (2007)

Vol. 1897: R. Doney, Fluctuation Theory for Lévy Processes, Ecole d'Été de Probabilités de Saint-Flour XXXV-2005. Editor: J. Picard (2007)

Vol. 1898: H.R. Beyer, Beyond Partial Differential Equations, On linear and Quasi-Linear Abstract Hyperbolic Evolution Equations (2007)

Vol. 1899: Séminaire de Probabilités XL. Editors: C. Donati-Martin, M. Émery, A. Rouault, C. Stricker (2007)

Vol. 1900: E. Bolthausen, A. Bovier (Eds.), Spin Glasses (2007)

Vol. 1901: O. Wittenberg, Intersections de deux quadriques et pinceaux de courbes de genre 1, Intersections of Two Quadrics and Pencils of Curves of Genus 1 (2007)

Vol. 1902: A. Isaev, Lectures on the Automorphism Groups of Kobayashi-Hyperbolic Manifolds (2007)

Vol. 1903: G. Kresin, V. Maz'ya, Sharp Real-Part Theorems (2007)

Vol. 1904: P. Giesl, Construction of Global Lyapunov Functions Using Radial Basis Functions (2007)

Vol. 1905: C. Prévôt, M. Röckner, A Concise Course on Stochastic Partial Differential Equations (2007)

Vol. 1906: T. Schuster, The Method of Approximate Inverse: Theory and Applications (2007)

Vol. 1907: M. Rasmussen, Attractivity and Bifurcation for Nonautonomous Dynamical Systems (2007)

Vol. 1908: T.J. Lyons, M. Caruana, T. Lévy, Differential Equations Driven by Rough Paths, Ecole d'Été de Probabilités de Saint-Flour XXXIV-2004 (2007)

Vol. 1909: H. Akiyoshi, M. Sakuma, M. Wada, Y. Yamashita, Punctured Torus Groups and 2-Bridge Knot Groups (I) (2007)

Vol. 1910: V.D. Milman, G. Schechtman (Eds.), Geometric Aspects of Functional Analysis. Israel Seminar 2004-2005 (2007)

Vol. 1911: A. Bressan, D. Serre, M. Williams, K. Zumbrun, Hyperbolic Systems of Balance Laws. Cetraro, Italy 2003. Editor: P. Marcati (2007)

Vol. 1912: V. Berinde, Iterative Approximation of Fixed Points (2007)

Vol. 1913: J.E. Marsden, G. Misiołek, J.-P. Ortega, M. Perlmutter, T.S. Ratiu, Hamiltonian Reduction by Stages (2007)

Vol. 1914: G. Kutyniok, Affine Density in Wavelet Analysis (2007)

Vol. 1915: T. Bıyıkoğlu, J. Leydold, P.F. Stadler, Laplacian Eigenvectors of Graphs. Perron-Frobenius and Faber-Krahn Type Theorems (2007)

Vol. 1916: C. Villani, F. Rezakhanlou, Entropy Methods for the Boltzmann Equation. Editors: F. Golse, S. Olla (2008)

Vol. 1917: I. Veselić, Existence and Regularity Properties of the Integrated Density of States of Random Schrödinger (2008)

Vol. 1918: B. Roberts, R. Schmidt, Local Newforms for GSp(4) (2007)

Vol. 1919: R.A. Carmona, I. Ekeland, A. Kohatsu-Higa, J.-M. Lasry, P.-L. Lions, H. Pham, E. Taflin, Paris-Princeton Lectures on Mathematical Finance 2004. Editors: R.A. Carmona, E. Çinlar, I. Ekeland, E. Jouini, J.A. Scheinkman, N. Touzi (2007)

Vol. 1920: S.N. Evans, Probability and Real Trees. Ecole d'Été de Probabilités de Saint-Flour XXXV-2005 (2008)

Vol. 1921: J.P. Tian, Evolution Algebras and their Applications (2008)

Vol. 1922: A. Friedman (Ed.), Tutorials in Mathematical BioSciences IV. Evolution and Ecology (2008)

Vol. 1923: J.P.N. Bishwal, Parameter Estimation in Stochastic Differential Equations (2008)

Vol. 1924: M. Wilson, Littlewood-Paley Theory and Exponential-Square Integrability (2008)

Vol. 1925: M. du Sautoy, L. Woodward, Zeta Functions of Groups and Rings (2008)

Vol. 1926: L. Barreira, V. Claudia, Stability of Nonautonomous Differential Equations (2008)

Vol. 1927: L. Ambrosio, L. Caffarelli, M.G. Crandall, L.C. Evans, N. Fusco, Calculus of Variations and Non-Linear Partial Differential Equations. Cetraro, Italy 2005. Editors: B. Dacorogna, P. Marcellini (2008)

Vol. 1928: J. Jonsson, Simplicial Complexes of Graphs (2008)

Vol. 1929: Y. Mishura, Stochastic Calculus for Fractional Brownian Motion and Related Processes (2008)

Vol. 1930: J.M. Urbano, The Method of Intrinsic Scaling. A Systematic Approach to Regularity for Degenerate and Singular PDEs (2008)

Vol. 1931: M. Cowling, E. Frenkel, M. Kashiwara, A. Valette, D.A. Vogan, Jr., N.R. Wallach, Representation Theory and Complex Analysis. Venice, Italy 2004. Editors: E.C. Tarabusi, A. D'Agnolo, M. Picardello (2008)

Vol. 1932: A.A. Agrachev, A.S. Morse, E.D. Sontag, H.J. Sussmann, V.I. Utkin, Nonlinear and Optimal

Control Theory. Cetraro, Italy 2004. Editors: P. Nistri, G. Stefani (2008)

Vol. 1933: M. Petkovic, Point Estimation of Root Finding Methods (2008)

Vol. 1934: C. Donati-Martin, M. Émery, A. Rouault, C. Stricker (Eds.), Séminaire de Probabilités XLI (2008)

Vol. 1935: A. Unterberger, Alternative Pseudodifferential Analysis (2008)

Vol. 1936: P. Magal, S. Ruan (Eds.), Structured Population Models in Biology and Epidemiology (2008)

Vol. 1937: G. Capriz, P. Giovine, P.M. Mariano (Eds.), Mathematical Models of Granular Matter (2008)

Vol. 1938: D. Auroux, F. Catanese, M. Manetti, P. Seidel, B. Siebert, I. Smith, G. Tian, Symplectic 4-Manifolds and Algebraic Surfaces. Cetraro, Italy 2003. Editors: F. Catanese, G. Tian (2008)

Vol. 1939: D. Boffi, F. Brezzi, L. Demkowicz, R.G. Durán, R.S. Falk, M. Fortin, Mixed Finite Elements, Compatibility Conditions, and Applications. Cetraro, Italy 2006. Editors: D. Boffi, L. Gastaldi (2008)

Vol. 1940: J. Banasiak, V. Capasso, M.A.J. Chaplain, M. Lachowicz, J. Miękisz, Multiscale Problems in the Life Sciences. From Microscopic to Macroscopic. Będlewo, Poland 2006. Editors: V. Capasso, M. Lachowicz (2008)

Vol. 1941: S.M.J. Haran, Arithmetical Investigations. Representation Theory, Orthogonal Polynomials, and Quantum Interpolations (2008)

Vol. 1942: S. Albeverio, F. Flandoli, Y.G. Sinai, SPDE in Hydrodynamic. Recent Progress and Prospects. Cetraro, Italy 2005. Editors: G. Da Prato, M. Röckner (2008)

Vol. 1943: L.L. Bonilla (Ed.), Inverse Problems and Imaging. Martina Franca, Italy 2002 (2008)

Vol. 1944: A. Di Bartolo, G. Falcone, P. Plaumann, K. Strambach, Algebraic Groups and Lie Groups with Few Factors (2008)

Vol. 1945: F. Brauer, P. van den Driessche, J. Wu (Eds.), Mathematical Epidemiology (2008)

Vol. 1946: G. Allaire, A. Arnold, P. Degond, T.Y. Hou, Quantum Transport. Modelling, Analysis and Asymptotics. Cetraro, Italy 2006. Editors: N.B. Abdallah, G. Frosali (2008)

Vol. 1947: D. Abramovich, M. Mariño, M. Thaddeus, R. Vakil, Enumerative Invariants in Algebraic Geometry and String Theory. Cetraro, Italy 2005. Editors: K. Behrend, M. Manetti (2008)

Vol. 1948: F. Cao, J-L. Lisani, J-M. Morel, P. Musé, F. Sur, A Theory of Shape Identification (2008)

Vol. 1949: H.G. Feichtinger, B. Helffer, M.P. Lamoureux, N. Lerner, J. Toft, Pseudo-Differential Operators. Quantization and Signals. Cetraro, Italy 2006. Editors: L. Rodino, M.W. Wong (2008)

Vol. 1950: M. Bramson, Stability of Queueing Networks, Ecole d'Eté de Probabilités de Saint-Flour XXXVI-2006 (2008)

Vol. 1951: A. Moltó, J. Orihuela, S. Troyanski, M. Valdivia, A Non Linear Transfer Technique for Renorming (2009)

Vol. 1952: R. Mikhailov, I.B.S. Passi, Lower Central and Dimension Series of Groups (2009)

Vol. 1953: K. Arwini, C.T.J. Dodson, Information Geometry (2008)

Vol. 1954: P. Biane, L. Bouten, F. Cipriani, N. Konno, N. Privault, Q. Xu, Quantum Potential Theory. Editors: U. Franz, M. Schuermann (2008)

Vol. 1955: M. Bernot, V. Caselles, J.-M. Morel, Optimal Transportation Networks (2008)

Vol. 1956: C.H. Chu, Matrix Convolution Operators on Groups (2008)

Vol. 1957: A. Guionnet, On Random Matrices: Macroscopic Asymptotics, Ecole d'Eté de Probabilités de Saint-Flour XXXVI-2006 (2009)

Vol. 1958: M.C. Olsson, Compactifying Moduli Spaces for Abelian Varieties (2008)

Vol. 1959: Y. Nakkajima, A. Shiho, Weight Filtrations on Log Crystalline Cohomologies of Families of Open Smooth Varieties (2008)

Vol. 1960: J. Lipman, M. Hashimoto, Foundations of Grothendieck Duality for Diagrams of Schemes (2009)

Vol. 1961: G. Buttazzo, A. Pratelli, S. Solimini, E. Stepanov, Optimal Urban Networks via Mass Transportation (2009)

Vol. 1962: R. Dalang, D. Khoshnevisan, C. Mueller, D. Nualart, Y. Xiao, A Minicourse on Stochastic Partial Differential Equations (2009)

Vol. 1963: W. Siegert, Local Lyapunov Exponents (2009)

Vol. 1964: W. Roth, Operator-valued Measures and Integrals for Cone-valued Functions and Integrals for Cone-valued Functions (2009)

Vol. 1965: C. Chidume, Geometric Properties of Banach Spaces and Nonlinear Iterations (2009)

Vol. 1966: D. Deng, Y. Han, Harmonic Analysis on Spaces of Homogeneous Type (2009)

Vol. 1967: B. Fresse, Modules over Operads and Functors (2009)

Vol. 1968: R. Weissauer, Endoscopy for GSP(4) and the Cohomology of Siegel Modular Threefolds (2009)

Vol. 1969: B. Roynette, M. Yor, Penalising Brownian Paths (2009)

Vol. 1970: R. Kotecký, Methods of Contemporary Mathematical Statistical Physics (2009)

Vol. 1971: L. Saint-Raymond, Hydrodynamic Limits of the Boltzmann Equation (2009)

Vol. 1972: T. Mochizuki, Donaldson Type Invariants for Algebraic Surfaces (2009)

Vol. 1973: M.A. Berger, Lectures on Topological Fluid Mechanics (2009)

Vol. 1974: F. den Hollander, Random Polymers: École d'Été de Probabilités de Saint-Flour XXXVII – 2007 (2009)

Recent Reprints and New Editions

Vol. 1702: J. Ma, J. Yong, Forward-Backward Stochastic Differential Equations and their Applications. 1999 – Corr. 3rd printing (2007)

Vol. 830: J.A. Green, Polynomial Representations of GL_n, with an Appendix on Schensted Correspondence and Littelmann Paths by K. Erdmann, J.A. Green and M. Schoker 1980 – 2nd corr. and augmented edition (2007)

Vol. 1693: S. Simons, From Hahn-Banach to Monotonicity (Minimax and Monotonicity 1998) – 2nd exp. edition (2008)

Vol. 470: R.E. Bowen, Equilibrium States and the Ergodic Theory of Anosov Diffeomorphisms. With a preface by D. Ruelle. Edited by J.-R. Chazottes. 1975 – 2nd rev. edition (2008)

Vol. 523: S.A. Albeverio, R.J. Høegh-Krohn, S. Mazzucchi, Mathematical Theory of Feynman Path Integral. 1976 – 2nd corr. and enlarged edition (2008)

Vol. 1764: A. Cannas da Silva, Lectures on Symplectic Geometry 2001 – Corr. 2nd printing (2008)

LECTURE NOTES IN MATHEMATICS   🐎 Springer

Edited by J.-M. Morel, F. Takens, B. Teissier, P.K. Maini

Editorial Policy (for the publication of monographs)

1. Lecture Notes aim to report new developments in all areas of mathematics and their applications - quickly, informally and at a high level. Mathematical texts analysing new developments in modelling and numerical simulation are welcome.

   Monograph manuscripts should be reasonably self-contained and rounded off. Thus they may, and often will, present not only results of the author but also related work by other people. They may be based on specialised lecture courses. Furthermore, the manuscripts should provide sufficient motivation, examples and applications. This clearly distinguishes Lecture Notes from journal articles or technical reports which normally are very concise. Articles intended for a journal but too long to be accepted by most journals, usually do not have this "lecture notes" character. For similar reasons it is unusual for doctoral theses to be accepted for the Lecture Notes series, though habilitation theses may be appropriate.

2. Manuscripts should be submitted either online at www.editorialmanager.com/lnm to Springer's mathematics editorial in Heidelberg, or to one of the series editors. In general, manuscripts will be sent out to 2 external referees for evaluation. If a decision cannot yet be reached on the basis of the first 2 reports, further referees may be contacted: The author will be informed of this. A final decision to publish can be made only on the basis of the complete manuscript, however a refereeing process leading to a preliminary decision can be based on a pre-final or incomplete manuscript. The strict minimum amount of material that will be considered should include a detailed outline describing the planned contents of each chapter, a bibliography and several sample chapters.

   Authors should be aware that incomplete or insufficiently close to final manuscripts almost always result in longer refereeing times and nevertheless unclear referees' recommendations, making further refereeing of a final draft necessary.

   Authors should also be aware that parallel submission of their manuscript to another publisher while under consideration for LNM will in general lead to immediate rejection.

3. Manuscripts should in general be submitted in English. Final manuscripts should contain at least 100 pages of mathematical text and should always include

   - a table of contents;
   - an informative introduction, with adequate motivation and perhaps some historical remarks: it should be accessible to a reader not intimately familiar with the topic treated;
   - a subject index: as a rule this is genuinely helpful for the reader.

   For evaluation purposes, manuscripts may be submitted in print or electronic form (print form is still preferred by most referees), in the latter case preferably as pdf- or zipped ps-files. Lecture Notes volumes are, as a rule, printed digitally from the authors' files. To ensure best results, authors are asked to use the LaTeX2e style files available from Springer's web-server at:

   ftp://ftp.springer.de/pub/tex/latex/svmonot1/ (for monographs) and
   ftp://ftp.springer.de/pub/tex/latex/svmultt1/ (for summer schools/tutorials).

Additional technical instructions, if necessary, are available on request from:
lnm@springer.com.

4. Careful preparation of the manuscripts will help keep production time short besides ensuring satisfactory appearance of the finished book in print and online. After acceptance of the manuscript authors will be asked to prepare the final LaTeX source files and also the corresponding dvi-, pdf- or zipped ps-file. The LaTeX source files are essential for producing the full-text online version of the book (see
http://www.springerlink.com/openurl.asp?genre=journal&issn=0075-8434 for the existing online volumes of LNM).

The actual production of a Lecture Notes volume takes approximately 12 weeks.

5. Authors receive a total of 50 free copies of their volume, but no royalties. They are entitled to a discount of 33.3% on the price of Springer books purchased for their personal use, if ordering directly from Springer.

6. Commitment to publish is made by letter of intent rather than by signing a formal contract. Springer-Verlag secures the copyright for each volume. Authors are free to reuse material contained in their LNM volumes in later publications: a brief written (or e-mail) request for formal permission is sufficient.

Addresses:
Professor J.-M. Morel, CMLA,
École Normale Supérieure de Cachan,
61 Avenue du Président Wilson, 94235 Cachan Cedex, France
E-mail: Jean-Michel.Morel@cmla.ens-cachan.fr

Professor F. Takens, Mathematisch Instituut,
Rijksuniversiteit Groningen, Postbus 800,
9700 AV Groningen, The Netherlands
E-mail: F.Takens@rug.nl

Professor B. Teissier, Institut Mathématique de Jussieu,
UMR 7586 du CNRS, Équipe "Géométrie et Dynamique",
175 rue du Chevaleret,
75013 Paris, France
E-mail: teissier@math.jussieu.fr

For the "Mathematical Biosciences Subseries" of LNM:

Professor P.K. Maini, Center for Mathematical Biology,
Mathematical Institute, 24-29 St Giles,
Oxford OX1 3LP, UK
E-mail: maini@maths.ox.ac.uk

Springer, Mathematics Editorial, Tiergartenstr. 17,
69121 Heidelberg, Germany,
Tel.: +49 (6221) 487-259
Fax: +49 (6221) 4876-8259
E-mail: lnm@springer.com